Subduction Zones
Part II

Edited by
Larry J. Ruff
Hiroo Kanamori

1989

Birkhäuser Verlag
Basel · Boston · Berlin

Reprint from Pure and Applied Geophysics
(PAGEOPH), Volume 129 (1989), No. 1/2

Editors' addresses:

Prof. L. J. Ruff
Dept. of Geological Sciences
University of Michigan
Ann Arbor, MI 48109
USA

Prof. H. Kanamori
Seismological Laboratory
California Institute of Technology
Pasadena, CA 91125
USA

Library of Congress Cataloging in Publication Data

Subduction zones / edited by Larry J. Ruff, Hiroo Kanamori.
p. cm.
»Reprint from Pure and applied geophysics (PAGEOPH), volume 128 (1988), no. 3/4« – –T. p. verso.
Bibliography: p.
Includes index.
ISBN 0-8176-1928-3 (U.S. : v. 1)
1. Plate tectonics. I. Ruff, Larry J., 1952– . II. Kanamori, H. (Hiroo), 1936– . III. Pure and applied geophysics.
QE511.4.S93 1989 551.1'36 – – dc 19 88-39469

(Revised for vol. 2)
Subduction zones.
Reprint from Pure and applied geophysics (PAGEOPH), volume 128 (1988), no. 3/4 and volume 129 (1989), no. 1/2.
Includes bibliographies and indexes.
1. Plate tectonics. I. Ruff, Larry J., 1952– . II. Kanamori, H. (Hiroo), 1936– . III. Pure and applied geophysics.
QE511.4.S93 1989 551.1'36 88-39469
ISBN 0-8176-2272-1 (U.S. : v. 2)

CIP-Titelaufnahme der Deutschen Bibliothek

Subduction zones / ed. by Larry J. Ruff ; Hiroo Kanamori. –
Reprint. – Basel ; Boston ; Berlin : Birkhäuser.
NE: Ruff, Larry J. [Hrsg.]
Reprint
Pt. 2 (1989)
Aus: Pure and applied geophysics ; Vol. 129
ISBN 3-7643-2272-1 (Basel ...) brosch.
ISBN 0-8176-2272-1 (Boston) brosch.

Printed in Germany on acid-free paper

ISBN 3-7643-2272-1
ISBN 0-8176-2272-1

Contents

PAGEOPH, Vol. 129, Nos. 1/2 (1989) 0033–4553/89/020001–05$1.50 + 0.20/0

Introduction to Subduction Zones

LARRY J. RUFF[1] and HIROO KANAMORI[2]

Subduction zones present many façades to those that observe them. From obvious features to obscure yet important processes, there are many aspects of subduction zones to observe and explain. Notable examples of obvious features are volcanoes, earthquakes, mountain belts, and deep sea trenches; while on the other hand, the unseen process of sediment subduction can produce long-term effects on the Earth's geochemical reservoirs. Indeed, most disciplines of the earth sciences have reasons to study subduction zones, and consequently all disciplines can make contributions to our understanding of subduction zones.

Each discipline has its own view of the subduction process. To volcanologists, subduction zones are the sites of volcanoes. The active volcanic belt is a ubiquitous feature of subduction zones; volcanoes may be built on continental or oceanic lithosphere. In addition, volcanism is associated with back-arc spreading. There is a considerable range in the composition of volcanic products in subduction zones. Many subduction zone volcanoes also represent great hazards to local populations, and violent eruptions may even cause global-scale variations in the atmosphere. To seismologists, subduction zones generate earthquakes, and are the most important tectonic setting for seismicity (see Figure 1). To geologists, subduction zones produce quite a diversity of geologic terranes and rock deformation, the settings vary from accretionary prisms to back-arc thrust regimes. To geodicists, subduction zones are the sites of some of the largest deviations in the gravity field as well as regions of dramatic horizontal strain and vertical uplift and subsidence.

But to plate tectonicists, subduction zones are relatively simple: subduction zones are convergent plate boundaries where oceanic lithosphere is consumed (see Figure 2). To embellish this "definition" from the perspective of global dynamics: subduction zones are regions where the cold upper thermal boundary layer of the convecting Earth sinks back into the hot mantle. The above definitions would seem to explain both the kinematic and dynamic *role* of subduction zones. But do these definitions explain why subduction zones are located where they are? what the

[1] Dept. of Geological Sciences, University of Michigan, Ann Arbor, MI 48109, U.S.A.
[2] Seismological Laboratory, California Institute of Technology, Pasadena, CA 91125, U.S.A.

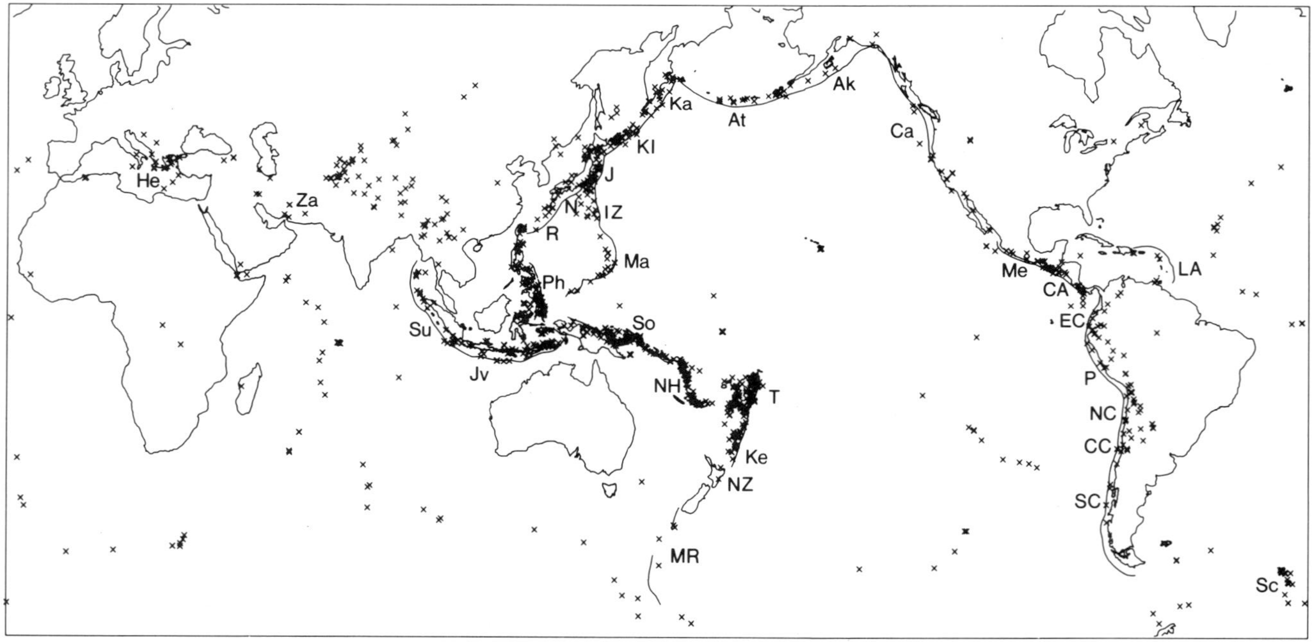

Figure 1

Global seismicity and subduction zones. The mercator projection extends from 65°S to 65°N, while the left and right margins are at 20°W. The continents and major islands are outlined, and major trench axes are also shown. Events in the Harvard CMT catalog from 1981 to 1986 are plotted as the crosses. Subduction zone segments are identified as follows together with the interacting plate pair (first named plate subducts beneath the second, ? where uncertain): He, *Hellenic arc*, Africa-Eurasia?; Za, *Zagros*, Arabia-Eurasia; Su and Jv, *Sumatra and Java*, India/Australia-Eurasia; So, *Solomon*, Solomon-Pacific?; NH, *New Hebrides*, India/Australia-Fiji/Pacific?; T, Ke, and NZ, *Tonga, Kermadec, and New Zealand* (*North Island*), Pacific-India/Australia; MR, *Macquarie Ridge*, India/Australia-Pacific; Ph, *Philippines* (*eastern margin*), Philippine-Eurasia?; R, *Ryukyu*, Philippine-Eurasia; N, *Nankai* (*southern Honshu*), Philippine-Eurasia; Ma and IZ, *Marianas and Izu-Bonin*, Pacific-Philippine; J, *Japan* (*northern Honshu*), Pacific-Eurasia? KI, *Kurile Islands* (*and Hokkaido*), Pacific-North America?; Ka, *Kamchatka*, Pacific-North America?; At and Ak, *Aleutians and Alaska*, Pacific-North America; Ca, *Cascades* (*Oregon-Washington-British Columbia*), Juan de Fuca-North America; Me, *Mexico*, Cocos-North America; CA, *Central America*, Cocos-Caribbean; EC, P, NC, CC, and SC, *Ecuador-Colombia, Peru, northern Chile, central Chile, and southern Chile*, Nazca-South America; Sc, *Scotia*, South America-Scotia?; LA, *Lesser Antilles*, South America? North America?-Caribbean. In addition to the above listed segments, there are several smaller subduction zones in the complex region of Southeast Asia. Most of the above-listed subduction zones receive some attention in this special issue.

geometry of subduction zone segments will be? the physiography of island arcs? the generation of island arc volcanoes? back-arc spreading? the diversity of earthquake occurrence? style of accretionary prisms?—clearly not. While plate tectonics provides the underlying kinematic role of subduction zones, it does not explain the great variety of phenomena associated with subduction zones. Although earthquakes, volcanoes, trenches, and global recycling might be second-order phenomena within the plate tectonics scenario, they are still first-order problems within earth sciences. Researchers in their various fields must go beyond plate tectonics and global dynamics to make progress in explaining the great diversity of phenomena associated with subduction zones.

The diverse contributions in this special issue reflect the many façades of subduction zones. The style of the contributions also varies from review papers to speculative accounts of obscure processes. The special issue is divided into two volumes with a total of four different sections: in Volume I (1) global-scale reviews, (2) accretionary prism processes, and in Volume II, (3) tectonics and subduction initiation, and (4) earthquake occurrence. While the size of this special issue requires the division into two volumes, there is considerable linkage between the two parts. Indeed, several papers could appear in more than one section. In the global-scale review section of Volume I: CLOOS and SHREVE present a two-part comprehensive description of accretionary prism and their model which explains several features; VASSILIOU and HAGER treat the problem of Wadati-Benioff zone seismicity and slab penetration with extensive modeling of stress regimes; WORTEL and VLAAR also treat Wadati-Benioff zone seismicity from the perspective of pressure-temperature controls on seismic/aseismic behavior; HSUI reviews some of the contributions of fluid mechanics to various processes associated with subduction zones; and MCLENNAN discusses the role of subduction zones in global geochemical cycling. Taken together, these global-scale discussions clearly indicate that many important present-day characteristics of subduction need to be better understood, not to mention the properties of subduction zones through time. In the accretionary prism processes section of Volume I: BRAY and KARIG examine the physical processes occurring in accretionary prisms, they focus on the dewatering process; SHI and WANG also present physical models of the accretionary prism with emphasis on heat flow; then COCHRANE *et al.* show the results of seismic experiments along the Cascades subduction zone to identify the deformational styles. Volume II is more focused on large-scale tectonics and/or seismicity. In the tectonics and subduction initiation section: CLOETINGH *et al.* treat the question of passive margin vulnerability to subduction initiation; SENO and TAKANO look at the remarkably complex triple junction interaction in central Honshu; MCCANN and HABERMANN present a model for the influence of ocean floor topography on segmentation of island arc platforms; while RUFF *et al.* present a seismotectonic study of the Macquarie Ridge complex and discuss the nature of subduction initiation. In the earthquake occurrence section of Volume II: HEATON and

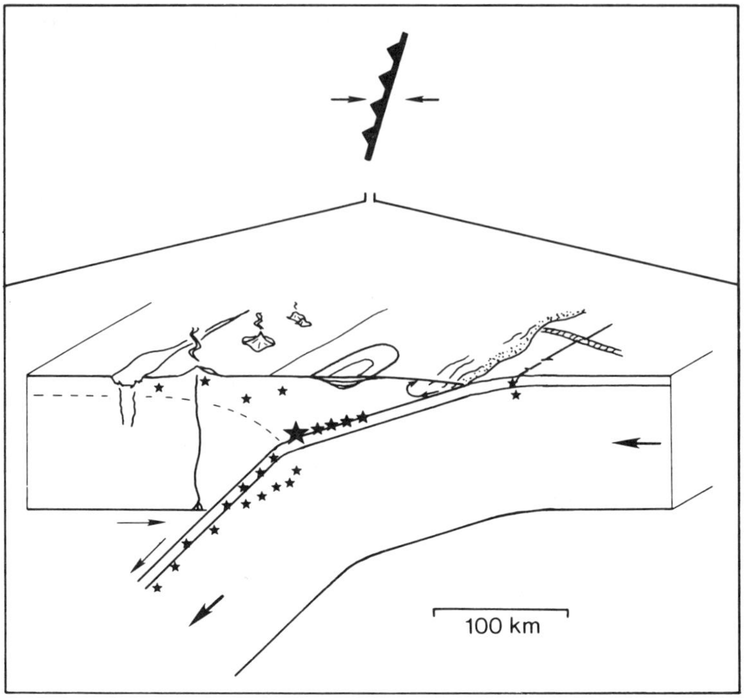

Figure 2
Two views of subduction zones. The map view at top depicts the role of subduction zones in plate tectonics; they are convergent plate boundaries where oceanic lithosphere is consumed. Upon closer examination (bottom), subduction zones display many features and processes of interest to many branches of earth science. Several of the important shallow features of subduction zones are represented in the cross-section: the oceanic lithosphere deforms as it subducts with normal faulting in the outer-rise and downward kinks at the trench axis and at a depth of ≈ 40 km; seismicity is represented by the stars; large underthrusting earthquakes (large star) occur on the plate contact interface, but only in a certain depth range that extends down to about 40 km; below this depth, seismicity occurs within the downgoing oceanic crust and upper mantle which also experience phase changes; "double Benioff zones" have been observed in the intermediate depth range; the deepest extent of the Wadati-Benioff zone varies from a hundred to seven hundred kilometers; there is also scattered seismicity in the outer-rise and in the upper plate above the "critical" isotherm (dashed line); the accretionary prism is a dynamic system between the trench axis and coastline with faulting, material flow, sediment subduction, underplating, and basin formation; dewatering and sediment metamorphism also occur in this environment; a volcanic belt occurs above the apex of the corner flow where the asthenosphere beneath the upper plate is viscously coupled to the downgoing plate; back-arc spreading occurs behind the volcanic arc. The cross-section is highly stylized—not all subduction zones display these features. An essential characteristic of subduction zones is their variability, thus it is important to add the third dimension to the stylized subduction zone. All of the above discussed features vary along arc strike, and furthermore, subduction zones evolve with time. Most of the above-mentioned characteristics and processes are addressed in this special issue.

HARTZELL give a detailed assessment of the anticipated strong motions from a hypothetical future great earthquake along the Cascades subduction zone; SMITH *et al.* present a comprehensive geophysical study of the North Island (New Zealand) subduction complex and earthquake occurrence; MORIYAMA *et al.* propose that a zone of unusual seismic coupling exists in the Izu-Bonin arc; and RUFF presents a "comparative subductology" approach to the speculative connection between sediment subduction and great earthquake occurrence.

It is worth repeating that the diversity of contributions in this special issue reflects the diversity of interesting and unexplained aspects of subduction. Of the many façades that subduction zones present to us, the kinematic role in plate tectonics is the best understood aspect. Many of the primary geological, geochemical, and geophysical characteristics of subduction zones still require explanation. It is our hope that this issue will help spark further scientific exchange and interaction throughout the earth sciences—the surest path to new knowledge about subduction zones.

PAGEOPH, Vol. 129, Nos. 1/2 (1989) 0033–4553/89/020007–19$1.50 + 0.20/0

On the Initiation of Subduction Zones

SIERD CLOETINGH,[1,2] RINUS WORTEL[1] and N. J. VLAAR[1]

Abstract—Analysis of the relation between intraplate stress fields and lithospheric rheology leads to greater insight into the role that initiation of subduction plays in the tectonic evolution of the lithosphere. Numerical model studies show that if after a short evolution of a passive margin (time span a few tens of million years) subduction has not yet started, continued aging of the passive margin alone does not result in conditions more favorable for transformation into an active margin.

Although much geological evidence is available in supporting the key role small ocean basins play in orogeny and ophiolite emplacement, evolutionary frameworks of the Wilson cycle usually are cast in terms of opening and closing of wide ocean basins. We propose a more limited role for large oceans in the Wilson cycle concept. In general, initiation of subduction at passive margins requires the action of external plate-tectonic forces, which will be most effective for young passive margins prestressed by thick sedimentary loads. It is not clear how major subduction zones (such as those presently ringing the Pacific Basin) form, but it is unlikely they form merely by aging of oceanic lithosphere. Conditions likely to exist in very young oceanic regions are quite favorable for the development of subduction zones, which might explain the lack of preservation of back-arc basins and marginal seas.

Plate reorganizations probably occur predominantly by the formation of new spreading ridges, because stress relaxation in the lithosphere takes place much more efficiently through this process than through the formation of new subduction zones.

Key words: Initiation of subduction, passive to active margin transition, preservation of back-arc basins, emplacement of ophiolites, mechanics of plate boundary formation, Wilson cycle, plate reorganizations.

Introduction

Initiation of oceanic lithosphere subduction is a key element in plate-tectonic schemes for the evolution of the lithosphere. The mechanisms that initiate the formation of new subduction zones are, however, not fully understood (DICKINSON and SEELY, 1979; FLINN, 1982; TURCOTTE and SCHUBERT, 1982; KOBAYASHI and SACKS, 1985). This is in part because, with a possible exception in the western Pacific (OKAL *et al.*, 1986; KROENKE and WALKER, 1986), there are no obvious

[1] Vening Meinesz Laboratory, Institute of Earth Sciences, University of Utrecht, P.O. Box 80.021, 3508 TA Utrecht, The Netherlands.

[2] Present address: Department of Sedimentary Geology, Institute of Earth Sciences, Free University, P.O. Box 7161, 1007 MC Amsterdam, The Netherlands.

present-day examples of the formation of new major trench systems. DICKINSON and SEELY (1979) summarize the possible mechanisms for initiation of oceanic lithosphere subduction and distinguish two classes: (1) plate rupture within an oceanic plate or at a passive margin and (2) reversal of the polarity of an existing subduction zone, possibly following a collision of an island arc with a passive margin (see also MITCHELL, 1984). A third mechanism is initiation of subduction by inversion of transform faults into trenches (UYEDA and BEN-AVRAHAM, 1972).

We deal here with mechanisms for the formation of new consuming plate boundaries rather than those for polarity reversal. Geological evidence for margins with widely different ages (e.g., DEWEY, 1969; COHEN, 1982) support the notion that passive continental margins in particular might be potential sites for the formation of consuming plate boundaries. Passive margins may, therefore, be expected to play a central role in the Wilson cycle of the opening and closing of oceans (WILSON, 1966).

The Role of Passive Margins in the Wilson Cycle

Several authors (TURCOTTE and SCHUBERT, 1982; HYNES, 1982) advocate a model based on the development of a major oceanic basin with old and, hence, cold and gravitationally unstable (VLAAR and WORTEL, 1976; OXBURGH and PARMENTIER, 1977) oceanic lithosphere at its continental margins before the basin can be closed. This seems to be more inspired by an actualistic comparison with the present size of the Atlantic Ocean rather than by a consideration of more pertinent geological observations. The same may be said of arguments for spontaneous foundering and subduction of oceanic lithosphere older than 200 Ma (HYNES, 1982), made on the basis of the temporary absence of oceanic lithosphere older than 200 Ma and the increase with age of the gravitational instability of oceanic lithosphere. The transformation of passive into active margins by spontaneous foundering of old unstable lithosphere is inhibited by the great strength of oceanic lithosphere (KIRBY, 1980; MCADOO *et al.*, 1985). We, therefore, focused on the state of stress at passive margins to investigate whether the stresses generated there are sufficiently high to induce lithospheric failure and transformation into an active margin.

Models for Passive to Active Margin Transition

Our numerical model studies (CLOETINGH, 1982; CLOETINGH *et al.*, 1983) have shown that in general, flexure induced by sediment loading dominates the stress state at passive margins. In that work we have demonstrated that, owing to the continuing accumulation of sediments at passive margins, the induced stress level increases with the age of the margin. An important rheological feature (BODINE *et al.*, 1981; CLOETINGH *et al.*, 1982) implemented in our thermo-mechanical models for the evolution of passive margins (CLOETINGH *et al.*, 1982; 1984) is that the strength of the lithosphere at the margin also increases with age. Finite-element stress calculations were made for a passive margin in different stages of evolution incorporating two different sediment loading models and the depth-dependent rheology inferred from extrapolation of rock-mechanics data (after GOETZE and EVANS, 1979).

As a *reference* model we adopted a sediment loading model in which the maximum thickness of the sedimentary wedge at passive margins followed a square-root-of-age relation (TURCOTTE and AHERN, 1977; WORTEL, 1980), reaching a maximum of 9.4 km at 200 Ma. This model constitutes a fair average of the sediment loading histories and resulting thicknesses observed at passive margins (SOUTHAM and HAY, 1981). In the second model, the *full load* model, the entire loading capacity of oceanic lithosphere is taken up by the sediments. This model is more appropriate for the huge sediment accumulations found at deltas, which clearly exceed the thicknesses depicted by the reference model. In this model the maximum thickness of the sedimentary wedges reaches 16 km at 200 Ma. For both the reference and full load models, sediment loading extends from the continental shelf to the continental rise, with maximum sediment thickness in the outer shelf/continental slope region (see Figure 1). Lithospheric strength profiles were adopted that combine the effect of pressure on brittle behavior (BYERLEE, 1968, 1978) and temperature- and stress dependence on ductile deformation from Goetze's flow laws for dry olivine (GOETZE, 1978; GOETZE and EVANS, 1979) with an assumed strain-rate of $10^{-18}\,s^{-1}$ (characteristic for sedimentary basin development). Such a rheology provides a good first-order description of the mechanical behavior of oceanic lithosphere, and is consistent with seismotectonic studies (WIENS and STEIN, 1983) and observations of the response of oceanic lithosphere to tectonic processes (MCADOO *et al.*, 1985). Similar to BODINE *et al.* (1981) we defined the depth at which the strength is 500 bar as the lower boundary of the *mechanically strong upper part of the lithosphere (MSL)*. Both its thickness and its maximum strength increase according to a square-root-of-age function from a few kilometers and a few kilobars, respectively, for young lithosphere to values of approximately 50 km and 10 kilobar for old oceanic lithosphere. Investigation of gravity anomalies at passive margins (KARNER and WATTS, 1982) has shown that the mechanical properties of oceanic lithosphere at passive margins are not essentially different from the rheological properties of "standard" oceanic lithosphere as inferred from studies of seamount loading. It should be noted that the models discussed here deal only with gross features of passive margin evolution, pertinent to the problem under consideration.

Figure 2 shows the stress field for the case of a reference load on a 100 Ma old passive margin. Differential stresses $\sigma_H - \sigma_V$ are greatest at the transition from oceanic lithosphere to rift-stage lithosphere. In Figure 3 the stress maximum for a 100 Ma old passive margin is displayed as a function of depth, down to the base of

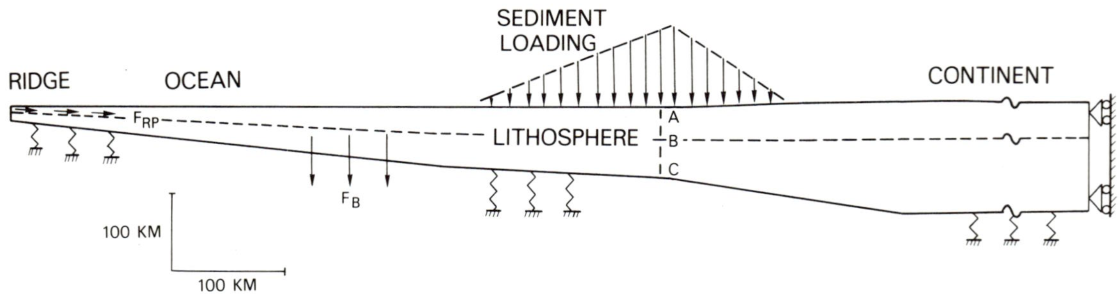

Figure 1

Model features geometry, rheology and system of forces acting on a passive margin in the interior of a plate not involved in collision or subduction processes. The bottom of the mechanically strong part of the lithosphere (MSL) is indicated by a broken horizontal line. F_{RP} is the push exerted by the oceanic ridge, F_B is the negative buoyancy associated with the cooling of the oceanic lithosphere when it moves away from the spreading center. The position of the sedimentary wedge corresponds to sedimentary loading on outer shelf, slope and rise.

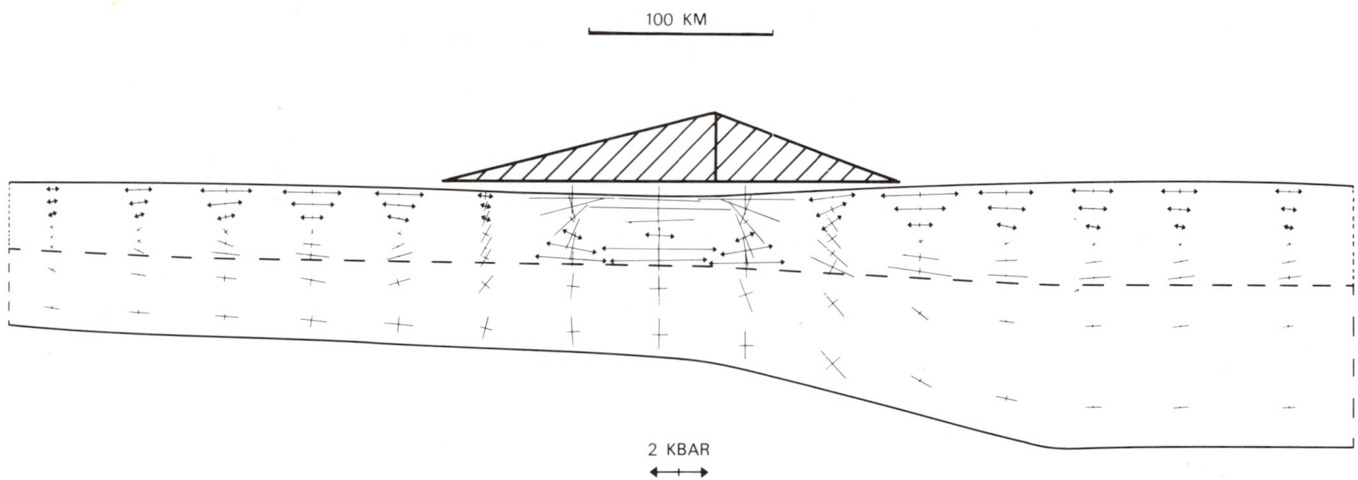

Figure 2
Stresses calculated for a passive margin at an age of 100 Ma based on the reference model of sediment loading. Flexure caused by sediment loading forms the dominant deformation mode at the margin. Principal stresses denoted by arrows are plotted in the undeformed configuration. Stresses are plotted only for the parts of the lithosphere where there is significant deformation. Symbols (↔) and (—+—) denote tension and compression, respectively.

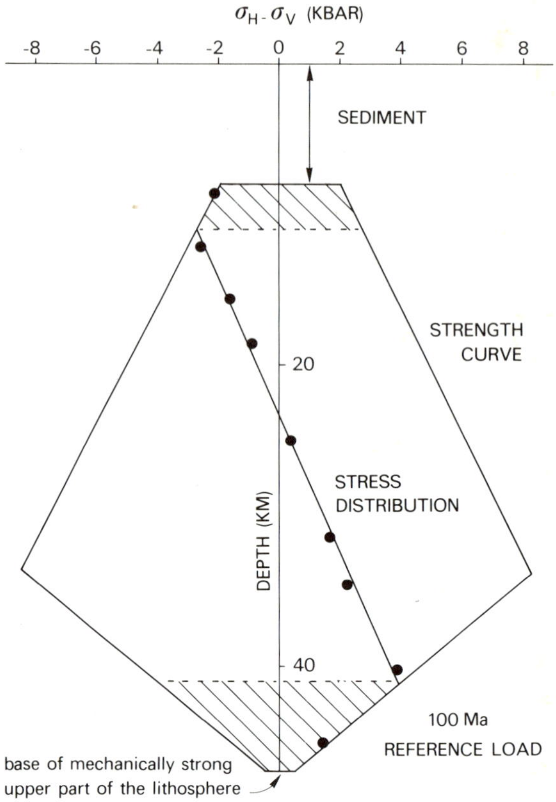

Figure 3
Comparison of stresses generated at 100 Ma old passive margin, under reference loading with lithospheric strength. Strength curve envelope and results of stress calculation (solid dots) are plotted as a function of depth down to the base of MSL at the point of maximum flexure (see also Figure 2). The stress distribution is given by the line inside the strength envelope connecting the solid dots. Differential stresses ($\sigma_H - \sigma_V$) are plotted against depth. Sign convention for the stresses: tension positive, compression negative. Zero-strength is assumed for the sediments. Hatched areas in the upper and lower part of the MSL denote failure by brittle fracture and ductile flow, respectively.

the MSL. The lowermost and uppermost regions of the MSL fail due to the high differential stresses developed here. However, the main part of the MSL remains in the elastic state as stresses are too low to result in rupture of the lithosphere. To induce complete lithospheric failure, the integrated force $F = \int \sigma \, dz$ must exceed the integrated depth-dependent strength $A = \int \sigma_y \, dz$, where σ is the stress distribution, σ_y the differential strength profile and z the depth. The ratio of the maximum stress generated to the maximum strength is essentially independent of lithospheric age since load, strength and thickness of MSL all exhibit the same square-root-of-age behavior due to the thermal evolution of the lithosphere.

The situation is very much different when the full loading capacity is taken up

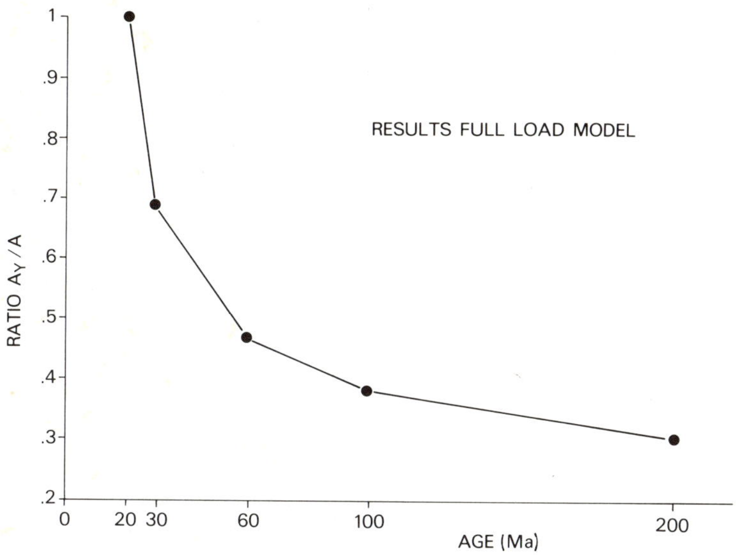

Figure 4
Ratio of A_y (the hatched area within the strength envelope, see Figure 3) to A (total area of the envelope) as a function of lithospheric age for the full load model where the entire loading capacity of the lithosphere is taken up by passive margin sedimentation. For ages below 20 Ma lithospheric failure and consequently initiation of subduction is induced.

by the sediments. The surplus of sediment added to the reference load promotes high stresses most effectively when deposited on a young (weak) passive margin. This is demonstrated in Figure 4 where we have plotted the ratio of A_y, corresponding to the area of the strength envelope in failure, to the total area A, corresponding to the integrated strength of the MSL. Figure 4 shows that full loading on passive margins with ages below 20 Ma leads to complete failure of the lithosphere. For ages in excess of 20 Ma the relative amount of failure decreases rapidly with age. Therefore, if subduction has not commenced at a passive margin during a youthful stage, continued aging alone will not result in conditions more favorable for the initiation of subduction. Similarly, stress levels required to be produced by external forces in order to create a new consuming plate boundary in this manner will increase with the growth of the oceanic basin as well. These findings unravel the enigma that gravitationally unstable oceanic lithosphere at the margin of the Atlantic is not subject to subduction (see KOBAYASHI and SACKS, 1985).

The Role of External Forces

In the preceding section the flexural stresses induced by sediment loading were shown to be of the order of several kilobars. In previous work (WORTEL and

CLOETINGH, 1981, 1983; CLOETINGH and WORTEL, 1985, 1986) we have demonstrated that the *regional* stress field caused by plate-tectonic forces (FORSYTH and UYEDA, 1975; CHAPPLE and TULLIS, 1977) is dominated by concentration of slab-pull forces. For passive continental margins located in the interiors of plates not involved in subduction or collision processes, the local stress field induced by sediment loading dominates the stresses induced by the remaining plate-tectonic forces, which in general is of the order of a few hundred bars (e.g., RICHARDSON *et al.*, 1979). We have shown that, under such conditions, the evolution of passive margins to maturity will not in itself lead to an initiation of subduction.

In general, external forces in addition to sediment loading are required to cause rupture of the lithosphere. A comparison could be made of the force required to induce complete lithospheric tensional failure and the estimates of age-dependent slab-pull and ridge-push forces (ENGLAND and WORTEL, 1980; DAVIES, 1983). However, as noted by CHAPPLE and TULLIS (1977), such comparisons do not take into account that the level of the regional stress field can be considerably magnified due to the geometrical shape of the plate boundaries and the complexity of boundary forces along lithospheric plates. More specifically, we have demonstrated the key importance of incorporating plate-tectonic forces due to variations in the age of the downgoing lithosphere at convergent boundaries in lithospheric stress modelling (WORTEL and CLOETINGH, 1981, 1985; CLOETINGH and WORTEL, 1985). These findings have important implications for the formation of new plate boundaries. In fact, we have demonstrated in our model study of the paleo-stress field of the Farallon plate (WORTEL and CLOETINGH, 1981) that the level of the tensional stress field caused by concentration of plate-tectonic forces was of sufficiently high magnitude to cause the inception of the Galapagos spreading center and fragmentation of the Farallon plate into the Cocos plate and the Nazca plate at 25–30 Ma ago (see Figure 5). CLOETINGH and WORTEL (1985, 1986) showed that tensional stresses induced by slab-pull forces associated with subduction of old oceanic lithosphere at various segments of the convergent boundaries of the Indo-Australian plate are transmitted into the interior of the plate, where they affect lithosphere of both oceanic and continental character. This is, for example, the case for the zone of concentrated extensional seismicity in young oceanic lithosphere in the equatorial Indian Ocean (STEIN *et al.*, 1987) and for eastern Australia, an area characterized by recent volcanic activity, probably associated with a regional tensional stress regime (DUNCAN and MCDOUGALL, 1988). Preferential rifting of continental lithosphere might occur due to its relative weakness compared to oceanic lithosphere (VINK *et al.*, 1984; STECKLER and TEN BRINK, 1986; SHUDOFSKY *et al.*, 1987). Thermo-mechanical models show that in this case stresses of the order of a few kilobars are required (HOUSEMAN and ENGLAND, 1986; CLOETINGH and NIEUWLAND, 1984). Fragmentation of plates under the influence of slab-pull forces is therefore seen to provide a mechanism for rifting of both oceanic and continental lithosphere. Due to its great strength, rifting of old oceanic

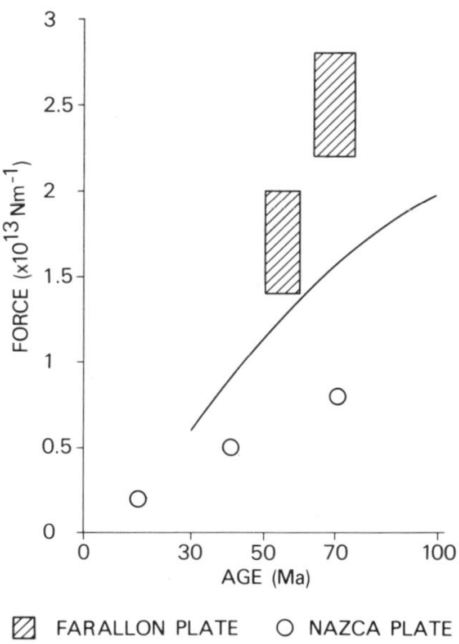

Figure 5

The force required to induce tensional failure in oceanic lithosphere plotted as a function of lithospheric age (solid line). Boxes indicate stress levels prior to the break-up of the Farallon plate into the Cocos and Nazca plates, calculated by WORTEL and CLOETINGH (1981). Circles indicate (intraplate) stress levels in the present-day Nazca plate (WORTEL and CLOETINGH, 1983). The values given are representative of large parts of the Farallon plate and Nazca plate, respectively.

lithosphere is not expected to occur on a large scale. This is in agreement with the observation that, with two exceptions (STEIN and COCHRAN, 1985; MAMMERICKS and SANDWELL, 1986) evidence for rifting of old oceanic lithosphere is absent.

A different situation arises in a tectonic setting dominated by compressional stresses, where even higher stress levels are required to exceed lithospheric strength. An example of this is found in the northeastern Indian Ocean, where an exceptionally high level of compressional stresses (Figure 6) is induced by the focusing of resistive forces associated with the Himalayan collision and subduction of young oceanic lithosphere in the northern part of the Sunda arc (CLOETINGH and WORTEL, 1985; 1986). The response of the oceanic lithosphere in the northeastern Indian Ocean to the high level of in-plane compressional stress approaching lithospheric strength is, however, not one of lithosphere failure, but one of buckling (WEISSEL *et al.*, 1980; MCADOO and SANDWELL, 1985; ZUBER, 1987). Figure 7 shows that the estimates of the forces required to buckle the ocean floor in the northeastern Indian Ocean given by MCADOO and SANDWELL (1985) are in excellent agreement with the calculated stress levels shown in Figure 6. As noted by CLOETINGH *et al.* (1983) the folding of the oceanic lithosphere caused by regional compressional stresses amplified by sediment loading might be a preparatory stage

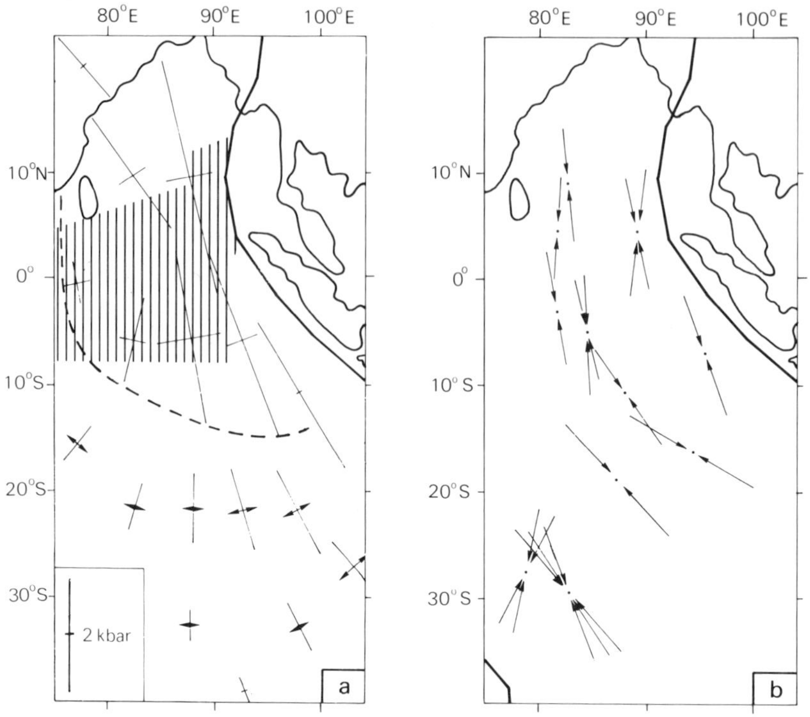

Figure 6
Regional stress field in the northeastern Indian Ocean. a (left) Calculated stress field after CLOETINGH and WORTEL (1986). The dashed line is the southern limit of the observed deformation in the northeastern Indian Ocean (GELLER *et al.*, 1983). Dashed area indicates the location of the proposed (WIENS *et al.*, 1985) diffuse plate boundary separating an Indian plate from an Australian plate. Plotted are principal horizontal non-hydrostatic stresses averaged over a uniform elastic plate with a reference thickness of 100 km. Note the high level of the stress field associated with the unique dynamic situation of the Indo-Australian plate. b (right) The orientation of maximum horizontal compressive stress inferred from an earthquake focal mechanism study by BERGMAN and SOLOMON (1985).

for the initiation of a new subduction zone in the northeastern Indian Ocean. More recently, strong independent evidence for the presence of a nascent diffuse plate boundary separating an Indian plate from an Australian plate has been presented by WIENS *et al.* (1985, 1986) based on an inversion of plate velocities and a study of the seismo-tectonics of the area. Hence, it seems that in the Indo-Australian plate stress levels induced by plate-tectonic forces are of sufficient magnitude to induce the inception of a new consuming plate boundary, even in old oceanic lithosphere. It is important to note that this process is the expression of the unique present-day dynamic situation of the Indo-Australian plate, which reflects for a major part the dramatic impact of the Eurasia-India collision, which is probably one of the most important tectonic events during the last 250 Ma of Earth history. For these

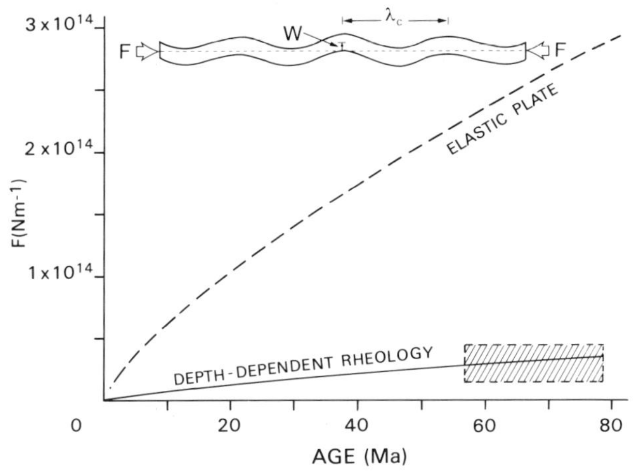

Figure 7

Buckling load versus age for oceanic lithosphere with a depth-dependent rheology inferred from rock-mechanics studies (GOETZE and EVANS, 1979) (solid curve) and for fully elastic oceanic lithosphere (dashed curve) (after MCADOO and SANDWELL, 1985). Box indicates the stress levels calculated for the area in the northeastern Indian Ocean (CLOETINGH and WORTEL, 1985, 1986, see Figure 6) where folding of oceanic lithosphere under the influence of compressional stresses has been observed (MCADOO and SANDWELL, 1985; GELLER *et al.*, 1983).

reasons, the mechanism underlying the formation of a nascent plate boundary in the central Indian Ocean is likely the exception rather than the rule for the creation of new subduction zones.

Closure of Small Oceanic Basins

Therefore, initiation of subduction of old oceanic lithosphere requires extremely high levels of intraplate stress to be induced by external forces, probably in association with unique plate-tectonic settings such as presently encountered in the northeastern Indian Ocean. As demonstrated by our modelling of stresses at passive margins, the level of stresses to be produced by plate-tectonic forces drops dramatically with decreasing age of the lithosphere, favoring initiation of young oceanic lithosphere subduction at the margins of small oceanic basins. These findings lead us to propose the modification of the classical sequence of the Wilson cycle concept shown in Figure 8, in which closure of newly rifted basins by initiation of subduction of young lithosphere occurs. This has implications for tectonics and volcanism which differ appreciably from those associated with deep subduction zones (VLAAR, 1983; CLOETINGH *et al.*, 1984; VLAAR and CLOETINGH, 1984). For example, a notable feature of shallow-angle subduction of young oceanic lithosphere is the induction of a compressional regime in the overriding plate, contrasting with the

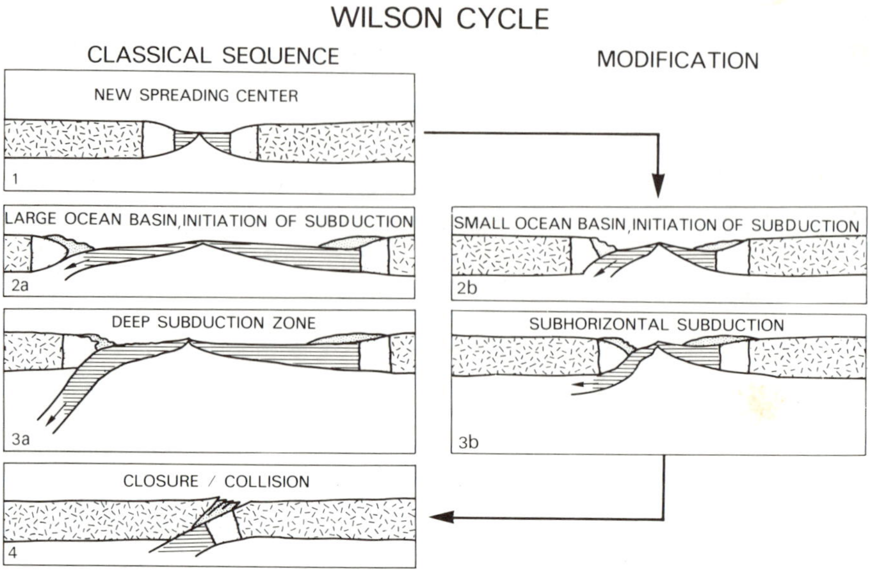

Figure 8

Scenarios for the Wilson Cycle. Left: classical scenario, in which rifting is followed by the formation of a large ocean basin. Initiation of subduction involves old oceanic lithosphere, which results in a deep subduction zone. Right: preferred scenario for Wilson cycle, in which closure of the newly rifted basin occurs through initiation of subduction of young oceanic lithosphere leading to either shallow-angle subduction or to obduction of gravitationally stable buoyant lithosphere (VLAAR, 1983). Implications for tectonics and volcanism strongly differ from those associated with deep subduction zone. Patterns indicate passive-margin sediments and continental, rift-stage and oceanic lithosphere.

general association of high-angle subduction of old oceanic lithosphere with an extensional regime in the overriding plate (ENGLAND and WORTEL, 1980). Closure of small oceanic basins explains the frequently observed absence of island arc volcanism in Wilsonian orogenies (e.g., TRUMPY, 1982) and the emplacement of ophiolites, fragments of young gravitationally stable oceanic lithosphere on the adjacent continent during the transformation of a passive margin, features that do not conform to the standard concepts of subduction of oceanic lithosphere and the long time span of the classical form of the Wilson cycle. An interesting analogy for the destruction of passive margins of small oceanic basins exists in the form of marginal basins, whose evolution is characterized by a short time span between their creation and collapse (less than 20 Ma, TAYLOR and KARNER, 1983). The modelling described in the present paper has demonstrated that the conditions likely to exist in these very young oceanic regions are quite favorable for the development of new subduction zones. These findings might have important consequences for evaluating the movement and accretion of terranes and other continental crust fragments.

As noted by CHURCH and STEVENS (1971), much of the geological evidence in collision orogens points to closing of small ocean basins rather than large oceans of the scale of the present-day Atlantic. Usually, precise data on the timing of the closing events are absent, which inhibits a quantitative comparison with our model results. Nevertheless, much of the geological evidence is clearly contradictory to activation of passive margins in a late stage of their evolution. As observed by MCWILLIAMS (1981), not all (Proterozoic) Pan-African and older mobile belts mark the sites of major ocean closure. Rather they are formed without the destruction of vast amounts of oceanic lithosphere. On the base of a survey of the tectonic framework of central and western Europe, ZWART and DORNSIEPEN (1978) pointed out that the Variscan orogens are also due to collision and closing of short-lived oceans of minor size. Investigations of Alpine orogeny (TRUMPY, 1982) have provided strong evidence for closure of small oceanic basins in an early stage after opening as have studies of the evolution of the Appenines (WINTERER and BOSELLINI, 1981). In fact, the Alpine basins, being characterized by young oceanic lithosphere, transcurrent faulting, extensive sediment loading and a compressive tectonic regime were ideally suited for the transformation of passive into active margins.

Frequently, however, only indirect and debatable arguments are available for the estimation of the size of the proposed oceans. This applies in particular to the "type locality" of the Wilson cycle, the Iapetus Ocean, which is of unknown width (WINDLEY, 1977). The latter situation is to a large extent caused by the orientation of the spreading, which has hampered reliable paleomagnetic estimates for the width of the Iapetus Ocean. Paleomagnetic evidence, however, certainly does not rule out that large Paleozoic oceans were consumed (e.g., MCELHINNY and VALENCIO, 1981). In fact, the best evidence for the Iapetus Ocean has come from the occurrence of ophiolite belts (ZWART and DORNSIEPEN, 1978). Several authors, (e.g., DEWEY, 1976; NICOLAS and LEPICHON, 1980; SPRAY, 1984) favour obduction of thermally young immature oceanic lithosphere to account for the short time gap documented between ophiolite formation and ophiolite emplacement.

Similarly, flat-angle subduction (VLAAR, 1983) during closure of a small ocean basin provides an important element in the dynamical evolution of foreland thrust belts (STOCKMAL *et al.*, 1986). Our findings also shed light on the far greater degree of basement fragmentation observed under foreland basins in collisional settings compared to their Andean analogs (ALLEN *et al.*, 1986). Evolutionary frameworks of the Wilson cycle in terms of opening and closing of wide oceans seem sometimes to be more inspired by an actualistic comparison with the present size of the Atlantic Ocean, than by a consideration of more pertinent geological observations. During its evolution, the Atlantic Ocean passed through a transition from a narrow to a wide ocean basin, without the formation of a system of subduction zones at its margins. Apparently, optimal loading conditions for transformation of passive into active margins were not fulfilled at this stage, while further aging has not made the passive margins more susceptible to initiation of subduction. To rely too heavily on such an actualistic analogon might, therefore, be very misleading. Moreover, estimated widths of oceans inferred from palinspastic restorations are frequently automatically increased with an additional 1000 km inferred from the lengths of slabs consumed in modern circum-Pacific subduction zones, associated with subduction of old oceanic lithosphere (e.g., WILLIAMS, 1980). Such reconstructions exclude *a priori* the possibility of the closing of a young oceanic basin and might even lead to overlooking the eventual interesting consequences of the subduction of young lithosphere. Our model studies have shown that if subduction has not started after a short evolution of a passive margin, continued aging of the passive margin alone does not result in conditions more favorable for transformation into an active margin. That requires a concentration of external forces in a unique dynamic situation such as the one in which the present-day Indo-Australian plate is involved. Therefore, we propose a more critical appraisal of the role large oceans play in the Wilson cycle.

Discussion

We have demonstrated that evolution of passive margins to maturity (200 Ma) will not in itself lead to initiation of subduction. In general, external forces in addition to sediment loading are required to cause rupture of the lithosphere.

The action of external forces will be most effective when young passive margins are prestressed by thick sedimentary wedges. The compressive stresses necessary to sustain the further development of subduction zones involving young stable lithosphere (MCKENZIE, 1977; ENGLAND and WORTEL, 1980) may be provided by stress concentration of plate-tectonic forces during plate reorganizations (CLOETINGH and WORTEL, 1985) and are an order of magnitude smaller than the stresses required to rupture the lithosphere.

Existing weakness zones located within plates might be more suitable sites for initiation of subduction than passive margins. Spreading ridges (TURCOTTE *et al.*, 1977) and transform faults (UYEDA and BEN-AVRAHAM, 1972) have been suggested. A spreading ridge should be thought of as the lower bound (0 Ma) of a young passive margin, and forms a rupture zone inherent in the lithosphere. This view is consistent with the results of a survey of recently initiated subduction zones in the Pacific (KARIG, 1982) showing that zones initiated during the Neogene are frequently on the sites of transform faults or else are rejuvenated pre-existing subduction zones or geometrical adjustments (OKAL *et al.*, 1986; KROENKE and WALKER, 1986). However, it is not clear how the major subduction zones that presently ring the Pacific have been initiated. For example, it is widely held that the present Pacific coastline of Northern and Southern America has been the site of semi-continuous eastward subduction since the Late Proterozoic (WINDLEY, 1977).

The different plate configurations and thermal regimes at that time may have provided conditions more suitable for initiation of subduction.

Thus, in general, we do not expect initiation of oceanic lithosphere subduction at passive margins to play a leading role in the plate reorganizations such as documented by RONA and RICHARDSON (1978). In an oceanic plate attached to a subduction zone the pull acting on the subducting slab can be concentrated to a sufficient stress level to induce the formation of new spreading centers (WORTEL and CLOETINGH, 1981; 1983). Therefore, we conjecture that plate reorganizations occur primarily through the formation of new spreading ridges, since stress relaxation in the lithosphere occurs much more easily via this process than through the formation of new subduction zones. During the former process new subduction zones might be subsequently created at the sites of already present transform faults, when the new spreading direction has a component perpendicular to the direction of the transform fault.

Better understanding the mechanics of the formation of new plate boundaries in general, and the processes underlying the transition of passive margins into active margins in particular, will enhance our insight in the evolution of sedimentary basins. We have shown (CLOETINGH *et al.*, 1985; CLOETINGH, 1986, 1988) that the associated reorganizations of lithospheric stress fields are recorded in the stratigraphic record of sedimentary basins. Specific short-term fluctuations (time scales of a few Ma and longer) in apparent sea levels inferred from passive margins and intracratonic basins can now be associated quantitatively with particular plate-tectonic reorganizations. Conversely, the seismostratigraphic record of sedimentary basins might provide a new source of information on paleo-stress fields (CLOETINGH, 1986; LAMBECK *et al.*, 1987) and, hence, on the dynamics of plate reorganizations. Furthermore, the opening and closure of oceanic basins, with the associated changes in the area/age distribution of the ocean floor during the Wilson cycle, has been shown (HELLER and ANGEVINE, 1985) to be the main controlling feature on the occurrence of long-term (time scales of tens of Ma) sea level cycles. Further work on modelling of paleo-stress fields along the lines set out by WORTEL and CLOETINGH (1981) is required to more fully document differences in the roles of initiation of rifting and the initiation of subduction in the tectonic evolution of the plates' interiors.

Conclusions

Aging of passive margins will not in itself lead to a spontaneous initiation of subduction. In general, the formation of new subduction zones at passive margins requires a focusing of external plate-tectonic forces. The action of these external forces will be most effective when young passive margins are prestressed by thick sedimentary wedges. Conditions likely to exist in very young oceanic lithosphere are quite optimal for the development of new subduction zones, which might explain the lack of preservation of back-arc basins and marginal seas. It is not clear how major subduction zones, such as those presently ringing the Pacific, form. Probably, plate reorganizations primarily take place through the formation of new spreading ridges, because stress relaxation in the lithosphere occurs much more effectively via this process than through the formation of new subduction zones.

Acknowledgments

Partial support for this work was provided by NATO grant 0148/87. Comments and useful suggestions by an anonymous reviewer are appreciated.

References

ALLEN, P., HOMEWOOD, P., and WILLIAMS, G. (1986), *Foreland basins: an introduction*, Special Publication International Association of Sedimentologists *8*, 8–15.

BERGMAN, E. A. and SOLOMON, S. C. (1985), *Earthquake source mechanisms from bodywave inversion and intraplate tectonics in the northern Indian Ocean*, Physics Earth and Planetary Interiors *40*, 1–23.

BODINE, J. H., STECKLER, M. S., and WATTS, A. B. (1981), *Observations of flexure and the rheology of the oceanic lithosphere*, Journal of Geophysical Research *86*, 3695–3707.

BYERLEE, J. D. (1968), *Brittle-ductile transition in rocks*, Journal of Geophysical Research *73*, 4741–4750.

BYERLEE, J. D. (1978), *Friction of rocks*, Pure and Applied Geophysics *116*, 615–626.

CHAPPLE, W. M. and TULLIS, T. E. (1977), *Evaluation of the forces that drive the plates*, Journal of Geophysical Research *82*, 1967–1984.

CHURCH, W. R. and STEVENS, R. K. (1971), *Early Paleozoic ophiolite complexes of the Newfoundland Appalachians as mantle-oceanic crust sequences*, Journal of Geophysical Research *76*, 1460–1466.

CLOETINGH, S., *Evolution of passive continental margins and initiation of subduction zones*, Ph.D. Thesis, University of Utrecht, p. 111, 1982.

CLOETINGH, S. (1986), *Intraplate stresses: a new tectonic mechanism for fluctuations of relative sealevel*, Geology *14*, 617–620.

CLOETINGH, S., *Intraplate stresses: a new element in basin analysis*, In K. L. Kleinspehn and C. Paola (eds), *New Perspectives in Basin Analysis* (Springer-Verlag, New York, 1988) pp. 205–230.

CLOETINGH, S., MCQUEEN, H., and LAMBECK, K. (1985), *On a tectonic mechanism for regional sealevel variations*, Earth and Planetary Science Letters *75*, 157–166.

CLOETINGH, S. and NIEUWLAND, F. (1984), *On the mechanics of lithospheric stretching and doming: a finite element analysis*, Geologie en Mijnbouw *63*, 315–322.

CLOETINGH, S. and WORTEL, R. (1985), *Regional stress field of the Indian plate*, Geophysical Research Letters *12*, 77–80.

CLOETINGH, S. and WORTEL, R. (1986), *Stress in the Indo-Australian plate*, Tectonophysics *132*, 49–67.

CLOETINGH, S. A. P. L., WORTEL, M. J. R., and VLAAR, N. J. (1982), *Evolution of passive continental margins and initiation of subduction zones*, Nature *297*, 139–142.

CLOETINGH, S. A. P. L., WORTEL, M. J. R., and VLAAR, N. J. (1983), *State of stress at passive margins and initiation of subduction zones*, American Association of Petroleum Geologists Memoir *34*, 717–723.

CLOETINGH, S. A. P. L., WORTEL, M. J. R., and VLAAR, N. J. (1984), *Passive margin evolution, initiation of subduction and the Wilson cycle*, Tectonophysics *109*, 147–163.

COHEN, C. R. (1982), *Model for a passive to active continental margin transition: implications for hydrocarbon exploration*, American Association of Petroleum Geologists Bulletin *66*, 708–718.

DAVIES, G. F. (1983), *Subduction zone stresses: constraints from mechanics and from topographic and geoid anomalies*, Tectonophysics *99*, 85–98.

DEWEY, J. F. (1969), *Continental margins: a model for conversion of Atlantic-type to Andean type*, Earth and Planetary Science Letters *6*, 189–197.

DEWEY, J. F. (1976), *Ophiolite obduction*, Tectonophysics *31*, 93–120.

DICKINSON, W. R. and SEELY, D. R. (1979), *Structure and stratigraphy of fore-arc regions*, American Association of Petroleum Geologists Bulletin *63*, 2–31.

DUNCAN, R. A. and MCDOUGALL, I., *Time-space relationships for Cainozoic intraplate volcanism in eastern Australia, the Tasman Sea, and New Zealand*, In R. W. Johnson and S. R. Taylor (eds), *Australian Academy of Sciences*, in press.

ENGLAND, P. and WORTEL, R. (1980), *Some consequences of the subduction of young slabs*, Earth and Planetary Science Letters *47*, 403–415.

FLINN, E. A. (1982), *The International Lithosphere Program*, EOS Transactions American Geophysical Union *63*, 209–210.

FORSYTH, D. W. and UYEDA, S. (1975), *On the relative importance of the driving forces of plate motion*, Geophysical Journal of the Royal Astronomical Society *43*, 163–200.

GELLER, C. A., WEISSEL, J. K., and ANDERSON, R. N. (1983), *Heat transfer and intraplate deformation in the central Indian Ocean*, Journal of Geophysical Research *88*, 1018–1032.

GOETZE, C. (1978), *The mechanisms of creep in olivine*, Philosophical Transactions Royal Society of London, Series A 288, 99–119.

GOETZE, C. and EVANS, B. (1979), *Stress and temperature in the bending lithosphere as constrained by experimental rock mechanics*, Geophysical Journal of the Royal Astronomical Society *59*, 463–478.

HELLER, P. L. and ANGEVINE, C. L. (1985), *Sea-level cycles during the growth of Atlantic-type oceans*, Earth and Planetary Science Letters *75*, 417–426.

HOUSEMAN, G. A. and ENGLAND, P. C. (1986), *A dynamical model of lithosphere extension and sedimentary basin formation*, Journal of Geophysical Research *91*, 719–729.

HYNES, A. (1982), *Stability of the oceanic tectosphere—a model for early Proterozoic intracratonic orogeny*, Earth and Planetary Science Letters *61*, 333–345.

KARIG, D. E. (1982), *Initiation of subduction zones: implications for arc evolution and ophiolite development*, Geological Society of London Special Publication *10*, 563–576.

KARNER, G. D. and WATTS, A. B. (1982), *On isostasy at Atlantic-type continental margins*, Journal of Geophysical Research *87*, 2923–2948.

KIRBY, S. H. (1980), *Tectonic stresses in the lithosphere: constraints provided by the experimental deformation of rocks*, Journal of Geophysical Research *85*, 6353–6363.

KOBAYASHI, K. and SACKS, I. S. (1985), *Preface to structures and processes in subduction zones*, Tectonophysics *112*, p. 561.

KROENKE, L. W. and WALKER, D. A. (1986), *Evidence for the formation of a new trench in the western Pacific*, EOS Transactions American Geophysical Union *67*, 145–146.

LAMBECK, K., CLOETINGH, S., and MCQUEEN, H. (1987), *Intraplate stresses and apparent changes in sea level: the basins of northwestern Europe*, In C. Beaumont and A. J. Tankard (eds), *Sedimentary Basins and Basin-forming Mechanisms*, Canadian Society of Petroleum Geologists Memoir *12*, 259–268.

MAMMERICKS, J. and SANDWELL, D. (1986), *Rifting of old oceanic lithosphere*, Journal of Geophysical Research *91*, 1975–1988.

MCADOO, D. C., MARTIN, C. F., and POULOUSE, S. (1985), *Seasat observation of flexure: evidence for a strong lithosphere*, Tectonophysics *116*, 209–222.

MCADOO, D. C. and SANDWELL, D. T. (1985), *Folding of oceanic lithosphere*, Journal of Geophysical Research *90*, 8563–8568.

MCELHINNY, M. W. and VALENCIO, D. A. (1981), *Paleoreconstruction of the continents*, Geodynamics Series American Geophysical Union *2*, 1–194.

MCKENZIE, D. P. (1977), *The initiation of trenches: a finite amplitude instability*, Maurice Ewing Series American Geophysical Union *1*, 57–62.

MCWILLIAMS, M. O., *Palaeomagnetism and Precambrian tectonic evolution of Gondwana*, In A. Kröner (ed.), *Precambrian Plate Tectonics* (Elsevier, Amsterdam, 1981) pp. 649–687.

MITCHELL, A. H. G. (1984), *Initiation of subduction of post-collision foreland thrusting and back-thrusting*, Journal of Geodynamics *1*, 103–120.

NICOLAS, A. and LEPICHON, X. (1980), *Thrusting of young lithosphere in subduction zones with special reference to structures in ophiolite peridotites*, Earth and Planetary Science Letters *46*, 397–406.

OKAL, E. A., WOODS, D. F., and LAY, T. (1986), *Intraplate deformation in the Samoa-Gilbert-Ralik area: a prelude to a change of plate boundaries in the southwest Pacific?* Tectonophysics *132*, 69–77.

OXBURGH, E. R. and PARMENTIER, E. M. (1977), *Compositional and density stratification in oceanic lithosphere—causes and consequences*, Journal of the Geological Society of London *133*, 343–355.

RICHARDSON, R. M., SOLOMON, S. C., and SLEEP, N. H. (1979), *Tectonic stress in the plates*, Reviews of Geophysics and Space Physics *17*, 981–1019.

RONA, P. A. and RICHARDSON, E. S. (1978), *Early Cenozoic global plate reorganization*, Earth and Planetary Science Letters *40*, 1–11.

SHUDOFSKY, G., CLOETINGH, S., STEIN, S., and WORTEL, R. (1987), *Unusually deep earthquakes in East Africa: constraints on the thermo-mechanical structure of a continental rift system*, Geophysical Research Letters *14*, 741–744.

SOUTHAM, J. R. and HAY, W. W., *Global sedimentary mass balance and sealevel changes*, In C. Emiliani (editor), *The Oceanic Lithosphere*, *The Sea*, vol. 7 (Wiley, New York, 1981) pp. 1617–1684.

SPRAY, J. G. (1984), *Possible causes and consequences of upper mantle decoupling and ophiolite displacement*, Geological Society of London Special Publication *13*, 255–268.

STECKLER, M. S. and TEN BRINK, U. S. (1986), *Lithospheric strength variations as a control on new plate boundaries: examples from the northern Red Sea region*, Earth and Planetary Science Letters *79*, 120–132.

STEIN, C. A. and COCHRAN, J. R. (1985), *The transition between the Sheba ridge and Owen Basin: rifting of old oceanic lithosphere*, Geophysical Journal of the Royal Astronomical Society *81*, 47–74.

STEIN, S., CLOETINGH, S., WIENS, D. A., and WORTEL, R. (1987), *Why does near ridge extensional seismicity occur primarily in the Indian Ocean*, Earth and Planetary Science Letters *82*, 107–113.

STOCKMAL, G. S., BEAUMONT, C., and BOUTILIER, R. (1986), *Geodynamic models of convergent margin tectonics: transition from rifted margin to overthrust belt and consequences for foreland-basin development*, American Association of Petroleum Geologists Bulletin *70*, 181–190.

TAYLOR, B. and KARNER, G. D. (1983), *Evolution of marginal basins*, Reviews of Geophysics and Space Physics *21*, 1727–1741.

TRUMPY, R., *Alpine paleogeography: a reappraisal*, In K. Hsu (editor), *Mountain Building Processes* (Academic Press, London, 1982) pp. 149–156.

TURCOTTE, D. L. and AHERN, J. L. (1977), *On the thermal and subsidence history of sedimentary basins*, Journal of Geophysical Research *82*, 3762–3766.

TURCOTTE, D. L., HAXBY, W. F., and OCKENDON, J. R. (1977), *Lithospheric Instabilities*, Maurice Ewing, Series (American Geophysical Union) *1*, 63–69.

TURCOTTE, D. L. and SCHUBERT, G., *Geodynamics: Applications of Continuum Physics to Geological Problems* (Wiley, New York, 1982) 450 pp.

UYEDA, S. and BEN-AVRAHAM, Z. (1972), *Origin and development of the Philippine Sea*, Nature *240*, 176–178.

VINK, G. E., MORGAN, W. J., and ZHAO, W.-L. (1984), *Preferential rifting of continents: a source of displaced terranes*, Journal of Geophysical Research *89*, 10072–10076.

VLAAR, N. J. (1983), *Thermal anomalies and magmatism due to lithospheric doubling and shifting*, Earth and Planetary Science Letters *65*, 322–330.

VLAAR, N. J. and CLOETINGH, S. A. P. L. (1984), *Orogeny and ophiolites: plate tectonics revisited with reference to the Alps*, Geologie en Mijnbouw *63*, 159–164.

VLAAR, N. J. and WORTEL, M. J. R. (1976), *Lithospheric aging instability and subduction*, Tectonophysics *32*, 331–351.

WEISSEL, J. K., ANDERSON, R. N., and GELLER, C. A. (1980), *Deformation of the Indo-Australian plate*, Nature *287*, 284–291.

WIENS, D. A., DE METS, C., GORDON, R. G., STEIN, S., ANGUS, D., ENGELN, J. F., LUNDGREN, P., QUIBLE, D., STEIN, C., WEINSTEN, S., and WOODS, D. F. (1985), *A diffuse plate boundary model for Indian ocean tectonics*, Geophysical Research Letters *12*, 429–432.

WIENS, D. A. and STEIN, S. (1983), *Age dependence of oceanic intraplate seismicity and implications for lithospheric evolution*, Journal of Geophysical Research *88*, 6455–6468.

WIENS, D.A., STEIN, S., DE METS, C., GORDON, R., and STEIN, C. (1986), *Plate tectonic models for Indian Ocean "intraplate" deformation*, Tectonophysics *132*, 37–48.

WILLIAMS, H., *Structural telescoping across the Apalachian orogen and the minimum width of the Iapetus Ocean*, Geological Association of Canada Special Paper *20*, 421–440, 1980.

WILSON, J. T. (1966), *Did the Atlantic close and then re-open?* Nature *211*, 676–681.

WINDLEY, B. F., *The Evolving Continents* (Wiley, Chichester, 1977) 385 pp.

WINTERER, E. L. and BOSELLINI, A. (1981), *Subsidence and sedimentation of Jurassic passive continental margin Southern Alps, Italy*, American Association of Petroleum Geologists Bulletin *65*, 394–421.

WORTEL, R., *Age-dependent subduction of oceanic lithosphere*, Ph.D. Thesis, University of Utrecht, p. 147, 1980.

WORTEL, R. and CLOETINGH, S. (1981), *On the origin of the Cocos-Nazca spreading center*, Geology *9*, 425–430.

WORTEL, R. and CLOETINGH, S. (1983), *A mechanism for fragmentation of oceanic plates*, American Association of Petroleum Geologists Memoir *34*, 793–801.

WORTEL, R. and CLOETINGH, S. (1985), *Accretion and lateral variations in tectonic structure along the Peru-Chile Trench*, Tectonophysics *112*, 443–462.

ZUBER, M. T. (1987), *Compression of oceanic lithosphere: a model for intraplate deformation in the central Indian Ocean*, Journal of Geophysical Research *92*, 4817–4825.

ZWART, H. J. and DORNSIEPEN, U. F. (1978), *The tectonic framework of central and western Europe*, Geologie en Mijnbouw *57*, 627–654.

(Received February 2, 1988, revised February 26, 1988, accepted March 30, 1988)

PAGEOPH, Vol. 129, Nos. 1/2 (1989) 0033–4553/89/020027–14$1.50 + 0.20/0

Seismotectonics at the Trench-Trench-Trench Triple Junction off Central Honshu

TETSUZO SENO[1] and TAKASHI TAKANO[2]

Abstract—We study earthquakes in and near the TTT type triple junction off Boso peninsula, central Honshu, to elucidate the plate interaction in this area. The Pacific, North America (northeast Japan) and Philippine Sea plates meet at the junction of the Japan and Izu-Bonin Trenches, and the Sagami Trough. We determine focal mechanisms using WWSSN data. We also determine accurate focal depths by modeling body-waves. There is no serious trade-off between focal depth and source time function for the events treated in this study.

The earthquake mechanisms and their focal depths show two major modes of deformation of the Pacific slab at the junction. One mode is represented by nearly vertical normal faults with strikes perpendicular to the Bonin Trench. This mode of faulting is dominant in regions south of the junction and characteristically the southwest block is downthrown. The other mode is represented by nearly vertical normal faults that strike parallel to the Japan Trench and indicate the northwest block is downthrown. This latter mode is dominant in regions north of the junction. The former mode may represent the accommodation of the slab geometry to the change in dip angle between the northeast Japan and Izu-Bonin arcs; the Izu-Bonin slab has a larger dip than that of the northeast Japan slab. The latter mode shows that normal faults parallel to the trench strike, usually seen in trench axis-outer rise regions, continue to occur further landward of the trench axis in the area just north of the junction. This might be caused by the loading of the Philippine Sea slab which penetrates between northeast Japan and the Pacific slab north of the Sagami Trough.

Further north of these normal faults north of the junction, we find earthquakes which represent the relative motion between the Pacific and North American plates. This means that the Philippine Sea slab does not exist there. With the aid of earthquakes which represent the Philippine Sea-Pacific and Philippine Sea-North America motions located northwest of the normal faults, we can depict a possible area where the Philippine Sea slab exists north of the Sagami Trough.

Key words: TTT type triple junction, Sagami Trough, Pacific slab, Philippine Sea slab.

1. Introduction

The Sagami Trough, a westward extension of the Nankai Trough, meets with the Japan—Izu-Bonin Trenches at about 34°N, 142°E, 200 km west of central

[1] International Institute of Seismology and Earthquake Engineering, Building Res. Inst., Tsukuba, Ibaraki-ken 305, Japan.

[2] Department of Earth Sciences, Ibaraki University, Mito, Ibaraki-ken, Japan.

Honshu (Figure 1). The Philippine Sea plate (PHS hereafter) subducts beneath southwest Japan along the Nankai Trough and beneath northeast Japan (NEJ hereafter) along the Sagami Trough (e.g., NAKAMURA *et al.*, 1984; SENO, 1977). The relative plate motions around Japan (SENO *et al.*, 1987) are shown in Figure 1. The motion vectors are computed with the hypothesis that Japan is divided into

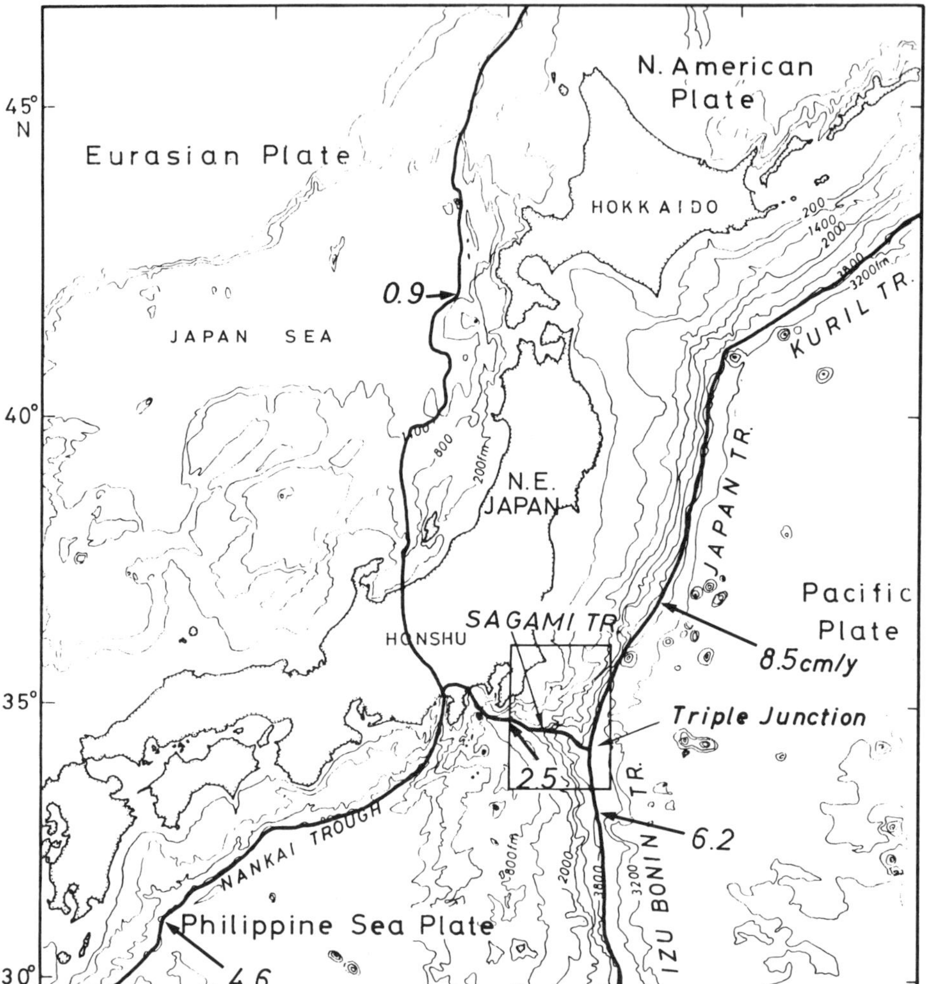

Figure 1

Plate boundaries and relative plate velocities around Japan. Northeast Japan is assumed as part of the North American plate (NAKAMURA, 1983; KOBAYASHI, 1983). Relative plate motions are calculated from SENO *et al.* (1987). Bathymetry is simplified from CHASE *et al.* (1971). Study area is indicated by the rectangle.

the Eurasian plate (southwest Japan) and the North American plate (NEJ) (KOBAYASHI, 1983; NAKAMURA, 1983). The Pacific plate (PAC) subducts beneath NEJ along the Japan Trench and beneath PHS along the Izu-Bonin Trench. The junction where the Sagami Trough meets the Japan—Izu-Bonin Trenches is thus a TTT (a) type triple junction (MCKENZIE and MORGAN, 1969).

This triple junction area has been surveyed extensively by the Japanese-French KAIKO project in 1984, including Seabeam, gravity, magnetism and seismic reflection. The Hydrographic Department, Maritime Safety Agency of Japan, also surveyed the area using Seabeam in 1984 (KATO *et al.*, 1985). The KAIKO project revealed the detailed morphology in the area (RENARD *et al.*, 1987; NAKAMURA *et al.*, 1987). The geometry of the boundary between the three plates is indicated by the broken line in Figure 2. There are two sedimentary basins, North Basin (NB) and South Basin (SB), in the inner lower trench slope near the junction (Figure 2). There are also bulges between these basins and the Izu-Bonin Trench, which are called North Nose (NN), Central High (CH) and South Nose (SN). The plate boundary between NEJ and PHS runs, from west to east, along the Boso canyon, through the northern edge of NB and near NN and CH, to where it connects with the Bonin Trench (SENO *et al.*, 1988).

In this paper, we obtain focal mechanisms and accurate focal depths of earthquakes to study the interaction between and deformation within the plates at the triple junction. Determination of accurate focal depths is essential to interpret focal mechanisms of earthquakes in terms of plate tectonics (e.g., STEIN and WIENS, 1986). This is especially true in this region because the interaction between the three plates is likely to be complicated. We determine accurate focal depths by matching waveforms of synthesized seismograms with those observed at teleseismic stations.

2. *Data and Method of Analysis*

We analyse earthquakes which occurred during the period from 1964 through 1981 in the area between 140.2°E and 142.2°E and 33.5°N and 36°N shown by the rectangle in Figure 1. We select earthquakes from the Bulletins of ISC (International Seismological Center) with magnitudes (m_b) 5.0 and larger to the south of 35°N and 5.3 and larger for the region north of 35°N. A smaller limit of the magnitude for the southern area is chosen because levels of seismicity in the area are lower than to the north. We also added to the above data set a large ($M_s = 6.7$) earthquake on September 18, 1984 at the triple junction and two events listed in the Harvard Centroid Moment Tensor Catalogue of 1982.

We determine focal mechanisms for these earthquakes using P wave first motions and S wave polarization angles from WWSSN (Worldwide Standardized Seismograph Network) records. Three mechanism solutions in the southernmost part of the area (B5, B9 and B13 in Figure 2) are from a companion paper which studies regions further south (MORIYAMA *et al.*, this volume). In total, we determined twenty-one new focal mechanisms for the area. Figure 3 shows the equal area plots of the *P*-wave first motions and *S*-wave polarization angles in the lower

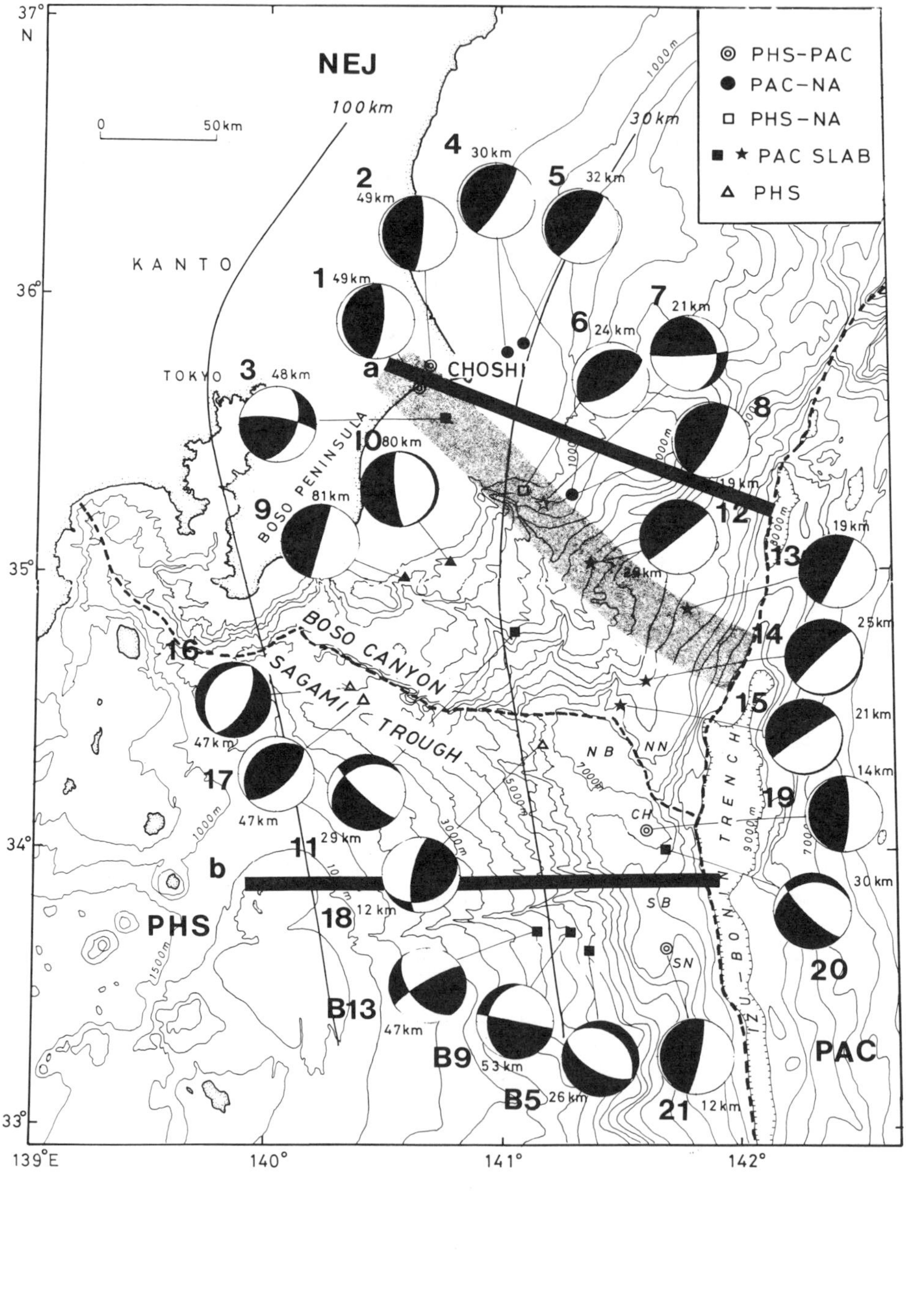

hemisphere. Nodal planes are also shown in these plots; they are well constrained by the data. Table 1 lists the parameters of the focal mechanism solutions.

We next determined accurate focal depths for the earthquakes with known mechanisms. We synthesized a long-period P-wave seismogram with various focal depths and found the focal depth at which the synthetic waveform best fits the observed one. Synthetics are generated with the algorithm of KROEGER and GELLER (1986), which incorporates all reflections and conversions from the shallow structure. We assumed a point source; calculated rays are convolved with the trapezoidal source time function, instrumental response and attenuation factor ($T_p^* = 1$ sec). We used the crustal structure shown by seismic refraction studies across the Japan trench inner slope along 39°45′N (MURAUCHI and LUDWIG, 1980). The crustal structure is varied according to the distance of the epicenter from the trench axis. We used the water depth measured directly at the epicenter in the bathymetric map (Figure 2).

Figure 4 shows some examples of the comparison between the synthesized and observed seismograms. The observed seismogram is indicated by the dotted line. We can constrain the source time function from the first half cycle of the waveform. No serious trade-off between the source depth and source time function occurs for small-size earthquakes as treated in this study (see STEIN and WIENS, 1986; SENO and GONZALEZ, 1987). The best-fit depth can be determined by matching the later phases, in particular, the second half cycle of the waveform. The best-fit depth is indicated by the arrows in Figure 4. We averaged the best fit focal depths at several stations for each event. Table 2 lists the focal depth, number of stations used, standard deviation and source time function. We also list the depths reported in the Bulletins of ISC for comparison.

3. *Interpretation and Discussion*

Figure 2 presents the focal mechanisms and the bathymetry of the area. Different symbols at epicenters indicate different tectonic interpretations of these

Figure 2

Fault plane solutions (lower hemisphere) plotted in the bathymetric map of the study area (HYDROGEOGR. DEPT., 1982; KATO *et al.*, 1985). Solid shading represents compressional quadrants. Three solutions, B5, B9 and B13 are from MORIYAMA *et al.* (1988). NB, North Basin; SB, South Basin; NN, North Nose; CH, Central High; SN, South Nose. Plate boundaries are indicated by the broken lines. The filled and double circles and open square at the epicenter denote interplate events interpreted to represent relative plate motions. The stars, filled squares, and filled triangles denote events which occurred within the Pacific plate. The open triangles denote events which probably occurred within the Philippine Sea plate. The estimated northern edge of the Philippine Sea slab is shaded. The thick lines indicate the projection lines for the cross-sections in Figure 5. The thin lines show the contours of the Pacific upper surface estimated from the regional seismicity map of YOSHII (1979).

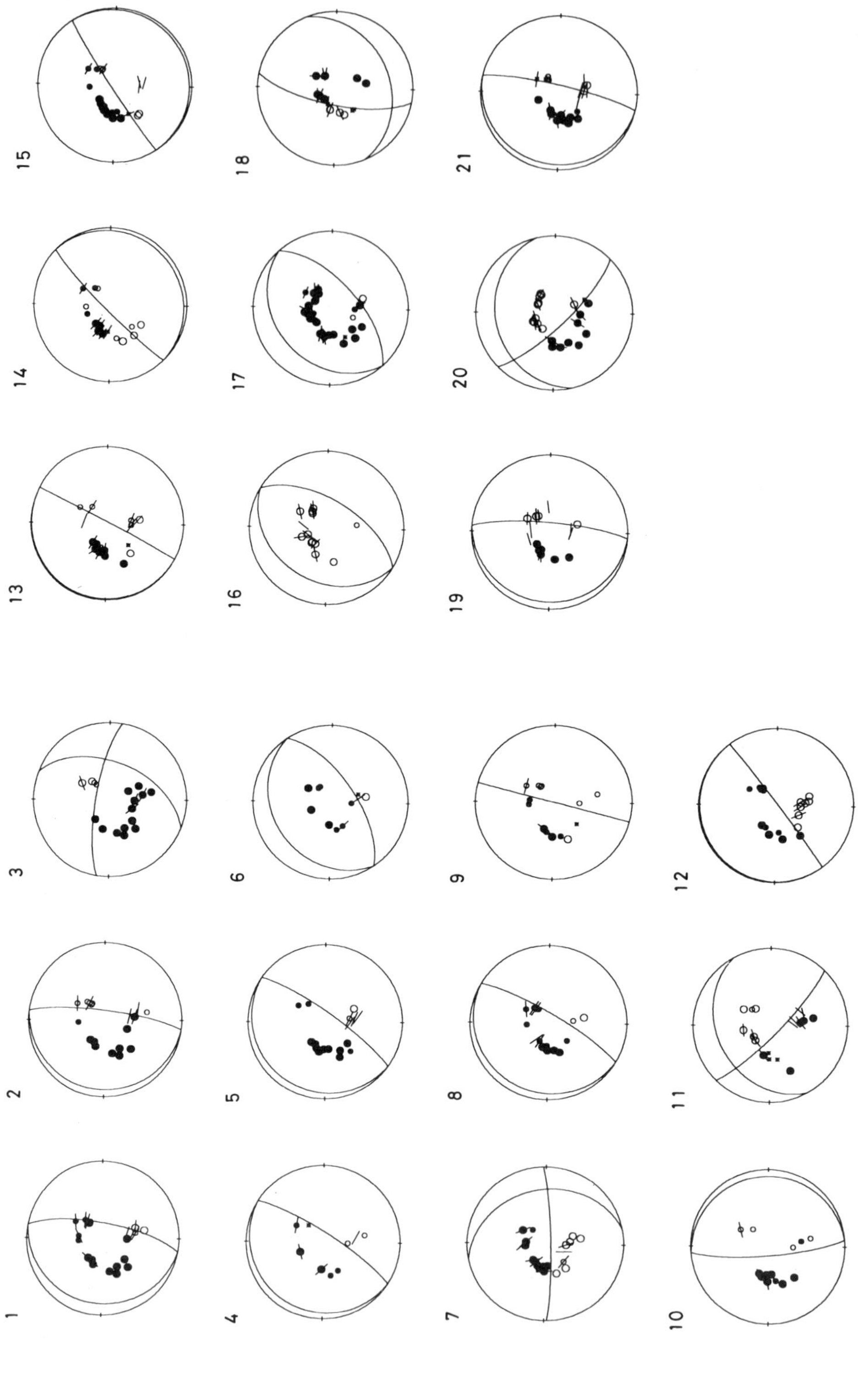
1
2
3
4
5
6
7
8
9
10
11
12
13
14
15
16
17
18
19
20
21

Table 1

Parameters of mechanism solutions of studied earthquakes.

No.	Date Y M D h m	Location Lat. (°N)	 Long. (°E)	m_b	Fault parameters plane A $\phi_{(°)}\delta$	 plane B $\phi_{(°)}\delta$	 T-axis $az_{(°)}pl$	 P-axis $az_{(°)}pl$
1	73 09 30 06 17	35.67	140.64	5.9	100 74	280 16	280 61	100 29
2	73 10 01 14 16	35.76	140.70	5.6	97 80	277 10	277 55	97 35
3	74 03 03 04 50	35.57	140.75	5.6	9 76	111 50	232 39	336 17
4	81 09 02 09 24	35.82	141.02	5.6	122 80	302 10	302 55	122 35
5	74 11 15 23 32	35.85	141.10	5.8	125 80	305 10	305 55	125 35
6	81 09 03 19 39	35.29	141.08	5.7	145 64	325 26	325 71	145 19
7	78 04 06 23 29	35.26	141.18	5.5	178 84	78 31	329 43	203 32
8	77 08 21 05 19	35.29	141.30	5.5	120 80	300 10	300 55	120 35
9	80 03 12 03 21	35.00	140.59	5.4	104 86	284 4	284 49	104 41
10	79 10 28 05 39	35.06	140.78	5.4	265 80	85 10	265 35	85 55
11	82 03 27 00 19	34.75	141.06	5.6	224 80	331 31	75 47	200 29
12	75 01 20 17 31	35.05	141.37	5.8	142 88	322 2	322 47	142 43
13	78 03 01 15 45	34.85	141.81	5.3	117 88	297 2	297 2	117 43
14	70 04 16 01 55	34.57	141.63	5.1	315 84	135 6	315 39	135 51
15	76 02 26 15 06	34.51	141.52	5.2	326 86	146 4	326 41	146 49
16	79 08 12 07 13	34.57	140.37	5.8	120 60	300 30	120 15	300 75
17	80 05 08 08 03	34.50	140.42	5.9	137 62	317 28	317 73	137 17
18	64 05 24 10 31	34.34	141.19	5.0	280 72	156 30	299 23	68 56
19	73 08 10 00 08	34.02	141.63	5.3	94 80	274 10	274 55	94 35
20	84 09 18 17 02	33.97	141.44	6.6	255 74	349 27	208 26	73 56
21	68 01 07 11 12	33.61	141.71	5.2	102 82	282 8	282 53	102 37

Epicentral locations and magnitudes are from ISC Bulletins except for No. 20 for which PDE data are listed. ϕ and δ denote dip direction and dip angle respectively.

events described below. Figures 5a and b show foci and modes of deformation in cross-sections perpendicular to the trench axis in the northern and southern areas. Note the trend of the trench axis is slightly different between the areas; in the north it is NNE-SSW and in the south almost N-S. We divide the focal mechanisms into those representing relative plate motions and those representing intraplate deformation.

◄

Figure 3

First motions of P-waves and S-wave polarization angles and nodal plane solutions in the equal area plot of the lower focal hemisphere. Number at the upper left is the event number in Table 1. The open and closed circles are downward and upward first motions of the vertical component of P-wave respectively. Large size symbols are from long-period records and small ones are from short-period records. Arrows indicate S-wave polarization angles. Stars indicate nodal first motions of P-wave.

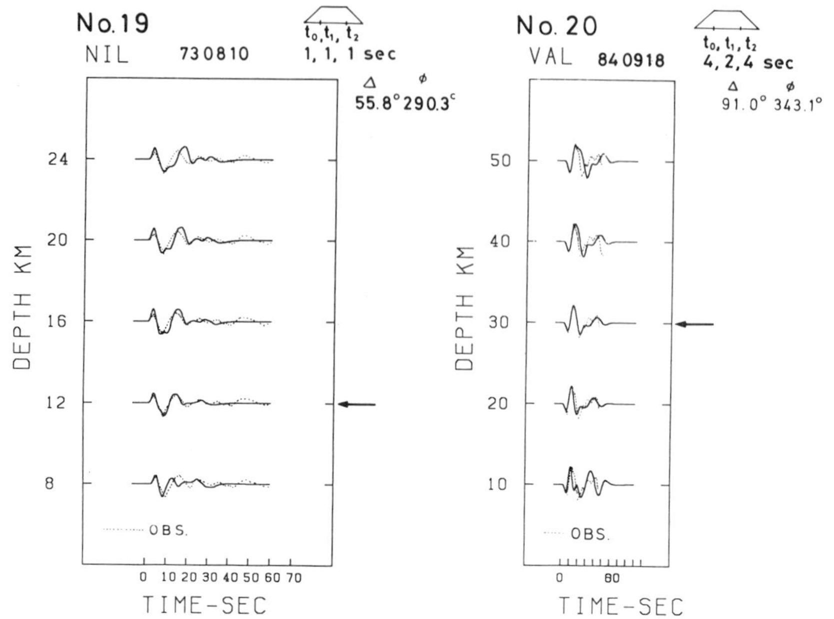

Figure 4
Comparison of synthesized seismograms at various depths with the observed seismogram. Examples at NIL station for Event 19 and at VAL station for Event 20 are shown. Arrows at the right indicate the best fit depth. t_0, t_1 and t_2 describe the source time function used for the calculation. Δ and ϕ denotes the epicentral distance and azimuth of the station.

3.1. Interplate Events

Thrust type mechanisms often represent differential motions between plates. In Figure 6, we compare slip vectors of the thrust type earthquakes except event 17 with the PAC-North America (NA), PAC-PHS and PHS-NA relative motions (Seno *et al.*, 1987). Events 4, 5 and 8, denoted by the solid circles in Figures 2 and 5, have slip vectors close to the PAC-NA motion. On the other hand events 1, 2, 19 and 21, denoted by the double circles, have slip vectors close to the PAC-PHS motion. Event 6 has a slip vector close to the PHS-NA motion.

In the cross-section of the northern area (Figure 5a), we can connect the foci of these thrust type earthquakes smoothly to the trench (8 km depth). We interpret these earthquakes (except event 6) as representing the differential motions between PAC and other plates (NA and PHS) at the upper surface of PAC. In Figure 2, the two contour lines represent the upper surface of PAC at 30 and 100 km depth, as estimated from the regional seismicity map (Yoshii, 1979). The upper surface of PAC shown in Figure 5a is consistent with these contours. In the southern section

Table 2

Depth determination and source time function.

No.	Date	Number of stations	Depth (km)	S.D. (km)	ISC Depth routine (km)	ISC Depth pP-P (km)	Source time func. t_0	t_1 (sec)	t_2
1	73 09 30	8	49	3.9	56	51	1	1	1
2	73 10 01	8	49	1.7	55	49	1	1	1
3	74 03 03	10	48	6.8	49	49	1	1	1
4	81 09 02	2	30	0	51	42	1	1	1
5	74 11 15	8	32	4.0	44	39	1	1	1
6	81 09 03	5	24	3.5	37	43	1.5	1	1.5
7	78 04 06	10	21	2.4	35	29	1	1	1
8	77 08 21	5	19	3.7	21	25	1	1	1
9	80 03 12	5	81	2.0	81	84	1	1	1
10	79 10 28	5	80	0	81	79	1	1	1
11	82 03 27	6	29	2.2	22	22	1	0	1
12	75 01 20	9	26	4.4	6	—	1	1	1
13	78 03 01	8	19	3.0	25	46	1	1	1
14	70 04 16	6	25	3.2	30	43	1	1	1
15	76 02 26	7	21	3.8	38	36	1	0	1
16	79 08 12	6	47	5.2	56	61	1	0	1
17	80 05 08	7	47	7.6	57	75	.5	0	.5
18	64 05 24	6	12	2.7	17	36	.4	0	.4
19	73 08 10	8	14	2.3	33	39	1	1	1
20	84 09 18	8	30	7.6	33*	—	4	2	4
21	68 01 07	8	12	5.3	36	51	1	1	1

*PDE S.D. denotes standard deviation.

(Figure 5b), we estimate the upper surface of PAC as the foci of events 19 and 21 are located on the surface and as consistent with the contours in Figure 2.

Events 1 and 2, interpreted as representing PAC-PHS motions, suggest that the PHS slab exists beneath Choshi at 50 km depth. Further to the west, beneath Kanto, the PHS slab is shown down to 50–100 km depth by the hypocenters of microearthquakes (ISHIDA, 1986; NOGUCHI, 1986). In contrast, offshore to the east, outside the coverage of microearthquake networks, the geometry of the PHS slab has not been revealed. In the present study, we find one event (event 6) representing the PHS-NA motion 50 km southeast of Choshi. Northeast of events 1, 2 and 6, we find events 4, 5 and 8 whose slip vectors can be interpreted as showing the PAC-NA motion. This suggests to us that the leading edge of the PHS slab exists between the location of events 1, 2 and 6 and that of events 4, 5 and 8.

South of the junction we find two events, events 19 and 21, representing the PAC-PHS motions. No PAC-PHS event is found offshore north of the Sagami Trough.

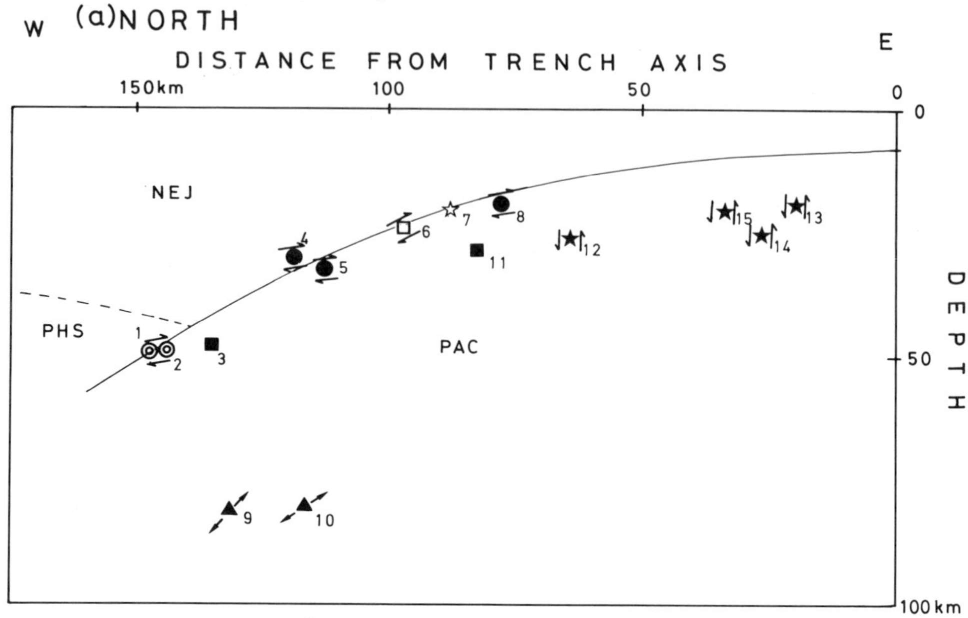

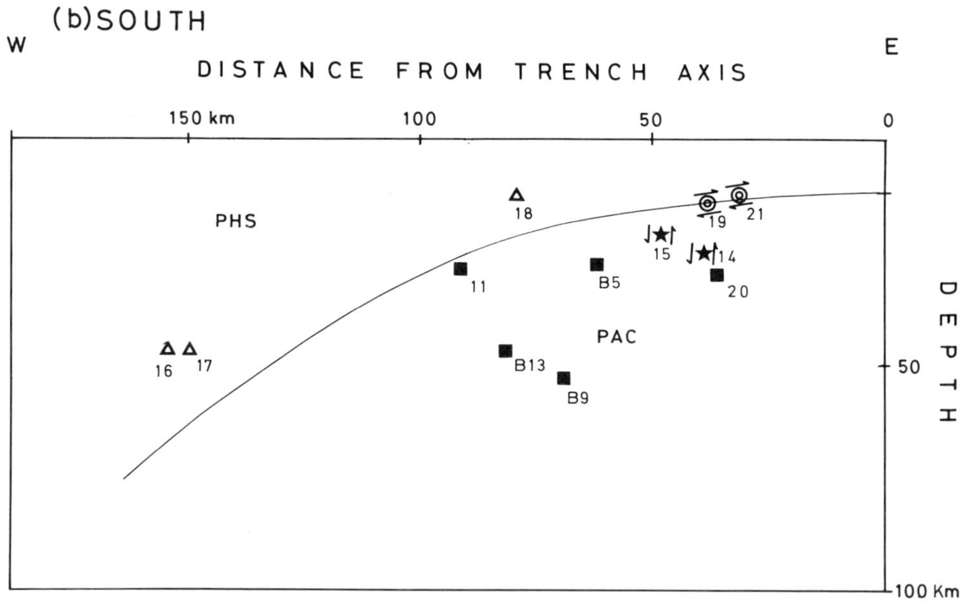

Figure 5

Vertical cross-section of the foci and mode of deformations of the studied earthquakes in the direction perpendicular to the trench axis; (a) in the north of 34.5°N and (b) in the south of 35°N along the lines shown in Figure 2. The thin solid line indicates the estimated upper surface of the Pacific slab. In (a), the Philippine Sea plate inferred at the line (a) in Figure 2 is indicated by the broken line.

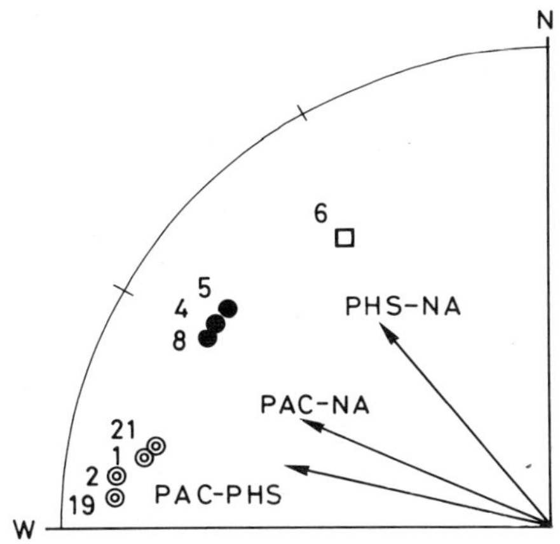

Figure 6
Comparison of the slip vectors of the thrust type events with the relative plate motions. The poles of the more vertical nodal planes are plotted in the equal area projection. The relative plate motions are calculated at 35°N, 141°E from SENO *et al.* (1987).

3.2. *Intraplate Events*

Other events except event 17 have focal mechanisms different from thrust type. Referring to the upper surface of PAC in Figure 5, Events 3, 9–15, B5, B9 and B13, and 20 are interpreted as events occurring within PAC. Events 16–18 are interpreted as events within PHS.

The events within PAC can be divided into three groups. The first group includes events 3, 11, 20 and B5–B13 (solid squares in Figures 2 and 5). The focal mechanisms of these events show nearly vertical nodal planes trending NW or WNW perpendicular to the arc trend. The sense of faulting is the SW block downthrown with respect to the NE block. This group of events is dominant, though not limited, to the south of the junction.

This group may represent the deformation of the PAC slab accommodating the larger dip angle of the slab beneath the Izu-Bonin arc than beneath NEJ shown by the intermediate and deep regional seismicity around Japan (YOSHII, 1979; UTSU, 1977). The PAC slab deforms to conform to the slab geometry at depth after subduction at the trench.

This type of deformation of the PAC slab is not restricted to the moderate or small size earthquakes. In 1953, a large earthquake ($M_s = 8.0$), called the 1953 Boso-Oki earthquake, occurred at the triple junction. ANDO (1971) showed that this event has the same focal mechanism and location as event 20 ($M_s = 6.7$). Furthermore the deformation is not restricted to the shallow part of the slab. In the deeper part of the slab beneath central Honshu, ITO and ANNAKA (1977) and ISACKS and MOLNAR (1971) obtained several similar mechanism solutions. This mode of deformation of the PAC slab is significant at all depths at the junction between the northeast Japan and Bonin arcs.

The second group is shown by events 12–15, denoted by the stars in Figures 2 and 5. These events show nearly vertical nodal planes trending NE, subparallel to the arc trend. The sense of faulting is the NW block downthrown with respect to the SE block. This group of events is dominant just north of the junction.

We interpret the deformation represented by the second group as reactivation of the normal faults seen in the outer trench slope-outer rise regions of the arc-trench system. Usually, outer-rise normal faults caused by bending of the oceanic plate prior to subduction are not seen landward of the trench axis (e.g., CHAPPLE and FORSYTH, 1979). We thus suggest a special reason for the development of the normal or nearly vertical faulting landward of the trench axis north of the triple junction area. We propose that the subduction of PHS at the junction beneath NEJ causes this reactivation of the normal faults. The PHS slab is forced into the interface between NEJ and PAC north of the junction, causing the PAC slab to bend below the PHS slab. If we accept this hypothesis, we can estimate the northern edge of the PHS slab, the shaded area in Figure 2, using the locations of the interplate PAC-PHS, PHS-NA and PAC-NA events and those of the second group of intraplate events.

The third group consists of events 9 and 10, denoted by the solid triangle in Figures 2 and 5. They have deep foci (80 km) and T axes parallel to the down-dip direction of the PAC slab; they would be down-dip tensional events often seen in subducting slabs (ISACKS and MOLNAR, 1971).

We interpret events 16–18 as intraplate deformation of PHS based on their locations with respect to the inferred PAC upper surface. Event 18 has a shallow focal depth (12 km) and a T axis dipping to the northeast. Thus this event may represent the bending of PHS prior to its subduction along the PHS-NEJ boundary near the North Basin.

The mechanisms of events 16 and 17 are difficult to understand. They have the same focal depths (47 km), but have the opposite mechanism solutions. This is further confirmed by the direct comparison of the P-waveforms; they coincide with each other even in the later phases if the polarity of the record of either of the events is reversed. Although event 17 has a slip vector subparallel to the PAC-NA motion, we cannot interpret this event as occurring at the PAC-NEJ interface, because PHS starts to subduct beneath NEJ just above the location of this event (Figure 2) and thus NEJ does not exist here above the PAC slab. Also we cannot interpret this representing the PAC-PHS motion because the slip vector is much deviated from that expected. Thus, although we believe this pair of events probably occurring within PHS, we leave them as caused by an unknown mechanism.

Event 7 is also difficult to interpret. It is located very close to the upper surface of PAC (Figure 5a). It has a nodal plane nearly vertical and perpendicular to the arc trend; however the sense of motion on this plane is opposite to that of the first group of intraplate events. The location of this event is on the estimated edge of the PHS slab (Figure 2). This event may be caused by complex interaction between the

PAC, NA and PHS plates at this edge. At present, we cannot provide any confident interpretation for this event.

4. Concluding Remarks

On the basis of the focal mechanisms and accurate focal depths of the earthquakes near the triple junction off central Honshu, we have attempted to elucidate the major modes of deformation within and the interactions between the plates in this TTT type triple junction area. The Pacific slab is deformed significantly near the junction. In the south of the junction, vertical faulting with the southwest block downthrown, with respect to the northeast block, is dominant. In the north of the junction, vertical faulting with the northwest block downthrown, with respect to the southeast block, is dominant. We interpret the former as representing the accommodation to the difference in the slab geometry at depth between the northeast Japan and the Izu-Bonin arcs. We interpret the latter as representing the downbending of the Pacific slab caused by the penetration of the Philippine Sea slab north of the Sagami Trough. Based on this interpretation, along with the locations of interplate earthquakes, we depict the possible northern edge of the Philippine Sea slab.

Interplate earthquakes between the Pacific and Northeast Japan plates occur north of the estimated edge of the Philippine Sea slab north of the Sagami Trough. Interplate earthquakes between the Pacific and Philippine Sea plates occur beneath Choshi and south of the triple junction. We find only one event representing the Philippine Sea-Northeast Japan motion offshore near the estimated edge of the Philippine Sea slab. To provide further understanding of the plate interaction and the slab geometry in this area, investigations including observations of seismicity and focal mechanisms using ocean bottom seismometers are necessary.

Acknowledgements

We thank the personnel of all the WWSSN stations who kindly sent us seismograms of the 1984 Boso-Oki earthquake. We are grateful to Seth Stein and two anonymous reviewers for critical reading of the manuscript. Discussions with the members of French-Japanese KAIKO project 1984, leg 2, and Katsuhiko Ishibashi and Shin'ichi Noguchi were also helpful.

REFERENCES

ANDO, T. (1971), *A fault model of the 1953 Boso-Oki earthquake*, Abstr. Seismol. Soc. Jpn No. 2, 49 (in Japanese).

CHAPPLE, W. M. and FORSYTH, D. W. (1979), *Earthquakes and bending of plates at trenches*, J. Geophys. Res. *84*, 6729–6749.

CHASE, T. E., MENARD, H. W., and MAMMERICKX, J. (1971), *Topography of the North Pacific*, Scripps Inst. Oceanography.

HYDROGEOGR. DEPT., Maritime Safety Agency of Japan (1982), Bathymetric Chart, Central Japan, no. 6313.

ISACKS, B. and MOLNAR, P. (1971), *Distribution of stresses in the descending lithosphere from a global survey of focal-mechanism solutions of mantle earthquakes*, Rev. Geophys. Space Phys. *9*, 103–174.

ISHIDA, M. (1986), *The configuration of the Philippine Sea and the Pacific plates as estimated from the high-resolution microearthquake hypocenters in the Kanto-Tokai district*, Japan. Rep. Nat. Res. Center Diast. Prev. *36*, 1–19 (in Japanese).

ITO, K. and ANNAKA, T. (1977), *Spatial clustering and focal mechanisms of the deep earthquakes in central Japan*, Zisin *30*, 201–212 (in Japanese).

KATO, S., NAGAI, T., TAMAKI, M., KONDO, T., TOMIYASU, Y., KATO, G., MUNEDA, K., and ASADA, A. (1985), *Submarine topography of the eastern Sagami Trough to the triple junction*, Rep. Hydrogr. Res. *20*, 1–24 (in Japanese).

KOBAYASHI, Y. (1983), *On the initiation of subduction of plates*, Earth Monthly *5*, 510–514 (in Japanese).

KROEGER, G. C. and GELLER, R. J. (1986), *An efficient method for synthesizing teleseismic body waves for sources in a vertically layered crust*, unpublished manuscript.

MCKENZIE, D. P. and MORGAN, W. J. (1969), *Evolution of triple junctions*, Nature *224*, 125–133.

MORIYAMA, T., TAJIMA, F., and T. SENO (1988), *An unusual zone of seismic coupling in the Bonin arc: the 1972 Hachijo-Oki earthquakes and related seismicity*, PAGEOPH, this issue.

MURAUCHI, S. and LUDWIG, W. J. (1980), *Crustal structure of the Japan Trench: The effect of subduction of ocean crust*, Init. Rep. Deep Sea Drilling Project, U.S. Government Printing Office, Washington, D.C. 56,57, 463–469.

NAKAMURA, K. (1983), *Possible nascent trench along the eastern Japan Sea as the convergent boundary between Eurasian and North American plates*, Bull. Earthq. Res. Inst. *58*, 711–722 (in Japanese).

NAKAMURA, K., SHIMAZAKI, K., and YONEKURA, N. (1984), *Subduction, bending and education. Present and Quaternary tectonics of the northern border of the Philippine Sea plate*, Bull. Soc. Geol. Fr. *26*, 221–243.

NAKAMURA, K., RENARD, V., ANGELIER, J., AZEMA, J., BOURGOIS, J., DEPLUS, C., FUJIOKA, K., HAMANO, Y., HUCHON, P., KINOSHITA, H., LABAUME, P., OGAWA, Y., SENO, T., TAKEUCHI, A., TANAHASHI, M., UCHIYAMA, A., and VIGNERESSE, J.-L. (1987), *Oblique and near collision subduction, Sagami and Suruga Troughs—Preliminary results of French-Japanese 1984 KAIKO cruise, Leg 2*, Earth Planet. Sci. Lett. *83*, 229–242.

NOGUCHI, S. (1986), *Geometry of the Philippine Sea plate and features of seismicity in Ibaraki Prefecture*, Earth Monthly *7*, 97–104 (in Japanese).

RENARD, V., NAKAMURA, K., ANGELIER, J., AZEMA, J., BOURGOIS, J., DEPLUS, C., FUJIOKA, K., HAMANO, Y., HUCHON, P., KINOSHITA, H., LABAUME, P., OGAWA, Y., SENO, T., TAKEUCHI, A., TANAHASHI, M., UCHIYAMA, A., and VIGNERESSE, J.- L. (1987), *Trench triple junction off central Japan—preliminary results of French-Japanese 1984 KAIKO cruise, Leg 2*, Earth Planet. Sci. Lett. *83*, 243–256.

SENO, T. (1977), *The instantaneous rotation vector of the Philippine Sea plate relative to the Eurasian plate*, Tectonophysics *42*, 209–226.

SENO, T. and GONZALEZ, D. G. (1987), *Faulting caused by earthquakes beneath the outer slope of the Japan Trench*, J. Phys. Earth *35*, 381–407.

SENO, T., MORIYAMA, T., STEIN, S., WOODS, D. F., DEMETS, C., ARGUS, D., and GORDON, R. (1987), *Redetermination of the Philippine Sea plate motion*, EOS, Trans. Amer. Geophys. Union *68*, 1474.

SENO, T., OGAWA, Y., TOKUYAMA, H., NISHIYAMA, E., and TAIRA, A. (1988), *Tectonic evolution of the triple junction off central Honshu for the past 1 m.y.*, Tectonophysics (in press).

STEIN, S. and D. A. WIENS (1986), *Depth determination for shallow teleseismic earthquakes: Methods and results*, Rev. Geophys. *24*, 806–832.

UTSU, T. (1977), *Seismology*, Kyoritsu Pub. Co., Tokyo, 286 pp. (in Japanese).

YOSHII, T. (1979), *Complication of geophysical data around the Japanese islands (I)*, Bull. Earthq. Res. Inst. *54*, 75–117 (in Japanese).

(Received September 26, 1986, revised January 29, 1988, accepted April 11, 1988)

PAGEOPH, Vol. 129, Nos. 1/2 (1989) 0033–4553/89/020041–29$1.50 + 0.20/0

Morphologic and Geologic Effects of the Subduction of Bathymetric Highs

WILLIAM R. MCCANN[1] and RAY E. HABERMANN[2]

Abstract—Morphologic and geologic observations suggest that subduction of bathymetric highs, such as aseismic ridges, chains of seamounts, and fracture zones, are important in the development of many forearc features and that those features form during relatively brief episodes of intense tectonism. A bathymetric high obliquely entering a subduction zone tends to compress sediments along its leading edge, resulting in arcward compression of the accretionary wedge. A landward deflection of the trench axis and a steepened inner wall result from this deformation. If a significant component of oblique slip occurs along the subduction zone, then along-strike movement of the accretionary wedge may also occur. Stresses resulting from subduction of bathymetric features with sufficient buoyancy or high relief extend farther landward than in the case of smaller, less buoyant features, inducing uplift of the leading edge of the overriding plate. Tectonic erosion of the base of the overriding plate and along-strike transport of arc material may also occur. The accelerated tectonism observed along several convergent margins can be attributed to the consumption of bathymetric irregularities on the seafloor rather than temporally abrupt changes in rates and directions of plate motions or other episodic events in the accretionary prism.

Key words: Subduction, tectonic erosion, uplift, aseismic ridges, subsidence, deformation.

Introduction

The forearc is that portion of the overriding plate between the outer rise and the volcanic front (SEELY, 1978). Forearc morphology depends on the age of the arc and seafloor being subducted, the rate and direction of plate convergence and the amount of sediment supplied to the subduction zone from either the overriding or downgoing plates (KARIG and SHARMAN, 1975; SEELY, 1978; CROSS and PILGER, 1982).

"Steady-state" subduction is achieved during the subduction of smooth, constant-age seafloor. The subduction of a bathymetric high represents a deviation from this state. Therefore, such events should be considered in examination of structures in forearcs.

VOGT (1973) and VOGT *et al.* (1976) discussed the response of subduction zones to the subduction of aseismic ridges. While their observations on the development of island arc cusps fits well for oceanic arcs with back-arc spreading, the effects of

---

[1] Department of Geology, University of Puerto Rico, Mayagüez, Puerto Rico 00708, U.S.A.
[2] NOAA/NGDC E/GC1 325 Broadway Boulder, CO 80303, U.S.A.

ridge subduction on continental margins or accretionary prisms remain obscure. In this paper we describe the effects that subduction of various types of bathymetric features have on crystalline rocks of overriding plates, and their accretionary prisms. Although along-strike transport of material is probably important as well (FORSYTHE and NELSON, 1985) we will not consider those effects here. In this paper we will concentrate on vertical tectonism. Bathymetric features cover a wide range of types including those slightly different from normal seafloor to large, nearly continental structures. Here, we concentrate on the subduction of ridges related to fracture zones and aseismic ridges generated at hot spots. They appear to modify the subduction process in an observable fashion without terminating it (VOGT *et al.*, 1976). Clearly, larger structures, such as oceanic plateaus, are best considered terranes that may later be accreted to continents (NUR and BEN-AVRAHAM, 1982; NUR, 1983). These accretion processes cause major changes in the character of the affected subduction zone.

Many authors note evidence of large-scale uplift or subsidence of forearcs (YONEKURA, 1983, Circum-Pacific; LUNDBERG, 1983, Costa Rica; VON HUENE *et al.*, 1980, Guatemala; KELLER, 1980, Japan; MORETTI and NGOKWEY, 1985; New Hebrides and Caribbean). These vertical motions are usually attributed to accretion (uplift) or tectonic erosion (subsidence) (HUSSONG *et al.*, 1976). Numerous models have been proposed to explain the cause of episodic accretion or erosion of material at the base of the accretionary prism or crystalline upper plate (MOORE *et al.*, 1979; KARIG and KAY, 1981; KARIG *et al.*, 1980; PAVLIS and BRUHN, 1983, HILDE, 1983). In this paper we propose a model which fits numerous observations and explains the episodic nature of these events without requiring rapid changes in convergence rates or directions. We recognize that episodic pulses of deformation (primarily collapse) of the accretionary prism may occur as a result of other causes. We suggest that the subduction of buoyant aseismic ridges is responsible for rapid erosion or intense deformation of accretionary prisms and the crystalline part of the overriding plates. Migration of the ridge-trench intersection with time produces a pulse deformation at the lower slope of the trench as well as uplift and later, subsidence of the inner wall. If tectonic erosion of the overriding plate has occurred, then the subsidence will be larger than the uplift earlier experienced as the ridge approached. This deformational pulse will migrate along the margin leaving a time transgressive record in the forearc sediments. The rate and direction of migration of such a deformation pulse may vary greatly from the direction and rate of relative plate motions, the governing factors being the strikes of the ridge, the trench and the direction of plate motion.

Deformation of Accretionary Prisms by the Consumption of Bathymetric Highs

In many subduction zones, sediments scraped off the downgoing plate or derived from the arc are deformed and thrust by imbricate thrust faults (Figure 1). Under

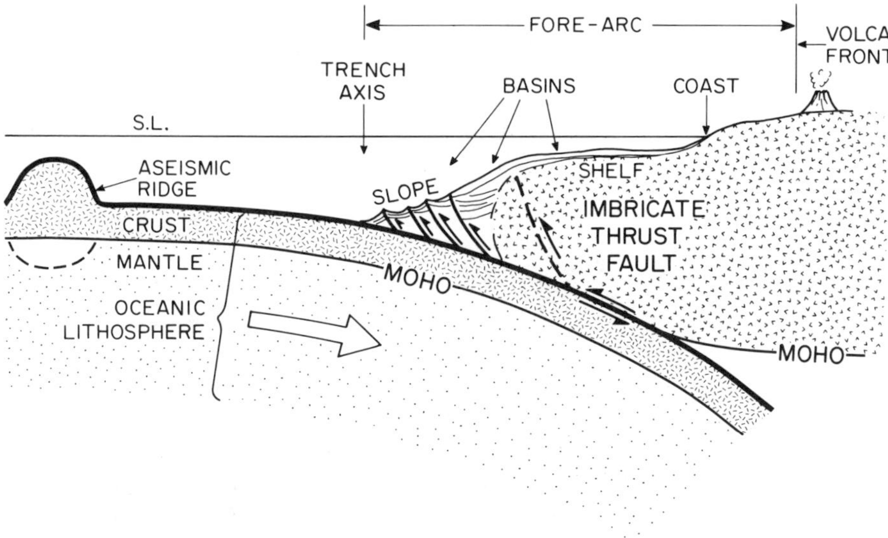

Figure 1
Main tectonic elements in a general subduction zone. Oceanic lithosphere descends beneath the upper plate, first disturbing sediments in slope region of forearc. Later, the downgoing plate meets the crystalline parts of the upper plate and, finally, oceanic plate enters the mantle beneath the crust of the upper plate. Aseismic ridge nearing the trench may or may not be compensated locally as shown by dashed line in mantle beneath it.

steady-state conditions (constant rate and direction of relative motion and subduction of relatively smooth, constant-depth seafloor) accretionary prisms tend toward a deformational equilibrium in which short-lived occurrences of major uplift, subsidence, compression or extension are observed. That is not to say, however, that accretion is or not occurring; the rate of accretion or erosion will depend upon the rate of sediment supply relative to the volume of sediment carried down the subduction zone (HILDE, 1983). Several accretionary prisms, however, do show periods of accelerated tectonism, which probably reflects nonsteady state conditions at present or in the past. It is those periods of accelerated tectonism in the accretionary wedge that we address first. Large, bathymetric features which appear to affect more landward structures on the thicker, stronger part of the overthrust plate will be discussed in later sections. A summary of our observations is listed in Tables 1 and 2 and some typical examples are discussed in the next sections.

Lesser Antilles

Normal seafloor entering the Puerto Rico and Northern Lesser Antillean Trenches is interrupted by a series of aseismic ridges on the North American plate (Figure 2). These topographic features lie close to the expected trend of fracture

Table 1

Bathymetric Ridges and Rises: Origin and Type

Feature	Location	Origin	Relief (km)	Compensation*	References
Amani Plateau	Ryukyu Is.	?	1.	Root	4, 19
Amlia Fracture Zone	Aleutian Is.	Fracture Zone	0.5–1.2	Flexure?	4
Carnegie Ridge	Ecuador	Hot spot	1.0–2.0	Root	11, 7
Cocos Ridge	Costa Rica	Hot spot	1.0–2.0	Root	4, 7
Daito Ridge	Ryukyu Is.	?	1.0	Root	4, 20
d'Entrecasteaux Fracture Zone	Vanuatu Is.	Subduction Zone?	1.6–2.8	?	11
Emperor Seamounts	Kamchatka	Hot spot	2.0	?	11, 16
Gulf of Alaska Seamounts	S. Alaska	Hot spot	1.0–1.5	?	4
Investigator Ridge	Sumatra	Fracture Zone	0.5	Flexure?	6
Iquique Ridge	N. Chile	Hot spot	0.5–1.0	Root?	11, 15, 17
Juan Fernandez Ridge	Chile	Hot spot?	1.0–2.0	?	11, 15, 14
Louisville Ridge	Tonga-Kermadec Is.	Fracture Zone	2.8–4.4	Flexure	11
Marcus Necker Ridge	Bonin Is.	?	2.0	Flexure?	4, 1
Magellean Seamounts	Mariana Is.	Mid Plate Volcanism?	1.0–2.0	Root, Flexure?	4, 1
Main-Barracuda Ridge	N.E. Caribbean	Fracture Zone	1.2	Flexure?	3, 12, 2
Mendaña Fracture Zone	Peru	Spreading Center	0.2–0.4	Root	18
Orozco Fracture Zone	Mexico	Fracture Zone	0.5?	Flexure?	4
Palau Kyushu Ridge	S. W. Japan	Remnant Arc	1.0	Root	4, 8, 9
Nazca Ridge	Peru	Hot spot	1.0–1.5	Root	11, 15, 14, 22
Roo Rise	Java	Continental Fragment?	1.0–1.5	Root	6, 13
Scarborough Seamounts	Philippines	?	2.0	?	21
Tehuantepec Ridge	S. Mexico	Fracture Zone	1.0	Flexure?	11
Tiburon Rise	Lesser Antilles	Fracture Zone	1.0	Flexure?	3, 12
Torres Rise	Torres Is.	?	3.2	?	11

References: 1, BODINE and WATTS (1979); 2, BIRCH (1970); 3, CASE and HOLCOMBE (1980); 4, CHASE *et al.* (1970); 5, CURRAY *et al.* (1977); 6, HAYES and EWING (1971); 7, HOLDEN and DIETZ (1972); 8, KARIG (1972), 9, LUDWIG *et al.* (1973); 10, MAILLET *et al.* (1983); 11, MAMMERICKX *et al.* (1973); 12, MCCANN and SYKES (1984); 13, NEWCOMB and MCCANN (1987); 14, PILGER and HANDSCHUMMACHER (1981); 15, PRINCE *et al.* (1980); 16, WATTS (1978); 17, CANDE (1983); 18, WARSI *et al.* (1983); 19, NISHIZAWA *et al.* (1983); 20, MURACHI *et al.* (1968); 21, HAYES and LEWIS (1984); 22, CUTLER (1977).

*Root indicates airy type compensation; Flexure indicates topography supported by elastic strength of lithosphere.

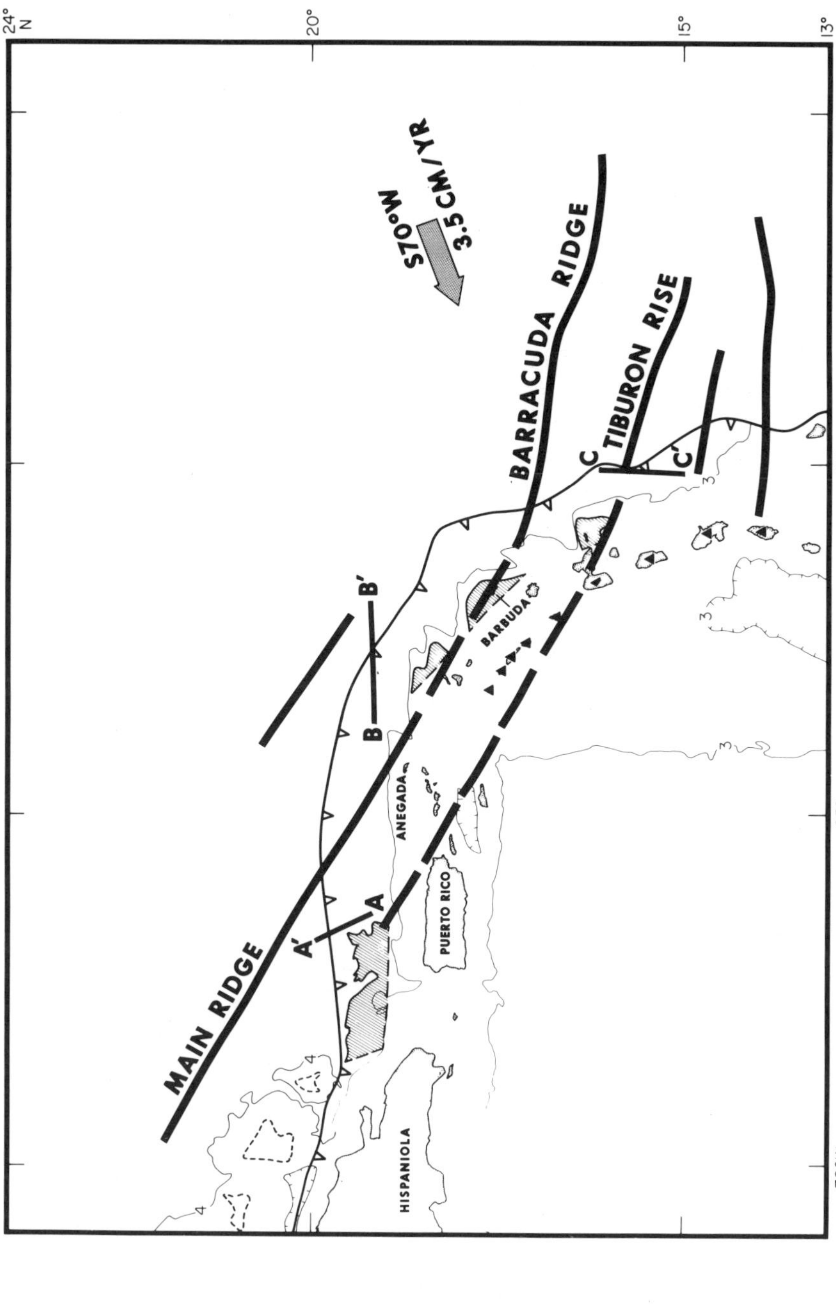

Figure 2

Tectonic framework of northeastern Caribbean. Aseismic ridges on the North American plate advance westerly toward the subduction zone. The northeastern corner of the Caribbean plate collides with ridges nearly broadside, unsubducted segments then advance along the westerly and southerly trending margins of the plate. Three seismic profiles are located at critical sections of the accretionary wedge; they elucidate the style of deformation observed before, during, and after the subduction of various ridges.

Table 2

Bathymetric Ridges and Rises: Effects on Subduction Zones

Feature	TA	IW	US	CO	References	Remarks
Amani Plateau	Sh, D	St, O	U?	U?	1, 19	Kiaki Island (forearc) uplifting
Amlia Fracture Zone	—	St, D	—	—	1, 11	Atka Basin truncated
Carnegie Ridge	Sh	St	St	U, T	14, 2	Trench slope break subaerial
Cocos Ridge	Sh, D?	St	St, U	U, T	1, 8, 2, 23	
Daito Ridge	Sh?	St?	St	U	25, 28	Upper Terrace absent
d'Entrecasteaux Fracture Zone	.A	St	St	U	1, 3, 24, 28	Santo Island occupies trench
Emperor Seamounts	Sh, D?	St	U?	U	1, 7	Volcanic chain offset, complex bathymetry on trench inner wall
Gulf of Alaska Seamounts	D	St?	—	—	1	May not have reached trench yet
Investigator Ridge	—	St, D	—	—	1, 17	Reactivated old faults in forearc
Iquique Ridge	—	—	—	—	5	May not have reached trench yet
Juan Fernandez Ridge	Sh	St	St	—	18	Complex bathymetry on outer wall of trench; large terrace on inner wall of trench
Louisville Ridge	Sh, D	St, D	St	U?	1, 20	Subaerial volcanism in wake of intersection
Marcus Necker Ridge	Sh, D	St, U	U	—	1, 22	Bonin Islands occupy forearc
Magellean Seamounts	Sh	U?	U?	—	1, 12, 9	"Seamounts" in forearc

Main-Barracuda Ridge	Sh?, O	St?	St?, U	U	15, 26	Forearc ridge/basins disrupted, shelf uplifted
Mendaña Fracture Zone	—	—	—	—	21	Disrupted shelf and slope basins
Orozco Fracture Zone	—	—	—	U, T	16	
Palau Kyushu Ridge	Sh, O	St, O	O?	—	1	
Nazca Ridge	Sh, St	St	St, O	U, T	1, 10, 13, 4, 6	No volcanism in wake of intersection, disruption of trench inner wall structures
Roo Rise	Sh, D	St, D, U	—	—	17	Forearc uplifted, sediments in forearc basin disrupted
Scarborough Seamounts	Sh, D	St, D	U	—	27	Up to 3 km uplift in forearc
Tehuantepec Ridge	Sh, D?	St	—	—	1	Effects may be related to change in age of seafloor
Tiburon Rise	Sh?, D	St?	St?, U	U	15	Forearc basin disrupted, shelf basin uplifted
Torres Rise	Sh, D	St	St, U	—	1, 3, 24	Torres islands lie on seaward edge of arc

Key to Headings and Entries: TA, Trench; IW, Trench Inner Wall; US Upper Slope; CO, Coastal Region; Sh, Shoals; St, Steep; U, Uplifted; O, Offset; D, Deflected landward; A, Absent.

References: 1, this paper; 2, BENTLEY (1974); 3, CARNEY and MACFARLANE (1982); 4, BROGGI (1946) and HSU (per. comm., 1987); 5, Cande (pers. comm., 1983); 6, CUTLER, (1977); 7, ERLICH, (1979); 8, FISCHER (1980); 9, FRYER and HUSSONG (1983); 10, JOHNSON and NESS (1981); 11, HOUSE and JACOB (1983); 12, HUSSONG and FRYER (1983); 13, KELLEHER and MCCANN (1976); 14, LONSDALE (1978); 15, MCCANN and SYKES (1984); 16, MCNALLY (pers. comm., 1984); 17, NEWCOMB and MCCANN (1987); 18, NISHENKO (1985); 19, SHIMIZAKI and NAKATA (1980); 20, VOGT *et al.* (1976); 21, WARSI *et al.* (1983); 22, HILDE (1983); 23, AMADEK *et al.* (1987); 24, TAYLOR *et al.* (1985); 25, WAGEMAN *et al.* (1970); 26, BRASIER and DONAHUE (1985); 27, HAYES and LEWIS (1985); 28, YONEKURA (1983).

zones created about 80 to 110 Ma ago when this seafloor was formed at the Mid-Atlantic Ridge. Magnetic anomalies, bathymetric trends and the pattern of deformed sediments of the inner wall of the trench strongly suggest that the northeastern corner of the Caribbean plate has recently overriden a segment of one of these ridges, with yet unsubducted segments lying along the northern and eastern margins of the plate boundary (McCann and Sykes, 1984). To the north lies the Main Ridge, presently intersecting the Virgin Islands-Puerto Rico subduction zone; to the east is the Barracuda Ridge, now entering the Lesser Antilles Trench off the Northern Lesser Antilles. As the strike of these features are not parallel to the vector of relative motion between the North American and Caribbean Plates, the intersections of these features with the subduction zone shift with time (Main Ridge to the west, Barracuda Ridge to the south). The deformation in this region is similar to that shown schematically in Figures 9D, E, and F.

The inner wall of the trench recently influenced by the Main-Barracuda Ridge shows few sub-bottom reflectors (Figure 3, profile B), indicating deformation of the accretionary prism. Also, a well-developed forearc ridge or basin does not exist in the accretionary wedge, in contrast to those portions of the subduction zone not yet effected by the passage of a ridge (Figure 3, profile A). The landward portions of the forearc overriding the ridges, such as near the islands of Anegada and Barbuda (Figure 2) show signs of recent uplift or reduced rates of regional subsidence (Horsefield, 1975; Brasier and Donahue, 1985). Both Barbuda and Anegada contain Pliocene to Quaternary Limestones and lie seaward of the normal chain of islands.

Westbrook (1982) noted similar effects on the subduction complex as a result of the consumption of ridges off the central and southern Lesser Antilles. He observed compressional features on the leading edge of each ridge. Profile C in Figure 3 shows an example of this deformation from the central Lesser Antilles. Here, the Tiburon Rise intersects the Lesser Antilles Trough, a forearc basin. Sediments closest to the ridge appear strongly tilted to the south.

Birch (1970) modelled gravity data across the Barracuda Ridge and concluded that the ridge is an upthrust block of crustal and mantle rocks and that the ridge does not have a low density root. The preceding observations, therefore, suggest that the effect that the subduction of an aseismic ridge has on the development of structures in the accretionary prism is not necessarily related to the buoyancy of the feature, but rather to its relief.

The rate of plate motion throughout this region is relatively constant, but the rate of convergence (i.e., portion of plate motion directed down-dip) varies greatly. Silver *et al.* (1983) observe variations in the development of accretionary wedge morphology near Sulawesi, Indonesia. They relate changes in rates of convergence to changes in basal shear stress. We do not feel it appropriate, however, to compare their observations to those in the northeastern Caribbean. The variation in convergence rate in the Caribbean occurs because of a change in the strike of the plate

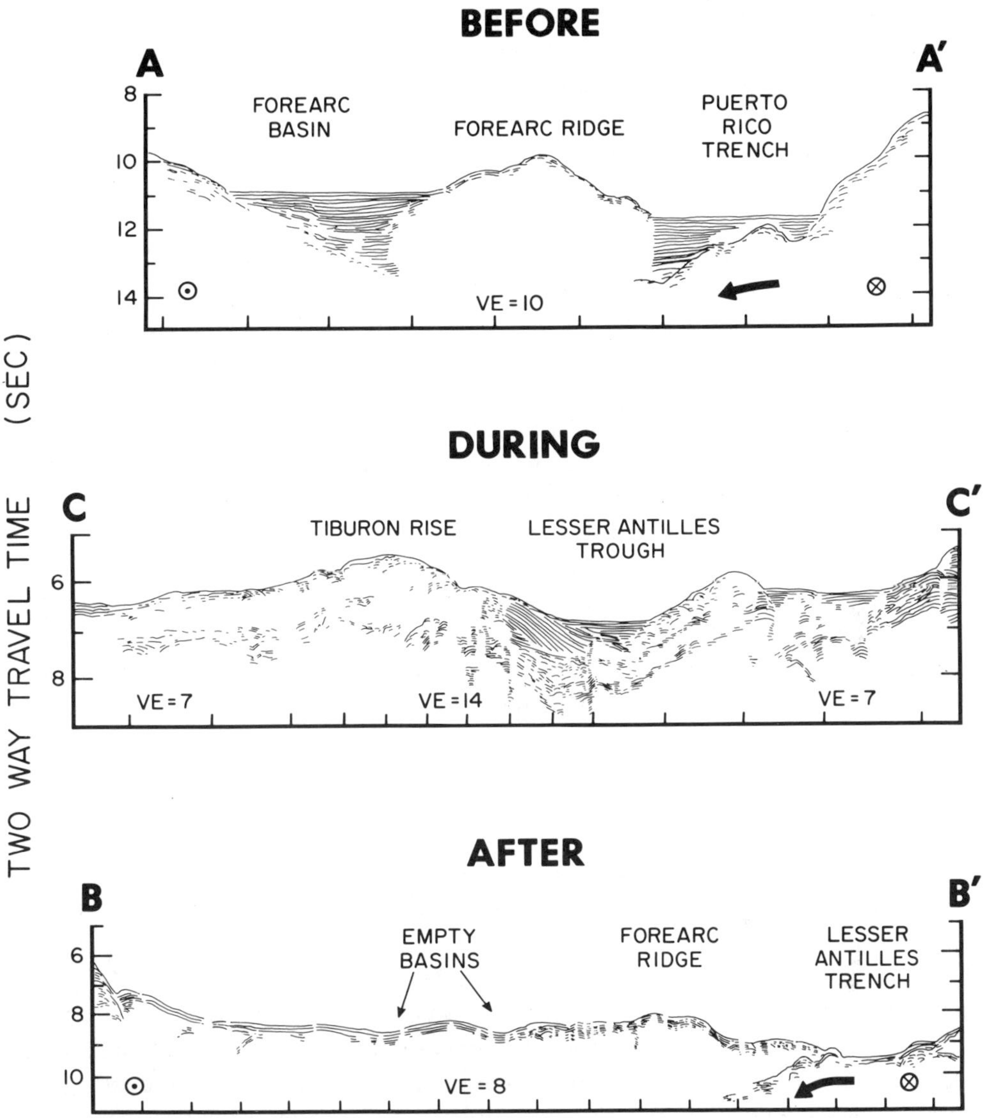

Figure 3

Line drawings of single channel profiler records illustrating the effects of the subduction of aseismic ridges. a) North of Puerto Rico lies a well defined forearc ridge/basin structure. These structures lie along a section of the arc that has not yet been affected by the westwardly advancing Main Ridge. Record from Conrad 19–03. b) Normal accretionary structures are absent in this Conrad 19–04 profile, showing a section of the arc recently disrupted by the passage of the Main-Barracuda Ridge. Shallow, coherent reflectors may represent either pelagic sediments or bubble pulse reflectors. Note the similar depths of the forearc ridge and basins and that the basins lack thick undeformed strata, unlike profile A-A′. c) Another section of Conrad 19–04 profile along the Lesser Antilles subduction zone. Tiburon Rise can be seen actively disrupting the strata in the Lesser Antilles Trough, a basin off the Central Lesser Antilles. Sediments are clearly tilted to the south, the direction of travel of the Tiburon Rise in this profile.

boundary relative to the slip vector, not because of a change in the rate of relative plate motion. Thus, we suggest that a change in basal shear affects the structures in the accretionary prism only in regions where the overall rate of plate motion changes, not just the component directed down the subduction zone.

Tonga-Kermadec

A linear bathymetric high with 2 to 3 km of relief, the Louisville Ridge, enters the subduction zone near the junction of the Tonga and Kermadec Trenches. The ridge lies along fracture zone trends resulting from Cretaceous spreading along the East Pacific Rise and appears to be an extension of the Eltanin Fracture Zone System (HAYES and EWING, 1971). Gravity modeling of the ridge topography indicates that it formed on crust older than 35 Ma and that the topography is supported by the flexural strength of the surrounding lithosphere, rather than by a low density root. The ridge may not be more buoyant than typical oceanic crust (WATTS *et al.*, 1980).

The intersection of the Louisville Ridge with the trench migrates southward at a rate of 6.5–8.0 cm/yr. Morphologic features associated with the past interactions of the ridge with the trench will, therefore, lie to the north of the present intersection. The bathymetry of the trench axis and inner wall display several features related to the subduction of the bathymetric high. First, the trench axis diverges from the linear trend of the Tonga-Kermadec trench system near its intersection with the Louisville Ridge (Figure 4, VOGT *et al.*, 1976), the offset gradually decreasing to the north. Second, the distance from the trench axis to the 1000 fathom (1 fathom = 1.8 meters) contour decreases from 100 km to 80 km south along the cusp, indicating a steepened inner wall north of the ridge-trench intersection. Finally, the inner wall of the trench, as defined by the 1000 and 3000 fathom contours, is unusually steep where the ridge descends beneath the forearc (hatched region, Figure 4), and the trench is deepest (> 5000 fathoms) just to the north of the steep wall. This deformation is similar to that shown in Figure 9C.

The deep trench just to the north of the intersection may result from the arcward underthrusting of trench sediments by the compressional effect of the ridge, the inability of sediments to pass to lower parts of the inner wall because of the presence of sediment dams, or ridges, formed by faulting in, or general uplift of, the steepened inner wall, or the elastic response of the lithosphere to the presence of the Louisville Ridge.

Sumatra

Seafloor entering the trench off the coast of Sumatra, western Indonesia, was created by a now extinct spreading system in the Indian Ocean (LIU, 1984). The largest offset in the spreading ridges, identified by magnetic anomalies near Sumatra,

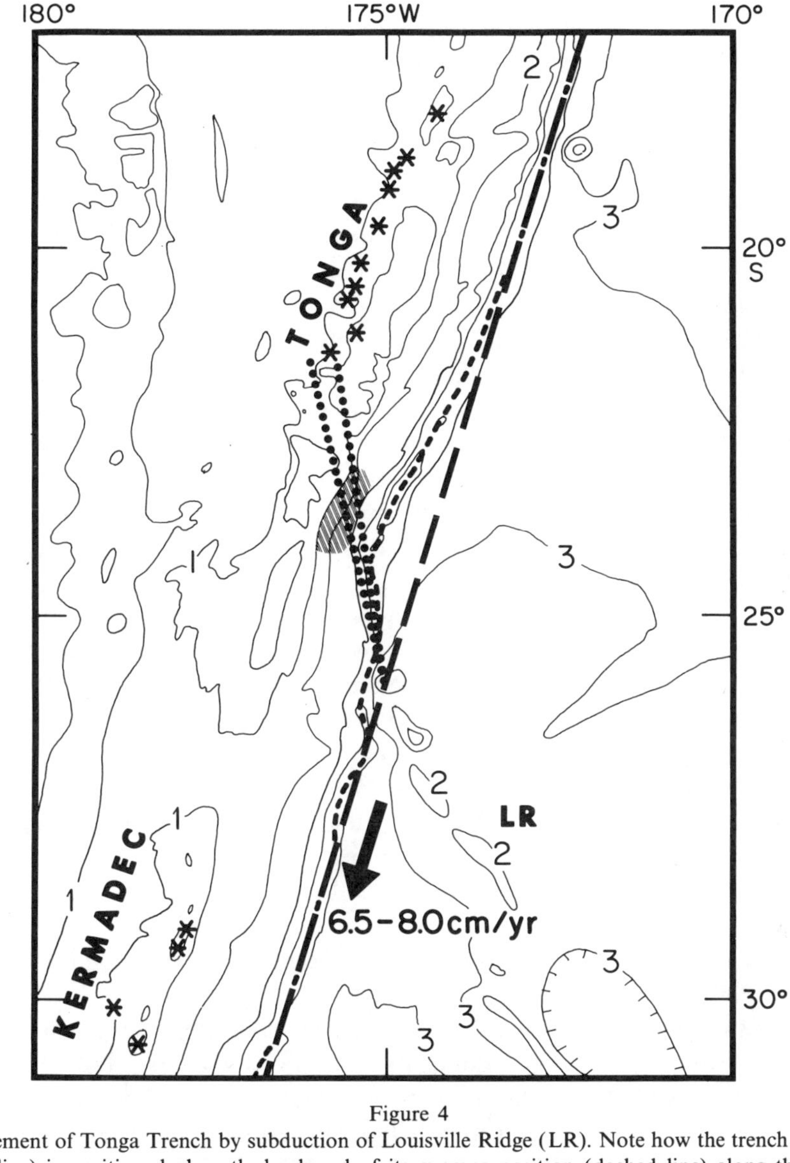

Figure 4
Displacement of Tonga Trench by subduction of Louisville Ridge (LR). Note how the trench axis (small dashed line) is positioned abruptly landward of its average position (dashed line) along the southern Tonga and Northern Kermadec arcs. Displacement is largest immediately to the north (i.e., in the wake) of the Louisville Ridge and gradually decreases to the north. Intersection of ridge and trench migrates to the south about 6.5–8.0 cm/yr. The distance between the 1000 and 3000 fathom contours is smallest (hatched) near the intersection, indicating a steepened inner wall. Asterisks are subaerial, quaternary volcanoes. Approximate map-view extension of Louisville Ridge into the subduction zone is shown by pair of dotted lines.

lies along the strike of the Investigator Ridge (JOHNSON *et al.*, 1976; LARSON *et al.*, 1978). The Investigator Ridge stands some 1000 to 1500 meters above the adjacent basement. One might expect to see a landward deflection of the trench axis and a steepening of the lower slope on the inner wall of the trench as in Tonga. However, the Investigator Ridge acts as a sediment dam, with thicker sedimentary sections to the west (MCDONALD, 1977). Sediments from this thick sedimentary section fill the trench as soon as the ridge passes, masking any simple morphologic expression of the compressional effect the ridge may have on the lower slope.

NEWCOMB and MCCANN (1987) show that the direction of slip between Java and the Indo-Australian Plate is due north. They also demonstrated that, because of the divergence between the slip direction of thrust earthquakes in Java and Sumatra, right-lateral slip along the Semanko fault in Sumatra is about 36 mm/yr. This means that the Sumatran forearc moves northwesterly with respect to the rest of Sumatra at the aforementioned rate. The Investigator Ridge strikes just slightly east of the north. Therefore, because the ridge lies along the slip vector between the Indo-Australian Plate and the core of Sumatra, its intersection with the Sumatran forearc will migrate along the arc at the same rate as motion along the Semanko fault, 36 mm/yr. Thus, the intersection of the Investigator Ridge with the trench migrates to the south.

Given the rate of 36 mm/yr, the Investigator Ridge passed beneath Nias Island 4.0–8.0 Ma (N, Figure 5). Detailed geologic and marine geophysical work near Nias Island show evidence of the passage of the Investigator Ridge.

Nias is one of several islands delineating the trench slope break of the Sumatran subduction zone. It is composed of Miocene to Recent sediments deposited in a near trench-slope environment. KARIG *et al.* (1980) report that the eastern edge of Nias marks a transition from deformed to undeformed sediments. Their results show that during a short interval in Pliocene time (5–2 Ma) the Banyak Islands, (B, Figure 5) north of Nias, experienced about 2 km of uplift by reverse faulting (KARIG *et al.*, 1980; MOORE *et al.*, 1980). Nias was strongly flexed in the later Pliocene and eastward directed overthrusting occurred in the sediments to the south, near Pini Island (P, Figure 5).

The thrusting near Pini Island is adjacent to the Pini Arch, an anticline composed of Oligocene to Recent sediments. NEWCOMB and MCCANN (1987) note that a free air gravity high trends from the Pini arch southward to the trench and then along the strike of the Investigator Ridge, suggesting that the two features are related.

Sediments to the east of Siberut Island have, in contrast to the strong pre-Quaternary deformation of the regions to the north, either experienced mild subsidence or the island has been uplifted in Late Tertiary-Quaternary time. Vertical movement on the east side of Siberut continues into recent time in contrast to the lack of tectonic activity east of Nias. KARIG *et al.* (1980) suggested that the episodic development of the arc directed overthrusting, flexure and uplift may have

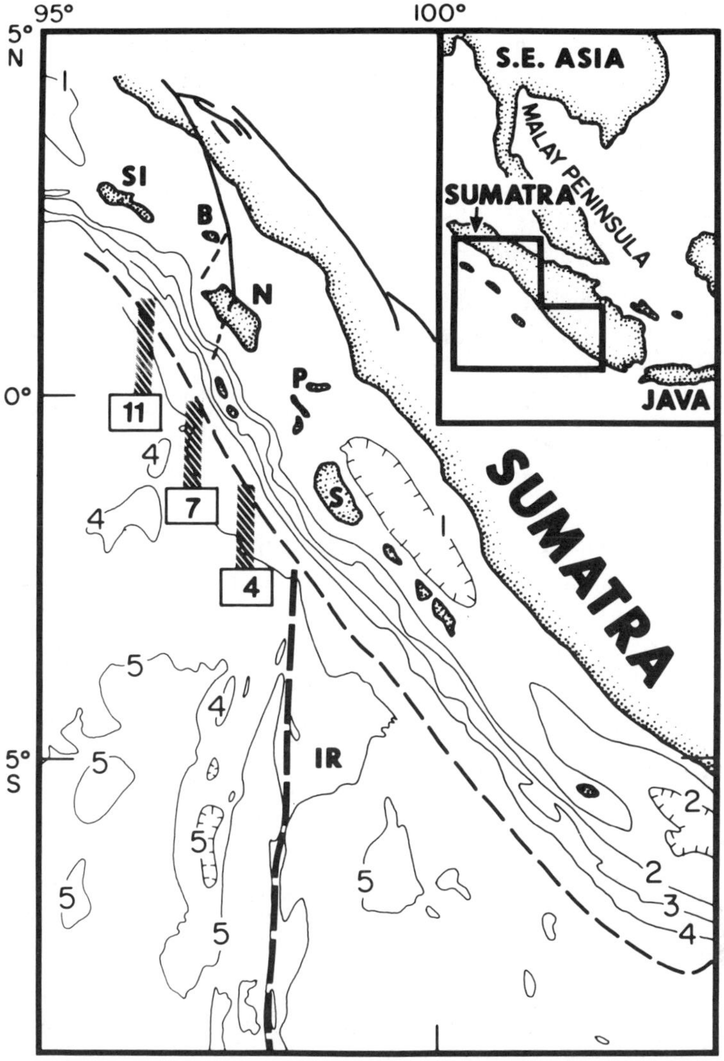

Figure 5
General bathymetry and Miocene-Recent position of intersection between Sumatra Trench (dashed line) and Investigator Ridge (IR). Numbers in boxes and hatched regions are times (in Ma) and positions of Investigator Ridge as it migrated along the Sumatra subduction zone at a rate of 36 mm/yr. Faults near the Banyak islands (B) and Nias (N) moved during the Pliocene, uplifting the Banyak islands some 2 km. Simulue (SI) may have been effected during an earlier period, the Pini islands (PI) lie along an arch in the overriding plate that may be related to the subduction of the Investigator Ridge; Siberut (S) is undergoing late Tertiary-Quaternary deformation. Contours are in meters.

resulted from an increase in the level of stress over the base of the accretionary prism. Given the present position of the Investigator Ridge, its rate of migration along the subduction zone, and the timing of the events in the vicinity of Banyak, Nias and Pini Island, we infer a relationship between ridge subduction and Pliocene-Recent deformation of the forearc. This deformation is similar to that shown in Figure 9D.

Tectonic Erosion and the Subduction of Aseismic Ridges in South America

Aseismic ridges associated with trends of fracture zones may be supported by the elastic strength of the surrounding lithosphere and, therefore, are not necessarily buoyant. Other aseismic ridges, in contrast, appear to be buoyant (supported by a low density root) and to have an effect on the overriding plate in addition to the accretionary prism. VOGT *et al.* (1976) showed that buoyant ridges resist subduction and can severely deform oceanic island arcs. When they intersect a continental plate, however, the effects may be quite different, causing a change in the dip of the downgoing seismic zone or a cessation of volcanism. Our observations as presented in Tables 1 and 2 suggest a correlation between morphologic and geologic changes of the leading edge of the continental plate and subduction of buoyant ridges. In this section we present evidence of such interactions along the northern portion of the Nazca-South America plate boundary.

Three major bathymetric features on the Nazca plate, the Nazca and Carnegie Ridges and the Mendaña Fracture Zone, enter the South American subduction zone between 5°N and 15°S. The structural makeup and origin of these features are known. The Carnegie and Nazca Ridges were formed at hotspots that lay on the axis of spreading centers. The Mendaña Fracture Zone is the extension of a transform fault that offset the now extinct Galapagos Rise in a right-lateral sense. For at least the last 16 Ma, the eastern portion of this fracture zone has been the locus of seafloor spreading (WARSI *et al.*, 1983).

Numerous basins along the western margin of South America, most of them submarine, have recorded vertical tectonic movements and deformational events. Here we examine the geologic record of the region in light of the passage of these bathymetric features.

Nature of the Bathymetric Ridges off South America

The Carnegie Ridge, an easterly trending bathymetric high (Figure 6), stands some 1.5 km above the surrounding seafloor. Crustal models of the ridge derived from refraction profiles indicate a thickened crustal section beneath the bulk of the ridge and suggest that its relief is supported by a low density root (HOLDEN and DIETZ, 1972). The Carnegie Ridge intersects the South American margin between 1°N and

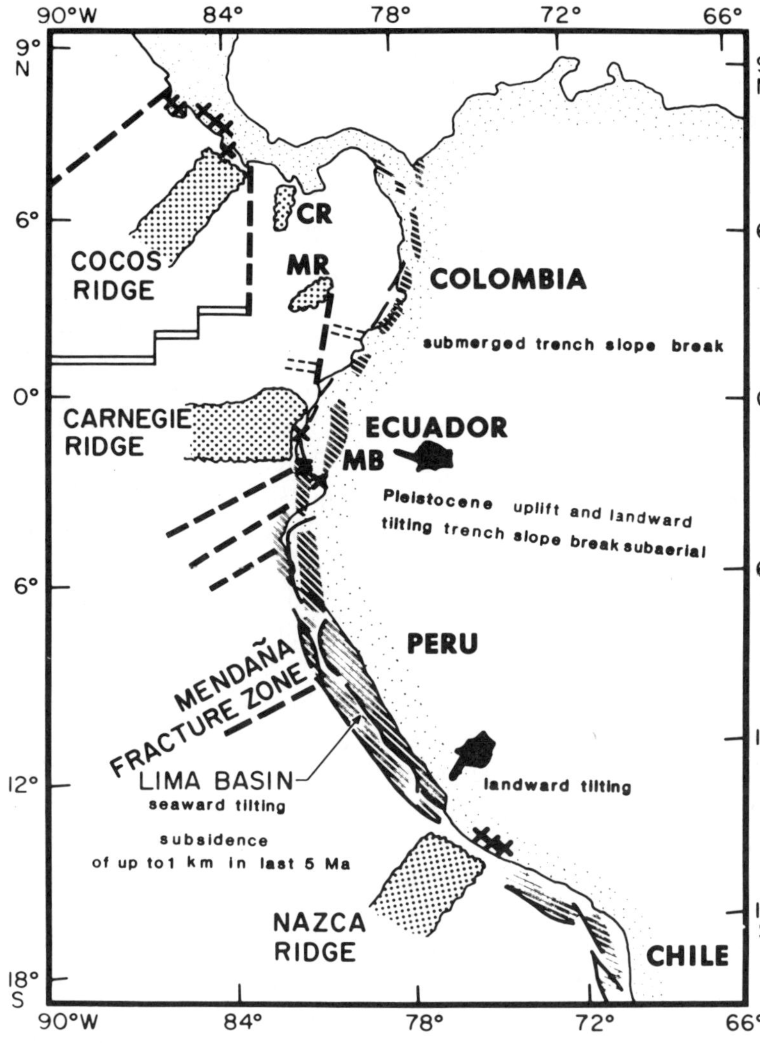

Figure 6
Major sedimentary and tectonic features on the western margin of South America. Hatched areas are shelf or slope basins of Miocene-Recent age, lines are sediments dams (upper-slope or shelf ridges). Crosses are regions with observed coastal terraces or other signs of uplift. Dashed lines are fracture zones or fracture ridges, double-dashed lines are extinct spreading centers.

3°S off the coast of Ecuador (Figure 6). The intersection of the Carnegie Ridge with the trench is nearly stationary because the ridge trends nearly parallel to the direction of relative motion between South America and the Nazca plate.

LONSDALE and MALFAIT (1974) and LONSDALE (1978) have estimated that the Carnegie Ridge first impinged upon the South American subduction zone 3 Ma ago, from convergence rates and marine geologic data. They inferred the location of the eastern edge of the ridge from the eastern terminus of its mirror image, the Malpelo Ridge.

The Nazca Ridge trends NE and intersects the Peruvian Trench near 15 degrees south. COUCH and WHISETT (1981), using marine gravity and refraction data, determined that the Nazca Ridge has a substantially thicker crust than surrounding seafloor and that its relief (1 km) is compensated by this low-density root. PILGER and HANDSCHUMACHER (1981) showed that the mirror image of the Nazca Ridge that was generated at the East Pacific rise is the Taumoto Ridge in the central Pacific. They estimated the length of the Nazca Ridge by projecting the mirror image of the Taumoto Ridge onto the Nazca Ridge. Cande (pers. comm., 1983), using new estimates of the poles of rotation, bathymetric data, and magnetic anomalies, concluded that the Line Islands were not a northerly extension of the Taumoto Ridge as assumed by PILGER (1983). Using the plate motions between the Nazca and South American plates derived by Cande, we find that the northernmost tip of the Nazca Ridge intersected the South American subduction zone 11 Ma ago at about the latitude of 9°S and that intersection migrated south to its present position (Figure 7a).

The Mendaña Fracture Zone is a 40 to 100 km wide zone of ridge-trough topography (WARSI *et al.*, 1983). Magnetic anomalies of Eocene/Oligocene age on either side of this feature show an age offset of 13 Ma, older seafloor being found on the southern side. This is consistent with the 200 m change in depth of the seafloor observed across the fracture zone (WARSI *et al.*, 1983). Ridges in the fracture zone rise some 200 to 400 meters above the surrounding seafloor. WARSI *et al.* (1983) described magnetic lineations and seafloor fabric parallel to the Mendaña fracture zone that suggest that it has been the site of slow seafloor spreading for the past 16 Ma. The axis of spreading enters the trench just south of 10°S and extends 400 km west of the trench.

We assume that the Mendaña fracture Zone had an easterly continuation equal to that of its mirror image, the Marquesas fracture zone (Cande, pers. comm., 1983). If this is the case, then it has been subducted beneath the Peruvian Trench from 6°S to 10°S in the last 20 Ma (Figure 7b). During much of that period it has been the locus of seafloor spreading.

Recent History of the Shelf and Slope Basins of South America

Numerous basins have been identified along the coastal region of Peru and Ecuador (Figure 6). In Ecuador, shelf basins are subaerial and bounded to the west by the trench slope break, which is onshore in this region (LONSDALE, 1978). Both the basins and trench slope break are submerged in northern Peru. During the Pleistocene an episode of accelerated landward tilting, uplift of the trench slope break, and rapid accumulation of sediments in the inner part of the Manabi basin (MB, Figure 6) occurred. Marine terraces, indicative of continuing uplift, are found

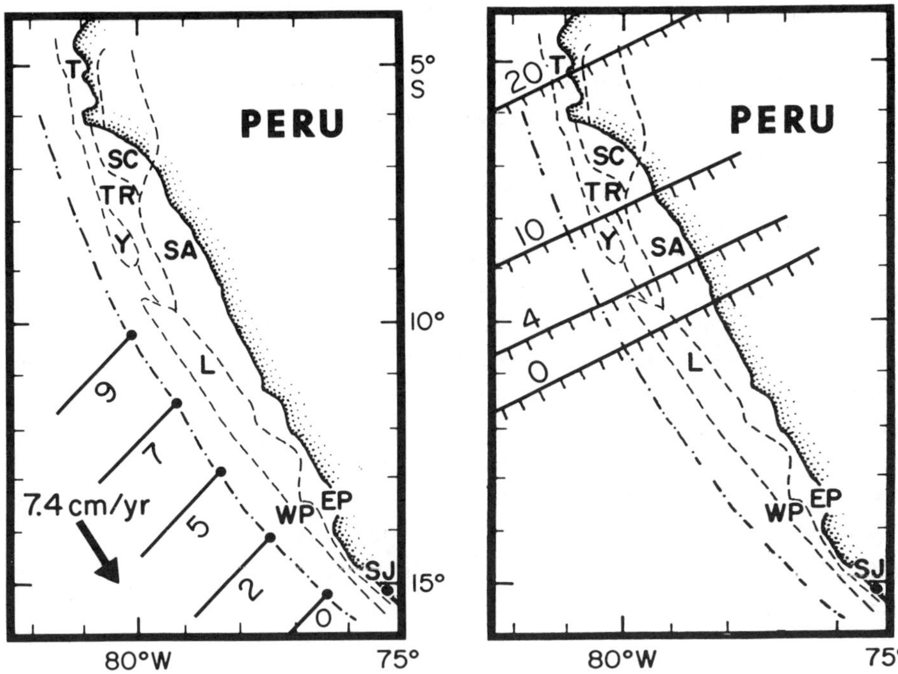

Fig. 7(a) Fig. 7(b)

Figure 7

Location of major forearc basins and bathymetric features along the margin of Peru (after THORNBURG and KULM, 1981). a) Earlier locations of Nazca Ridge (solid lines) and intersection with the Peru Trench (dot-dash line) shown as solid circles. Approximate age of intersection shown in Ma (Cande pers. comm., 1983). Rate of migration of ridge-trench intersection is also shown. Cande (pers. comm., 1983) estimates Nazca Ridge first impinged on trench near 8°S about 11 Ma ago. Lima Basin near 12°S is known to have begun subsidence of 1100 meters about 5.0 to 2.6 Ma ago, about the time when the Nazca Ridge passed by the region. We interpret this spatio-temporal coincidence as evidence of tectonic erosion of the base of the Peruvian forearc by the Nazca Ridge. SJ is San Juan, location of uplifted marine terraces. b) Locations of Mendaña Fracture Zone during the last 20 Ma are shown as solid lines with hatchures on older side. Basins in this part of the forearc experienced strong deformation during the period in which the fracture zone passed southward. Basins are: Talara, T; Sechura, SC; Trujillo, TR; Yaquina, Y; Salaverry, SA: Lima, L; West Pisco, WP; East Pisco, EP.

up to 300 meters above sea level along the coast of Ecuador, at the southern part of the intersection of the Carnegie Ridge with the trench. Shelf basins to the north lie on the coast or at sea and their trench slope breaks are, in general, submerged (Figure 6).

Along most of the Peruvian coast from 6°S to 15°S, north of the Nazca Ridge, there is a broad continental shelf over 100 km wide. It contains numerous sedimentary basins (THORNBURG and KULM, 1981; COULBOURN, 1981). The basins lie on continental crust and some of the forearc ridges, which act as sediments dams, are known to be composed of continental material (JONES, 1981; KULM *et al.*, 1981b).

The basins on the shelf or the upper slope show a distinct change in deformational style near 10°S. From 6° to 10°S the basins tend to be narrow and short; extensive faulting and folding deformed their sediments in Miocene time. Although uplift occurred in the early Miocene, the basins were later resubmerged. To the south of 10°S the basins are long and wide and differential uplift in the Miocene was followed by large-scale subsidence in the last 5 Ma. Much of the strata in the largest basin, the Lima Basin, is undeformed.

The Pisco Basins lie south of the Lima Basin, near the intersection of the Nazca Ridge with the trench. Sediments in those basins, unlike the strata in the Lima Basin, show recent landward tilting and upwarping (COUCH and WHISETT, 1981; JOHNSON and NESS, 1981). Numerous reflection records shown by JOHNSON and NESS (1981) and interpreted by THORNBURG and KULM (1981) show the shoaling and coalescence of the edges of the West Pisco Basin as one approaches the Nazca Ridge from the north until, opposite the ridge, thin sediments are pierced by underlying basement. They suggested that the north to south change in the character of the basin is related to the impingement of the Nazca Ridge on the continental margin. Near San Juan (SJ, Figure 7a) uplift apparently extends onland as a staircase of marine terraces rise to 250 m above sealevel (BROGGI, 1946).

Deformation of the South American Margin by the Subduction of Bathymetric Ridges

The timing and style of tectonic events in the basins from 5°S to 15°S in Miocene to Recent time appear to be intimately related to the subduction of the Mendaña Fracture Zone and the Nazca Ridge. In the region 5°S to 10°S, Miocene folding and faulting was followed by reduced sedimentation, suggesting sediment dams landward of the basin were produced during the faulting/folding episode. This intense deformation in the Miocene, not observed in the more southerly basins, is coincident in space and time with the subduction of the Mendaña Fracture Zone. Throughout much of the time when the Mendaña Fracture Zone was passing beneath the basins from 5°S to 10°S it was site of seafloor spreading. We suggest that there is a causal relationship between the subduction of rough seafloor associated with seafloor spreading on the Mendaña Fracture Zone and the deformation of the leading edge of the South American plate. This deformation is similar to that in the forearc of Figure 9F.

The Lima Basin, extending from 10°S to 13°S, shows little deformation of strata, but rather large-scale subsidence. Subsidence near 12°S started 5.0–2.6 Ma ago along the seaward edge of the basin and 1 Ma ago along the landward side (KULM *et al.*, 1981a,b). The time of initiation of this subsidence is coincident with the passage of the Nazca Ridge through this region. We suggest that the two events are related and that the subsidence of the Lima Basin results from erosion of the base of the overriding plate by the passage of the Nazca Ridge. Mild uplift preceding the subsidence reflects the upwarping of the overriding plate over the topography of the ridge itself and an isostatic response to the introduction of a buoyant feature into the subduction zone. Given a root 8 kilometers thick and a density contrast of .25 g/cc (COUCH and WHISETT, 1981), then an isostatic uplift of almost 600 meters is expected, in approximate agreement with the 250 meters of Quaternary uplift observed by BROGGI (1946). Later subsidence is then a combination of the passage of the ridge topography and the abrupt removal of some of the base of the overriding plate. Recent ODP drilling at site 686 in the West Pisco Basin indicates the initiation of subsidence some 1.5 Ma, just after the passage of the Nazca Ridge (SUESS and VON HUENE, 1987).

HSU and BLOOM (1985) and HSU *et al.* (1986) report the prominence of marine terraces near the intersection of the Nazca Ridge with the coast. He observed maximum uplift rates near the southern edge of the Ridge intersection with the coast.

LONSDALE (1978) has related the recent geologic record of Ecuador to the consumption of the Carnegie Ridge. This deformation is similar to that shown in Figure 9F. During the Pleistocene a major episode of accelerated landward tilting of the accretionary prism, uplift of the Coastal Range, and rapid accumulation of sediments in the inner part of the Manabi Basin occurred (Figure 6, MB). The Pleistocene marine terraces, at 300 m above sea level along the coast, suggest that uplift is continuing. The spatial and temporal coincidence of this phase of uplift, with the arrival of the Carnegie Ridge at the trench, strongly suggests a causal relationship between the two events.

Discussion

One expects the crystalline edge of the overriding plate to rest in a relatively inactive state on the downgoing plate under conditions of constant convergence rate and direction, constant age and smoothness of seafloor as well as constant sediment supply to the subduction zone. Of course, strong deformation occurs at the most seaward edge of the overriding plate, but the rest of the overthrust block is far removed from that tectonic activity. Disturbance from this equilibrium can come from a change in one or more of the factors assumed constant above (SHREVE and CLOOS, 1986). Clearly, changes in rates and directions of convergence will have an effect. We will not discuss them here as we wish to concentrate on the other parameters.

Subduction of Seafloor Relief and Deformation of Accretionary Prisms

The vast majority of the area of oceanic plates is relatively smooth. However, several distinct features such as ridges and troughs associated with ancient transform faults, aseismic ridges formed either by hot spot volcanism or by fragmentation of continental material, and seamounts distributed along chains or in a nonlinear fashion, have been noted. The subduction of any one of these features disturbs the equilibrium previously mentioned. This disturbance includes a change in the regional stress distributions caused by the relief, and possibly the buoyancy, of the incoming feature.

In a theoretical study of the shapes of accretionary wedges, DAVIS *et al.* (1983) term the seaward slope of the accretionary wedge, a and the landward dip of the basal décollement or downgoing oceanic plate, b. Wedges in a state of quasi-static equilibrium are described as critically tapered an $a + Rb = F$ where R and F are constants whose values are dependent upon rock strengths, densities and pore pressures. Subcritical wedges are those with $a + Rb < F$, supercritical wedges have $a + Rb > F$. Subduction of a bathymetric high leads to an increase in the dip of the basal décollement (increase in b) and, as overall taper is conserved, an equal but opposite change in the dip of the wedge (decrease in a). The parameter $R < 1$ for reasonable densities (DAVIS *et al.*, 1983), so the new geometry is in the subcritical domain. This leads to compressional deformation of the wedge as it builds up to a steeper, and more stable, critical taper. On passing down the back side of the bathymetric feature, the wedge becomes supercritical and deformation ceases; even after the feature passes deeper into the subduction zone the wedge is still supercritical, having a relatively steep slope resulting from the compressional event caused by the subduction of the feature.

Our observations of wedge geometries and deformation of accretionary prisms show remarkable agreement with this theory. We have noted deformation of accretionary prisms in the Caribbean, Sumatran and Tonga-Kermadec subduction zones. In Tonga-Kermadec we observed the steepened (supercritical) wedge in the arc segment recently intersected by the Louisville Ridge. In the other cases deformation of the accretionary wedge is observed, but the steepened nature of the prism, if it exists, is masked by other features such as change in sediment thickness on the incoming plate or complex shape of the subducting margin.

As the shear modulus of oceanic crust is greater than that of sediments, a bathymetric high standing above the strata blanketing a downgoing plate will present a different (higher) shear to the base of an accretionary prism. The increased shear stress will cause the prism to move into the subcritical domain, inducing renewed deformation. So two complimentary effects may induce deformation of accretionary prisms, change in wedge geometry and change in shear stress at the base of the wedge.

Effects of Relief on the Subduction Process

The bathymetric relief of a feature on the downgoing plate increases the vertical stress on the overriding plate as material is displaced. The relief also increases the horizontal stress as it thrusts material landward (force A, Figure 8b and Figures 9a–h). Stresses in the downgoing plate must also change as a result of this interaction. The load applied on the ridge by the overriding plate will tend to create or reactivate preexisting fractures, pushing the feature down into the downgoing plate. Horizontal stresses will tend to shear off the feature, possibly obducting it into the overriding plate.

Seismicity and focal mechanisms suggest that certain bathymetric features deform when subducted. FRANKEL *et al.* (1980), MCCANN and SYKES (1984), FISCHER and MCCANN (1984) studied seismicity in the vicinity of the Virgin

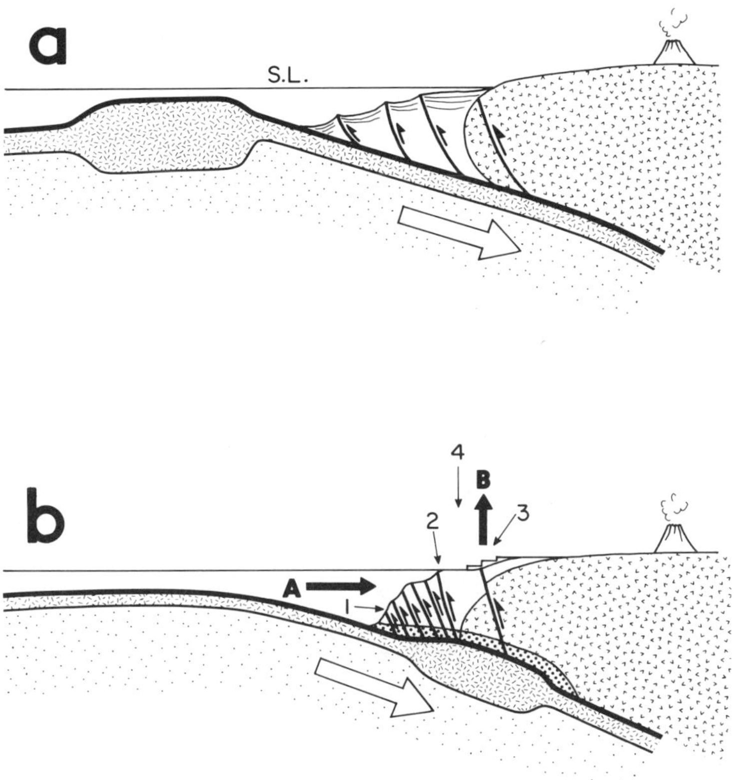

Figure 8
Cross section of the effects of ridge subduction. a) Subduction zone in quasi-equilibrium state as ridge approaches trench. b) Possible effects of subduction of aseismic ridge: 1—steepened inner wall of trench resulting from horizontal compression of accretionary prism, 2—deformation of sediments in accretionary prism by folding and/or thrusting, 3—development of coastal terraces as upper plate readjusts to relief of downgoing plate, 4—massive subsidence occurs if tectonic erosion at base of overriding plate occurs. S.L. is sea level, A represents compressive forces exerted on overriding plate by relief of ridge, B represents vertical forces related to buoyancy of downgoing ridge. Heavy dotted pattern in (b) is possible region of tectonic erosion.

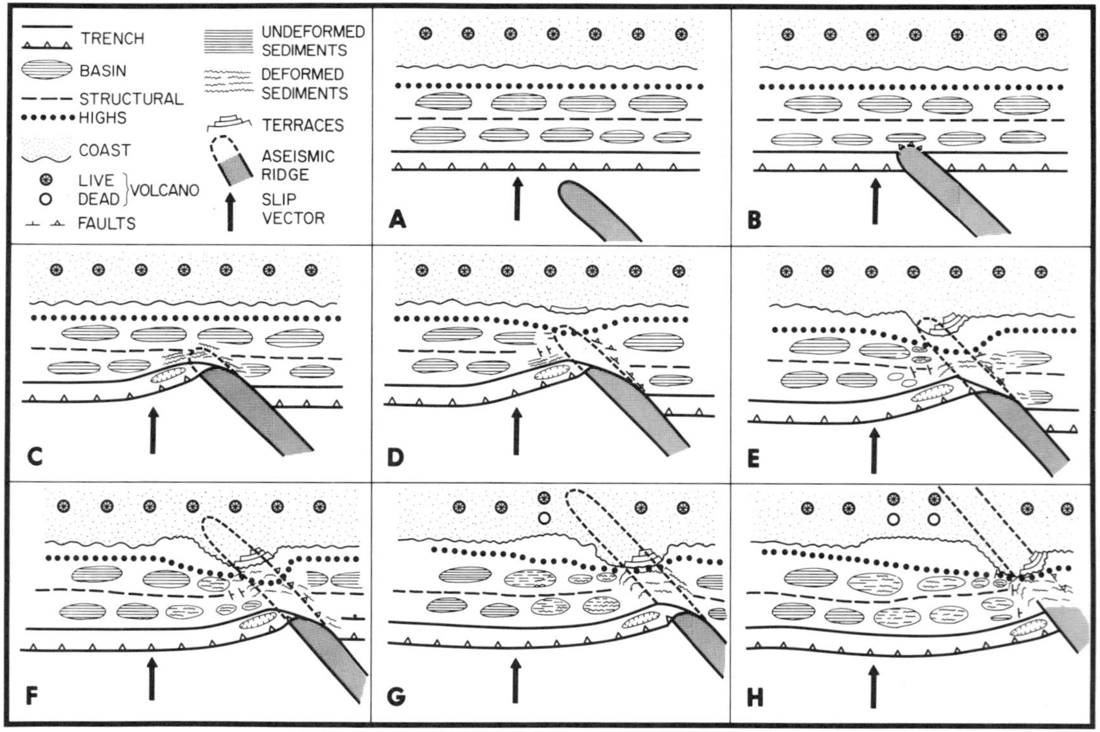

Figure 9

Map view of effects of subduction of bathymetric features. Subduction zone is generalized to contain characteristics of most convergent margins. Most prominent effects of ridge subduction are observed in accretionary wedge and adjacent sedimentary basins. Accretionary wedge is thrust landward, being made unstable by landward tilting. Basins may also be disrupted. Thrusting predominates on advancing flank of ridge, normal faulting occurs on trailing edge. Structural highs tend to be deflected landward if they are in the compressed wedge or migrate seaward if they are part of the uplifted, crystalline overriding plate. Coastline advances as upper plate is uplifted. Raised marine terraces are likely along leading edge of ridge, embayments or other signs of subsidence along trailing edge. If ridge is very buoyant, then ocean floor may underplate the upper plate, possibly causing a cessation of volcanism. If ridge is only buoyant enough to cause shallower angle of descent of lower plate, then volcanism is displaced landward.

Islands. There, the Main Ridge, a bathymetric high on the North American Plate, descends beneath the Caribbean Plate in a region of intense microearthquake activity. Much of this activity occurring within the downgoing plate, is diffusely distributed around the intersection of the Main Ridge with the crystalline part of the Caribbean Plate, and occurs in swarms. CHUNG and KANAMORI (1978) and EISSLER and KANAMORI (1982) studied earthquakes in the southwest Pacific occurring near the intersection of aseismic ridges with subduction zones. They observed strike-slip and normal fault mechanisms indicating reactivation of preexisting features and slab-detachment forces at work.

Much more is known about the reaction of the overriding plate to the subduction of bathymetric highs, because marine, geologic and seismic data are available. The wedge of accreted sedimentary material, if one exists, is most exposed to the presence of bathymetric features of the downgoing plate. This is because of its outboard position in the subduction zone and the lack of strength of the materials present there. The equilibrium state of this material is characterized by deformation, including folding and imbricate thrusting; in the more landward positions, sediments may accumulate in basins remaining relatively undeformed. The presence of a topographic feature will intensify the deformation of the sediments by uplifting and thrusting them landward and, depending upon the competency of the material underlying the sedimentary basin, deform some or all of the previously undeformed basin sediments (Figures 8 and 9). The uplifting is a simple response of the sediments to accommodate the presence of the topography; the landward thrusting is a response to the horizontal stresses exerted by the relief of the feature as it compacts the sedimentary wedge. The result of this interaction is a more deformed accretionary wedge with a steepened inner wall. The Caribbean, Sumatra and Tonga-Kermadec are examples of the deformation of accretionary wedges.

As the feature enters deeper into the subduction zone it interacts with the crystalline part of the overriding plate. The density, thickness and strength of the upper plate are greater than those of the sediments, so it is more difficult to produce observable effects. Nevertheless, the larger topographic features must have their topography accommodated, so that uplift of the overriding plate is likely to be observed (Figure 9D, E, and F). Bathymetric highs in general, have elevations of 1 to 2 km above surrounding seafloor. Uplifts along coasts rarely exceed a few hundred meters, so the net uplift in the overriding plate is about 10 percent of the relief of the feature itself. Some uplift presumably occurs through motion along faults within the overriding plate, although an elastic response cannot be ruled out. A large ($M_s = 7.5$) earthquake in 1974 occurred within the overriding wedge of the Caribbean plate near the intersection of the Barracuda Ridge with the subduction zone. STEIN *et al.* (1982), MCCANN *et al.* (1982) and MCCANN and SYKES (1984) suggested that this shock is related to intraplate deformation associated with the subduction of the ridge.

Uplift associated with consumption of bathymetric highs may be recognized by the presence of uplifted marine terraces. While deformation should occur in the vicinity of ridge-trench intersections, maximum rates of uplift should be observed on the leading edge of the interaction. That is, most rapid uplift will occur as the upper plate rides up to accommodate the incoming topography (Figure 9F). As the feature passes deeper into the subduction zone, any signs of uplift are likely to be destroyed as they resubmerge. Therefore, the only permanent record is the offshore deformation of sediments deposited on the blocks that experienced relative motion as the feature passed by that section of the upper plate.

Buoyancy and Tectonic Erosion

Several features we considered are not only characterized by relief on the seafloor, but also are isostatically compensated by low-density crustal roots. They are buoyant when compared to surrounding seafloor and, therefore, may resist subduction into the upper mantle. This resistance exhibits itself as a body force pushing the feature upward. These stronger vertical forces will be transmitted across the interface or zone of contact between the plates, into the overriding plate (force B in Figure 8). The increased normal force on the interface increases the drag on the upper plate by the lower one. Complimentary to this effect, as noted above, is the fact that bathymetric highs may stand above the strata deposited on the downgoing plate, and, therefore, may present a higher basal shear stress to the overriding plate. The high shear stresses at the base of the upper plate may induce tectonic erosion of the upper plate by shearing. In this case, parts of the upper plate may be carried down to deeper levels of the subduction zone, but it is unlikely that they will descend deeper than the Moho of the upper plate because this material, also being buoyant, resists entering the mantle. The net result of the erosion process is the removal of part of the seaward base of the upper plate with under-plating at the landward section. Concurrent with this process one expects isostatic uplift of the landward section and subsidence of the seaward one. This whole process should be a rather rapid one, occurring during the passage of an aseismic ridge which probably lasts only a few million years. Thus, the effects described can be differentiated from other, slower processes. The rapid subsidence of the Lima Basin is an example of such a rapid tectonic event.

The nature of the response of the upper-plate to ridge passage will depend on its previous history, the heterogeneity of its material and the strength of interplate coupling. Weakly coupled convergent margins are unlikely to experience erosion, as the necessary stresses are not present. Homogeneous, upper-plate material might more evenly distribute stresses, thus, resisting erosion. A segment having already experienced erosion may more easily accommodate (again) the additional relief. These last cases may be intimately tied with the strength of interplate coupling, a topic too broad for discussion here.

Sediment Supply to Subduction Zones

The larger features on the seafloor act as barriers to sediment flow, resulting in wide variations in the distribution of sediment on the seafloor. The passage of one of these features along a subduction zone may suddenly expose the arc to an increased or decreased horizontal load depending upon the distribution of sediments. This sudden change of load could have a dramatic effect on the leading edge of the upper plate. While we have ascribed the sudden acceleration of tectonic activity near northern Sumatra to the passage of the Investigator Ridge, we cannot rule out the possibility that the observed effects result from stresses exerted by the Bengal Fan, lying in the wake of the ridge. Other convergent margins presently consuming plates with thick sedimentary sequences show active deformation and uplift of the accretionary prism (Alaska, Southern Lesser Antilles).

Ridge Subduction and Arc Volcanism

The Louisville Ridge may be associated with a major change in the character of the Tonga arc, namely the restriction of volcanic activity to a narrow band from 18°–23°S. Several workers noted the relationship between the subduction of aseismic ridges beneath the South American margin and the lack of Quaternary volcanism (KELLEHER and MCCANN, 1976; BARAZANGI and ISACKS, 1976; NUR and BEN-AVRAHAM, 1981; PILGER, 1983). KELLEHER and MCCANN (1976) and PILGER (1983) ascribed the lack of volcanic in the southern Tonga arc to a modified subduction process caused by the subduction of the Louisville Ridge. A close look at the relationship between volcanism in the Tonga arc and the Louisville Ridge, however, suggests that the ridge may be associated with the activation, not the cessation of subaerial volcanism. The extension of the Louisville Ridge into the subduction zone reaches 100 km depth (the typical depth of a slab beneath a volcanic chain) beneath the southernmost volcanoes in the Tonga chain. Subduction beneath the Tonga arc may not, in general, result in subaerial volcanism. Perhaps the Louisville Ridge activates the production of volcanic material, or its subduction may sufficiently uplift the arc platform so that volcanism becomes subaerial and more readily observed in this sparsely populated region. This uplift may not result from the buoyancy of the ridge, as noted earlier the ridge may not be buoyant, but rather from compressional stresses exerted on the overriding plate by the downgoing topography arching the arc platform.

Conclusions

Our observations of recent tectonics of subduction zones led us to draw the following conclusions: 1) The subduction process tends to an equilibrium state that can be disturbed by changes in several factors, including relief of incoming seafloor and rate of sediment supply to the subduction zone. 2) Increased relief, or increased sediment supply will tend to compress the upper plate, thrusting sediments landward, possibly deforming ponds of undeformed sediments, possibly causing deformation within the crystalline part of the overriding plate and uplift along coastal sections of the upper plate. 3) Buoyant features will tend to cause stronger effects and may erode significant parts of the upper plate, resulting in large scale subsidence of the leading edge of the overriding plate. 4) These effects will tend to occur abruptly (i.e., within a few million years) as most of the irregularities on the seafloor are limited features.

Acknowledgements

Discussions with B. Perrin and S. Nishenko were helpful during the early stages of this study. Special thanks go to T. Boyd, G. Suarez, D. Davis and K. Jacob for enlightening and stimulating conversations and review comments. Figures were drafted by K. Nagao. This work was supported by the Division of Earth Sciences, National Science Foundation, grant EAR 82–13083.

References

Amadek, S., Tajima, F. and Weins, D. (1987), *Seismic rupture associated with subduction of the Cocos Ridge*, Tectonics *6*, 757–774.

Barazangi, M. and Isacks, B., (1976), *Spatial distribution of earthquakes and subduction of the Nazca Plate beneath South America*, Geology *4*, 686–692.

Bentley, L., (1974), *Crustal structure of the Carnegie Ridge, Panama Basin and Cocos Ridge*, unpubl. M.S. Thesis, Univ. of Hawaii.

Birch, F., (1970), *The Barracuda Fault Zone in the Western North Atlantic: Geological and Geophysical Studies*, Deep Sea Research *17*, 847–859.

Bodine, J. and Watts, A. (1979) *On Lithospheric flexure seaward of the Bonin and Marianas trenches*, Earth and Planet. Sci. Lett. *43*, 132–148.

Broggi, J., (1946), *Las Terrazas Marinas de la Bahia de San Juan en Ica*, Soc. Geol. Peru Bol. *19*, 21–23.

Brasier, M. and Donahue, J. (1985), *Barbuda—an Emerging Reef and Lagoon Complex on the Edge of the Lesser Antilles Island Arc.*, J. Geol. Soc. London *142*, 1101–1117.

Carney, J. and Macfarlane, A. (1982), *Geologic Evidence Bearing on the Miocene to Recent Structural Evolution of the New Hebrides Arc*, Tectonophysics *87*, 147–175.

Case, J. and Holcombe, T. (1980), *Geologic-tectonic map of the Caribbean Region*, U.S. Geol. Surv. Misc. Invest. Ser., MAP I–1100.

Chase, C., Menard, H., and Mammerickx, J., (1980), *Bathymetry of the North Pacific, Charts 1–6*, Scripps Inst. of Oceanography, La Jolla, CA.

Chung, W. and Kanamori, H. (1978), *Subduction Process of a Fracture Zone and Aseismic Ridges—the Focal Mechanism of the New Hebrides Earthquake of 1969 January 19 and Some Related Events*, Geophys. J. Roy. Astrom. Soc. *54*, 221–240.

Couch, R. and Whisett, R. (1981), *Structures of the Nazca Ridge and Continental Slope of Southern Peru*, Geol. Soc. Amer. Mem. *154*, 587–618.

COULBOURN, W. (1981), *Tectonics of the Nazca Plate and the Continental Margin of Western South America 18 to 23S*, Geol. Soc. Amer. Mem. *154*, 587–618.

CROSS, T. and PILGER, R. (1982), *Controls of Subduction Geometry, Location of Magmatic Arcs, and Tectonics of Arc and Back-arc Regions*, Geol. Soc. Amer. Bull. *93*, 545–562.

CURRAY, G., SHOR, G., RAITT, R., and HENRY, H. (1977), *Seismic refraction and reflection studies of crustal structure in the eastern Sunda and western Banda Arcs*, J. Geophys. Res. *82*, 2479–2489.

CUTLER, S. (1977), *Geophysical Investigation of the Nazca Ridge*, M Sc. Thesis, Univ. of Hawaii, Honolulu.

DAVIS, D., SUPPE, J., and DAHLEN, T. (1983), *Mechanics of fold-and-thrust belts and accretionary wedges*, J. Geophys. Res. *88*, 1153–1172.

EISSLER, H. and KANAMORI, J. (1982), *A large normal-fault earthquake at the junction of the Tonga Trench and the Louisville Ridge*, Phys. Earth Planet. Int. *29*, 161–172.

ERLICH, E. (1979), *Recent structures of Kamchatka and position of Quaternary volcanoes*, Bull. Volcanol. *42*, 13–42.

FISCHER, K. and MCCANN, W. (1984), *Velocity modeling and earthquake relocations in the northeast Caribbean*, Bull. Seism. Soc. Amer. *74*, 90–104.

FISCHER, K. (1980), *Recent tectonic movements of the Costa Rican Pacific Coast*, Tectonophysics *70*, t25–t33.

FORSYTHE, R. and NELSON, E. (1985), *Geological manifestations of Ridge collision: Evidence from the Golfo de Penas-Taitao Basin, Southern Chile*, Tectonics *4*, 477–495.

FRANKEL, A., MCCANN, W., and MURPHY A. (1980), *Observations from a seismic network in the Virgin Islands region: Tectonic structures and earthquake swarms*, J. Geophys. Res. *85*, 2669–2678.

FRYER, P. and HUSSONG, D. (1983), *Effects of subduction of seamounts on the Mariana and Izu-Bonin forearc regions* (*abstract*), Trans. Amer. Geophys. Union. *64*, 829.

HAYES, D. and EWING, M. (1971), *The Louisville Ridge—a possible extension of the Eltanin fracture zone*, Antartic Res. Ser. *15*, 223–228.

HAYES, D. and LEWIS S. (1985), *Structure and tectonics of the Manila Trench System, Western Luzon, Philippines*, Energy *10*, 263–279.

HILDE, T. (1983), *Sediment subduction versus accretion around the Pacific*, Tectonophysics *99*, 381–397.

HOLDEN, J. and DIETZ, R. (1972), *Galapagos Gore, NazCoPac triple junction and Carnegie Cocos Ridge*, Nature *235*, 266–269, 1972.

HOUSE, L. and K. JACOB (1983), *Earthquakes, plate subduction, and stress reversals in the Eastern Aleutian Arc*, J. Geophys Res. *88*, 9347–9374.

HORSEFIELD, W. (1975) *Quaternary vertical movements in the Greater Antilles*, Geol. Soc. Amer. Bull. *86*, 933–938.

HSU, T. and BLOOM, A. (1985), *Quaternary marine terraces and maximum tectonic uplift rate of the Peruvian coast at 15.5S latitude*, (abs) Geol. Soc. Amer., 614.

HSU, T., WEHMILLER, J., and BAUTIN, B. (1986), *Correlation of late pleistocene Peruvian coastal marine terraces 14.5 to 15.5S latitude by amino acid enantiometric ratios*, (abs) Geol. Soc. Amer. 641.

HUSSONG, D. and FRYER, P. (1983), *Forearc tectonics and the morphology of the Northern Mariana arc* (*abstract*), EOS, Trans. Amer. Geophys. Union *64*, 828.

HUSSONG, D. *et al.* (1976), *Crustal structure of the Peru–Chile trench*; *8–12S latitude*, in *The Geophysics of the Pacific Ocean Basin and its margins*, AGU, Washington D.C., 17–85.

JOHNSON, B., POWELL, C., and VEEVERS, J. (1976), *Spreading History of the Eastern Indian Ocean and Greater India's northward flight from Antarctica and Australia*, Geol. Soc. Amer. Bull. *87*, 1560–1566.

JOHNSON, S., and NESS, G. (1981), *Shallow structures of the Peru Margin 12–18S*, Geol. Soc. Amer. Mem. *154*, 525–544.

JONES, P. (1981), *Crustal Structures of the Peru continental margin and adjacent Nazca Plate, 9S latitude*, Geol. Soc. Amer. Mem. *154*, 423–444.

KARIG, D. (1972), *Remnant arcs*, Geol. Soc. Amer. Bull. *83*, 1057–1068.

KARIG, D. and SHARMAN, G. (1975), *Subduction and accretion in trenches*, Geol. Soc. Amer. Bull. *86*, 377–389.

KARIG, D., LAWRENCE, M., and CURRAY, J. (1980), *Structural framework of the forearc basin*, J. Geol. Soc. Lond. *127*, 1–15.

KARIG, D. and KAY, R. (1981), *Fate of sediments on the descending plate at convergent margins*, Phil. Trans. R. Soc. Lond. Ser. A, *A301*, 233–251.

KELLEHER, J. and MCCANN, W. (1976), *Buoyant zones, great thrust earthquakes, and unstable boundaries of subduction*, J. Geophys. Res. *89*, 4885–4895.

KELLER, G., (1980), *Benthic foraminifera and Paleobathymetry of the Japan Trench area, Leg 57, Deep Sea Drilling Project*, in Scientific Party Init. Rept. DSDP 56, 57 Pt. 2, Washington, D.C., 835–865.

KULM L., PRINCE, R., FRENCH, W., JOHNSON, S., and MASIAS, A. (1981a), *Crustal structure and tectonics of the central Peru Continental Margin*, Geol. Soc. Amer. Mem. *154*, 445–468.

KULM, L., SCHRADER, H., RESIG, J., THORNBURG, T., MASIAS, A., and JOHNSON, L. (1981b), *Late Cenozoic carbonates on the Peru continental margin: Lithostratigraphy, biostratigraphy and tectonic history*, Geol. Soc. Amer. Mem. *154*, 468–508.

LARSON, R., CARPENTER, G., and DIEBOLD, J. (1978), *A geophysical study of the Wharton Basin near the Investigator fracture zone*, J. Geophys. Res. *83*, 773–782.

LIU, C., CURRAY, J., and MCDONALD, J. (1984), *New constraints on the tectonic evolution of the eastern Indian Ocean*, Earth and Planet. Sci. Lett. *65*, 331–342.

LONSDALE, P., (1978), *Ecuadorian Subduction System*, Am. Assoc. Petrol. Geol. Bull. *62*, 2454–2477.

LONSDALE, P. and MALFAIT, B. (1974) *Abyssal dunes of foraminiferal sand on the Carnegie Ridge*, Geol. Soc. Amer. Bull. *85*, 1697–1712.

LUDWIG, W., MURACHI, S., DEN, N., BUHL, P., HOTTA, H., EWING, M., ASANUMA, T., YOSHII, T., and SAKAJIRI, N. (1973), *Structure of east China Sea–west Philippine Sea margin of southern Kyushu, Japan*, J. Geophys. Res. *78*, 2526–2536.

LUNDBERG, N. (1983), *Development of forearcs in intraoceanic subduction zones*, Tectonics *2*, 51–61.

MAILLET, P., MANZIER, M., SELO, M., and STROZER, D. (1983), *The d'Entrecasteaux Zone (southwest Pacific): A petrological and geochronological reappraisal*, Mar. Geol. *53*, 179–197.

MAMMERICKX, J., SMITH, S., TAYLOR, I., and CHASE, T. (1973), *Bathymetry of the Southwest Pacific, Charts 11–21*, Inst. Marine Res., Univ. of Ca., La Jolla, Ca.

MCCANN, W., DEWEY, J., MURPHY, A., and HARDING, J. (1982), *A large normal-fault earthquake in the overriding wedge of the Lesser Antilles subduction in the overriding wedge of the Lesser Antilles subduction zone: The earthquake of October 8, 1974*, Bull. Seismol. Soc. Amer. *72*, 2267–2284.

MCCANN, W. and SYKES, L. (1984), *Subduction of aseismic ridges beneath the Caribbean plate: Implications for the seismicity and seismic potential of the northeastern Caribbean*, J. Geophys. Res. *89*, 4493–4519.

MCDONALD, J., (1977), *Sediments and the structure of the Nicobar Fan, northeast Indian Ocean*, Ph.D. Thesis, Univ. of Ca. Scripps Inst. Ocean., San Diego, Ca.

MORETTI, I. and NGOKWEY, K. (1985), *Aseismic Ridge Subduction and Vertical Motion of Overriding Plate*, in *Caribbean Dynamics* (ed. A. Mascle) (Technip Paris) pp. 245–253.

MOORE, G., BILLMAN, H., HEHANUSSA, P., and KARIG, D. (1980), *Sedimentology and paleobathymetry of Neogene trench slope deposits, Nias Island, Indonesia*, J. Geol. *88*, 161–180.

MOORE, J. *et al.* (1979), *Progressive accretion in the Middle America Trench, southern Mexico*, Nature *281*, 638–642.

MURACHI, S. *et al.* (1968), *Crustal structure of the Philippine Sea*, J. Geophys. Res. *73*, 3143–3171.

NEWCOMB, K. and MCCANN W. (1987), *Seismic history and seismotectonics of the Sunda Arc*, J. Geophys. Res. *92*, 421–439.

NISHENKO, S., (1985), *Seismic potential for great interplate earthquakes along the Chilean margin of South America: a quantitative reappraisal*, J. Geophys. Res. *90*, 3589–3616.

NISHIZAWA, A., SUYEHIRO, K. and SHIMIZU, H. (1983), *Seismic Refraction Experiment at the Amami Plateau*, J. Phys. Earth *31*, 159–171.

NUR, A., and Z. BEN-AVRAHAM, (1981), *Volcanic gaps and the consumption of aseismic ridges in South America*, Geol. Soc. Amer. Mem. *154*, 929–940.

NUR, A., and BEN-AVRAHAM, Z. (1982), *Oceanic Plateaus, the Fragmentation of Continents, and Mountain Building*, J. Geophys. Res. *87*, 3644–3661.

NUR, A. (1983), *Accreted Terranes*, Rev. of Geophys. and Space Phys. *21*, 1779–1785.

PAVLIS, T. and BRUHN, R. (1983), *Deep-seated flow as a mechanism for the uplift of broad forearc ridges and its role in the exposure of high P/T metamorphic terranes*, Tectonics *2*, 473–497.

PILGER, R. (1983), *Plate reconstructions, aseismic ridges and low-angle subduction beneath the Andes*, Geol. Soc. Amer. Bull. *92*, 448–456.

PILGER, R. and HANDSCHUMACHER, D. (1981), *The fixed-hotspot hypothesis and origin of the Easter-Sala y Gomez-Nazca trace*, Geol. Soc. Amer. Bull. *92*, 448–456.

PRINCE, R., SCHWELLER, W., COULBOURN, T., SHEPERD, G., NESS, G., and MASIAS, A. (1980), *Bathymetry of the Peru-Chile continental margin and Trench*, Geol. Soc. Amer. Map Ser. MC-*34*.

SEELY, D., (1978), *The Evolution of structural highs bordering major forearc basins*, Bull. Am. Assoc. Petrol. Geol. *29*, 245–60.

SHIMIZAKI, K. and NAKATA, T. (1980), *Time-Predicatable recurrence model for large earthquakes*, Geophys. Res. Lett *7*, 279–282.

SHREVE, R. and CLOOS, M. (1986), *Dynamics of Sediment subduction, Melange formation, and prism accretion*, J. Geophys. Res. *91*, 10229–45.

SILVER, E., MCCAFFREY, R., and SMITH, R. (1983), *Collision, Rotation, and the initiation of subduction in the evolution of Sulawesi, Indonesia*, J. Geophys. Res. *88*, 9407–9418.

STEIN, S., ENGELN, J., WEINS, D., FUJITA, K., and SPEED, S. (1982), *Subduction seismicity and tectonics in the Lesser Antilles Arc*, J. Geophys. Res. *87*, 8642–8664.

SUESS, F., and VON HUENE, R. (1987), *Summary of Leg 112 Results*, Joides J. *13*, 31–43.

TAYLOR, F., JOUANNIC, C., and BLOOM, A. (1985), *Quaternary Uplift of the Torres Islands, Northern New Hebrides Frontal Arc: Comparison with Santo and Malekula Islands, Central New Hebrides Frontal Arc*, J. of Geology *93*, 419–438.

THORNBURG, T. and KULM, L. (1981), *Sedimentary basins of the Peru continental margin: Structure, stratigraphy and Cenozoic tectonics from 6 to 16 south latitude*, Geol. Soc. Am. Mem. *154*, 393–422.

VOGT, P. (1973), *Subduction of aseismic ridges*, Nature *241*, 189.

VOGT, P., LOWRIE, A., BRACEY, D., and HEY, R. (1976), *Subduction of aseismic oceanic ridges: Effects on shape, seismicity and other characteristics of consuming plate boundaries*, Geol. Soc. Amer. Spec. Paper *172*.

VON HUENE, R. *et al.* (1980), *The Deep Sea Drilling Project midtrench transect off Guatemala*, Geol. Soc. Amer. Bull. *91*, 421–432.

WAGEMAN, J., HILDE, T., and EMERY, K. (1970), *Structural Framework of East China Sea and Yellow Sea*, Amer. Assoc. Petrol. Geol. Bull. *54*, 1611–1643.

WARSI, E., HILDE, T., and SEARLE, R. (1983), *Convergence Structures of the Peru Trench Between 10S and 14S*, Tectonophysics *99*, 313–329.

WATTS, A. (1978), *An analysis of isostasy of the world's oceans. 1. Hawaiian-Emperor Seamount chain*, J. Geophys. Res. *83*, 5989–6004.

WATTS, A., BODINE, J., and RIBE, N. (1980), *Observations of flexure and the geologic evolution of the Pacific Ocean Basin*, Nature *283*, 532–537.

WESTBROOK, G. (1982), *The Barbados ridge complex: Tectonics of a mature forearc system*, in *Trench-Forearc Geology: Sedimentation and Tectonics on Modern and Ancient Active Plate Margins* (ed. Leggett, J.) Geol. Soc. London, London) pp. 275–290.

YONEKURA, N. (1983), *Late Quaternary Vertical Crustal Movements in and Around the Pacific as Deduced From Former Shoreline Data*, in *Geodynamics of the Western Pacific-Indonesian Region* (T. HILDE and S. UYEDA, eds.) AGU Geodynamic Series Vol. 11, Boulder, Colorado.

(Received September 26, 1986, accepted June 7, 1987)

PAGEOPH, Vol. 129, Nos. 1/2 (1989) 0033–4553/89/020071–59$1.50 + 0.20/0

Large Earthquakes in the Macquarie Ridge Complex: Transitional Tectonics and Subduction Initiation

LARRY J. RUFF,[1] JEFFREY W. GIVEN,[2] CHRIS O. SANDERS[3] and CHRISTINE M. SPERBER[1]

Abstract—While most aspects of subduction have been extensively studied, the process of subduction initiation lacks an observational foundation. The Macquarie Ridge complex (MRC) forms the Pacific-Australia plate boundary between New Zealand to the north and the Pacific-Australia-Antarctica triple junction to the south. The MRC consists of alternating troughs and rises and is characterized by a transitional tectonic environment in which subduction initiation presently occurs. There is a high seismicity level with 15 large earthquakes ($M > 7$) in this century. Our seismological investigation is centered on the largest event since 1943: the 25 MAY 1981 earthquake. Love, Rayleigh, and P waves are inverted to find: a faulting geometry of right-lateral strike-slip along the local trend of the Macquarie Ridge (N30°E); a seismic moment of 5×10^{27} dyn cm ($M_W = 7.7$); a double event rupture process with a fault length of less than 100 km to the southwest of the epicenter and a fault depth of less than 20 km. Three smaller thrust earthquakes occurred previous to the 1981 event along the 1981 rupture zone; their shallow-dipping thrust planes are virtually adjacent to the 1981 vertical fault plane. Oblique convergence in this region is thus accommodated by a dual rupture mode of several small thrust events and a large strike-slip event. Our study of other large MRC earthquakes, plus those of other investigators, produces focal mechanisms for 15 earthquakes distributed along the entire MRC; thrust and right-lateral strike-slip events are scattered throughout the MRC. Thus, all of the MRC is characterized by oblique convergence and the dual rupture mode. The "true" best-fit rotation pole for the Pacific-Australia motion is close to the Minster & Jordan RM2 pole for the Pacific-India motion. Southward migration of the rotation pole has caused the recent transition to oblique convergence in the northern MRC. We propose a subduction initiation process that is akin to crack propagation; the 1981 earthquake rupture area is identified as the "crack-tip" region that separates a disconnected mosaic of small thrust faults to the south from a horizontally continuous thrust interface to the north along the Puysegur trench. A different mechanism of subduction initiation occurs in the southernmost Hjort trench region at the triple junction. Newly created oceanic lithosphere has been subducted just to the north of the triple junction. The entire MRC is a "soft" plate boundary that must accommodate the plate motion mismatch between two major spreading centers (Antarctica-Australia and Pacific-Antarctica). The persistence of spreading motion at the two major spreading centers and the consequent evolution of the three-plate system cause the present-day oblique convergence and subduction initiation in the Macquarie Ridge complex.

Key words: Earthquakes, seismotectonics, subduction initiation, soft plate boundary.

[1] Department of Geological Sciences, University of Michigan, Ann Arbor, MI 48109, U.S.A.

[2] Sierra Geophysics, Kirkland, WA 98033, U.S.A.

[3] Seismological Laboratory, California Institute of Technology, Pasadena, CA 91125, U.S.A. (now at USGS, Menlo Park, CA 94025).

1. Introduction

Subduction zones consume oceanic lithosphere and are an indispensible part of plate tectonics. Unlike the oceanic lithosphere production system which can be linked as a nearly continuous, albeit sinuous, strand around the earth, subduction zones are a rather dissociated group and are found in several isolated corners of the world. While plate tectonics can predict that subduction zones are required along certain plate boundaries, it does not stipulate how subduction zones initiate and develop. The preservation of newly created oceanic lithosphere and the propensity for spreading centers to fragment continents leaves a wealth of geological information on the initiation and evolution of spreading. On the other hand, the subject of subduction initiation has little observational basis. To find such observations, we need to look at some muddled tectonic regimes. The Macquarie Ridge complex presents a natural laboratory for studies of subduction initiation.

2. Tectonics of the Macquarie Ridge Complex

The Macquarie Ridge complex is a complicated physiographic feature that trends approximately north-south between South Island, New Zealand and the Pacific-Antarctica spreading center. This feature consists of a sequence of troughs and ridges, with Macquarie Island as the only exposed expression. The seismically active Macquarie Ridge complex (hereafter: MRC) is crudely continuous with the Tonga-Kermadec-New Zealand seismic activity. The basic physiographic features and seismicity of the MRC are shown in Figure 1. The earthquake epicenters generally cluster about the bathymetric expression of the MRC. The high level of seismic activity, 15 events with $M_s > 7$, clearly suggests that the MRC is the active plate boundary between the Pacific and India plates. In the past, there has been some confusion over the nature of this plate boundary. Various authors have referred to the MRC as: a spreading ridge, a convergent plate boundary, a strike-slip margin, or some combination of the above. Global plate tectonic models place the rotation pole between the Pacific and India plates close to the MRC, therefore the relative plate motion can change rapidly along the MRC. The rotation pole from the global model RM2 (MINSTER and JORDAN, 1978) is plotted in Figure 1. The relative motions along the MRC predicted by this pole are depicted by the velocity vectors. Scattered intermediate-depth earthquakes beneath the southwestern corner of South Island provide evidence for convergence in this region (see SCHOLZ *et al.*, 1973).

Minster and Jordan treated the entire India-Australia region as one plate. Several observations suggest that there is some decoupling between India and Australia of a transform nature along the trend of the Ninety-East Ridge (STEIN and OKAL, 1978; WIENS *et al.*, 1985). Since the MRC is at the eastern edge of the

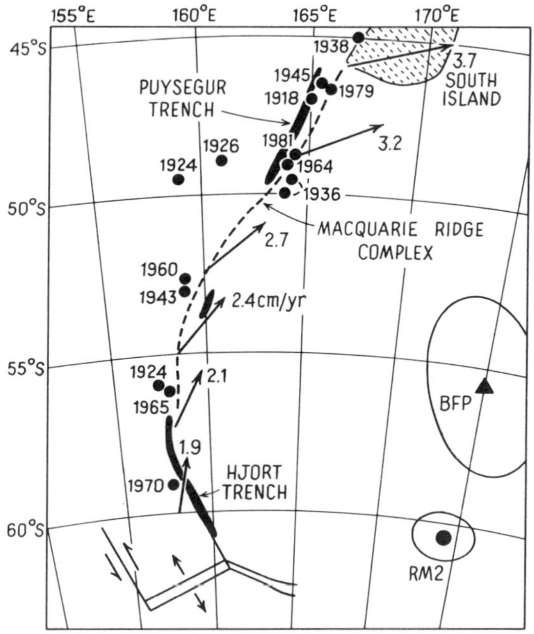

Figure 1
Simplified tectonics and seismicity of the Macquarie Ridge complex. Large earthquakes (magnitude greater than 7) are plotted as dots with the year of their occurrence. Two Pacific-Australia plate motion rotation poles and their error ellipses are plotted: the best-fitting pole (BFP) and the global model pole (RM2) from MINSTER and JORDAN (1978). The motion of the Australia plate relative to the Pacific plate along the Macquarie Ridge is shown by the vectors, based on the RM2 pole with velocity in cm/yr. The bathymetric rise of the Macquarie Ridge is schematically plotted as the dashed line, while the troughs are indicated by the solid regions. The spreading segment and Balleny transform fault between the Antarctica and Australia plates are also sketched. The Pacific-Antarctica-Australia triple junction is located at $61\frac{1}{2}$°S 161°E.

India plate, the interaction along this boundary describes the relative motion of the Pacific plate (PAC) and the eastern part of the India plate, which includes Australia. Henceforth, we shall refer to the Australia plate (AUS) rather than the India plate when discussing the interaction along the MRC. The RM2 pole for the Pacific-India motions will be used as the initial PAC-AUS pole. The termination of the MRC at the Pacific-Antarctica spreading center results in a triple junction at this point. Hence, we shall eventually consider the interaction with the Antarctica plate (ANT).

Since the Pacific-Australia rotation pole is located close to the MRC, changes in the rotation pole location over geological time will be reflected in the evolution of the MRC. The plate tectonic reconstruction of MOLNAR *et al.* (1975) indicates that some type of spreading activity formed the entire MRC since the Oligocene (38 Ma). The change from a tensional to compressional environment implies that the rotation pole has migrated southward since the Oligocene (also see WALCOTT, 1978). The seismicity of South Island and the uplift of the South Island Alps are also consistent with a southward migration of the rotation pole, mostly since the Miocene (SCHOLZ *et al.*, 1973). To summarize the basic geological setting, the MRC originally formed in an oceanic spreading center environment and subsequently served as a strike-slip plate boundary, with the northernmost part changing to a partly convergent boundary since the Miocene.

There are some inconsistencies in the geological setting that appear significant. In their comprehensive treatment of the magnetic lineations in this region, WEISSEL *et al.* (1977) noted that some oceanic crust appears to be missing; this crust was generated at the ANT-AUS spreading ridge segment at the PAC-ANT-AUS triple junction. It is noteworthy that this missing AUS plate crust is in the vicinity of the Hjort trench, a deep (maximum depth greater than 6 km) arcuate trough. WEISSEL *et al.* (1977) offered two explanations for the missing crust: (1) it was subducted at the Hjort trench, and (2) the plate boundary jumped to the present location at the Hjort trench. WEISSEL *et al.* then showed that the closure condition at the PAC-ANT-AUS triple junction implies subduction along the PAC-AUS boundary. RUFF and CAZENAVE (1985) addressed the possibility of subduction at the Hjort trench using SEASAT geoid anomalies and limited first motion data for the large 1924 earthquake; they concluded that subduction has probably occurred along the northern Hjort trench with Australia lithosphere subducting beneath the older Pacific lithosphere. If subduction has indeed occurred, note that newly created oceanic lithosphere has been subducted.

Another complication is the difference in the best-fitting and RM2 rotation poles (see Figure 1). The best-fitting pole is based only on the relative motion data along the PAC-AUS plate boundary, while the RM2 pole results from the global inversion of all data. As discussed in MINSTER and JORDAN (1978), the adjustment of the PAC-AUS pole from the BFP to the RM2 location is mostly due to the requirement of PAC-ANT-AUS triple junction closure. Notice that the BFP location implies spreading at the Hjort trench instead of convergence. Thus, we are faced with the question: is the Minster and Jordan BFP truly the best representation of PAC-AUS relative motions? It is instructive to look at the data source for the BFP. Since there are no recognized spreading ridge or transform fault segments between the Pacific and Australia plates, all the relative motion data comes from earthquake slip vectors. Since no great subduction zone earthquakes along the Kermadec trench were used in the Minster and Jordan study, there could be a bias in the rotation pole determination. Minster and Jordan did use the slip vectors from two earthquakes in the MRC, these two occurred in 1964 and 1965 and the first-motion focal mechanisms are from BANGHAR and SYKES (1969). These two slip vectors contributed significantly to the BFP location. The magnitudes of these two earthquakes are 6.9 and 6.6, hence they are not the most significant events in the MRC. One result of our study will be the addition of slip vectors from some of the largest earthquakes in the MRC.

The entire catalog of MRC earthquakes is plotted in Figure 2. Notice that the seismicity is not uniformly distributed throughout the MRC. The most striking cluster is around 50°S, with an apparent lack of seismicity both to the north and south of this cluster. The largest earthquake to have occurred since 1943, the 25 MAY 1981 event with $M_s = 7.6$, is located in this prominent cluster. Five earthquakes with $M > 7$ have occurred in the MRC since 1963. We have studied the largest of these earthquakes and have compiled the available focal mechanism information for other smaller earthquakes. Many large earthquakes occurred before 1963. Although it is difficult to study these earlier events, one aspect that can be

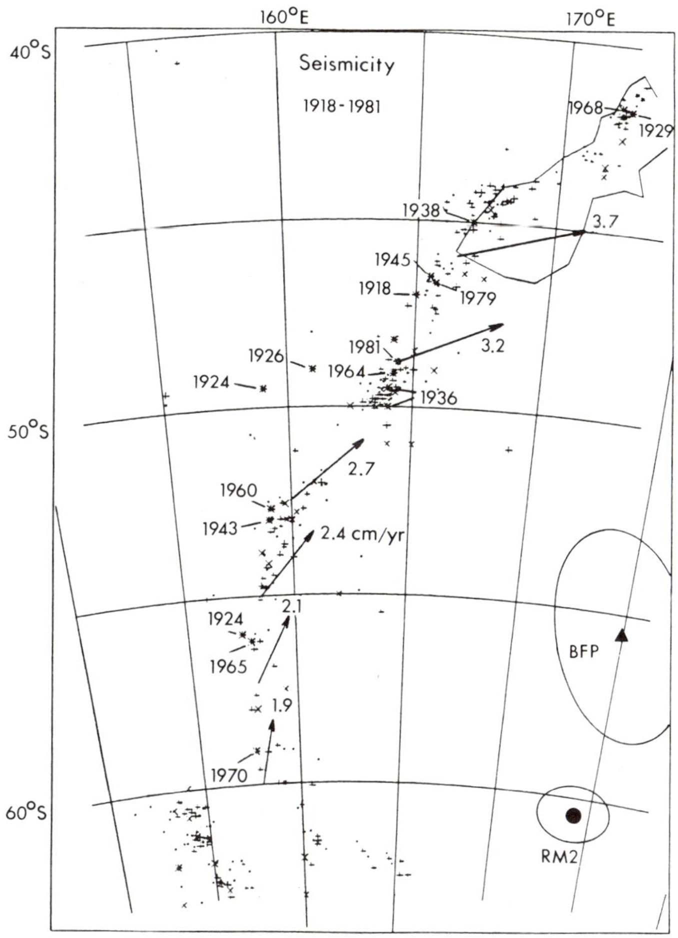

Figure 2
Macquarie Ridge complex seismicity, 1918 to 1981. All epicenters in the USCGS catalog are plotted, the magnitude threshold varies through this time period. The seismic activity is not uniform at any magnitude level. In particular, there is a concentration of events of all sizes in the region of the large 1981 earthquake from 49°S to 50°S. The BFP and RM2 rotation poles are plotted.

studied is their epicentral location. We have systematically checked the listed locations for large earthquakes that occurred between 1918 and 1963 for which the ISS catalog publishes arrival times. In addition, due to rather unique circumstances we have been able to determine focal mechanism constraints for two of these pre-1963 events. Based on this comprehensive study of large events in the MRC, we reconsider the tectonic setting. Despite the confusing seismicity, we find a consistent picture within the framework of rigid plate tectonics. We propose a model for subduction initiation based on the seismotectonics of the Puysegur trench region, and then speculate on global generalizations of the Macquarie Ridge complex tectonic environment.

3. Large Recent (Post-1963) Earthquakes

Focal mechanisms have been determined for only a few of the MRC events. As discussed above, only two of the smaller earthquakes were used to derive the RM2 relative plate motions. We have determined the focal mechanisms for the three largest earthquakes since 1963: the 1970 event in the Hjort trench region, and the 1979 and 1981 events in the Puysegur trench region. The 1979 and 1981 events are the two largest earthquakes in the time period from 1943 to the present. We begin with separate sections for the three earthquakes that we studied in detail, with emphasis on the large 1981 event. We then summarize the information available for all recent events.

3.1. The Great 25 MAY 1981 Strike-slip Earthquake

This event occurred in the seismicity cluster at the southern end of the Puysegur trench (Figure 3). The ISC hypocentral parameters for this event are: origin time, 5:25:9.4 (hr:min:s), 25 MAY 1981; epicenter, 48.55°S and 164.7°E; depth, 0 km (constrained). With values for the surface wave magnitude ranging from 7.4 to 7.7, this event is one of the largest earthquakes of 1981. Since it is also the largest magnitude earthquake in the MRC since 1943, any study of the mechanics of this region, and indeed of the Pacific-Australia plate interaction, must consider this event. An additional interesting aspect of this event is that the *P* wave first-motions indicate a strike-slip focal mechanism (see Figure 4), thereby implying that this event is one of the largest strike-slip earthquakes to be well-recorded. The focal mechanisms for two smaller earthquakes in the seismicity cluster are thrust mechanisms. Therefore, it is important to ascertain whether the 1981 earthquake was in fact a strike-slip event, as opposed to a precursory strike-slip event followed by a thrust main event. To obtain the overall average focal mechanism of this event, we have used the moment tensor inversion technique described in KANAMORI and GIVEN (1981). First we look at the aftershocks of this earthquake.

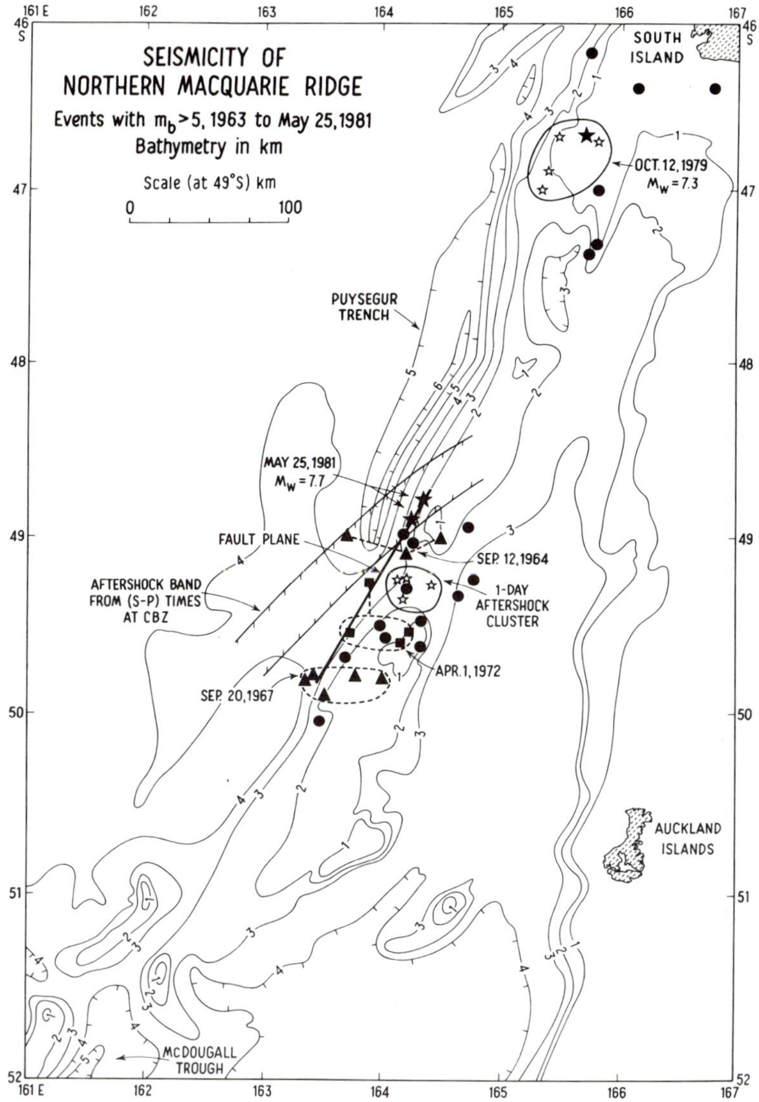

Figure 3

Detailed seismicity of the northern Macquarie Ridge, 1963 to 1981 mainshock. Epicenters and magnitudes are from the USCGS catalog. All events are larger than magnitude 5 except for three of the four events in the 1-day aftershock cluster of the 1981 event (open stars), these are the four largest aftershocks (29 or more stations reporting). All smaller aftershocks recorded at CBZ (Campbell Island) fall into the plotted band, based on relative *S-P* times. Two possible mainshock epicenters are plotted as solid stars. The bold line through the 1981 epicenters indicates the maximum possible extent of the 1981 rupture zone. For the 1979 earthquake, the mainshock and 1-day aftershocks are plotted as the solid and open stars, respectively, with the approximate fault extent indicated by the solid line. In the 1981 region, there are several earlier sequences of mainshock/aftershocks that are linked by broken lines: 12 SEP 1964, mainshock $M = 6.9$ (triangles); 20 SEP 1967, mainshock $M = 6.1$ (triangles); 1 APR 1972, mainshock $M = 6.3$ (squares). Focal mechanisms determined for some of these earlier events all indicate thrust faulting. Bathymetry is taken from maps published by the New Zealand D.S.I.R. Oceanographic Institute.

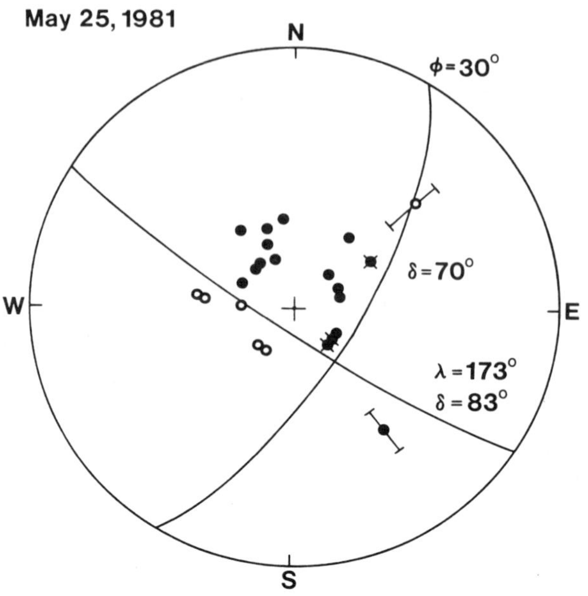

Figure 4
Preferred focal mechanism solution and WWSSN *P* wave first motions for the 1981 event. Lower-hemisphere equal-area projection with compressional and dilatational arrivals plotted as solid and open circles, respectively. This convention is used in all subsequent focal mechanisms. Crosses indicate near-nodal readings, and error bars indicate take-off angle uncertainty associated with close stations. Parameters for the nodal planes are strike, dip, and rake (ϕ, δ and λ), and are obtained from fault plane inversion.

3.1.1. Aftershock distribution. The fault area ruptured by an earthquake is typically estimated by the one-day aftershock area. The four largest one-day aftershocks (29 or more stations reporting times) of the 1981 event are plotted in Figure 3, with magnitudes between 4.5 and 5.1. This cluster lies about 50 km southwest of the epicenter and thus supports the selection of the NE-SW nodal plane as the fault plane. A rupture length of 50 km is much less than we would expect for a strike-slip earthquake of this magnitude. For example, the 1976 Guatemala earthquake ($M_s = 7.6$) had large surface displacements over a length of 250 km (PLAFKER, 1976). Even if we use all the one-day aftershocks for the 1981 event in the ISC catalog above magnitude 4, there are no epicenters north of the mainshock nor south of 50°S. The teleseismic one-day aftershocks for the 1981 event thus imply a rupture length of 100 km or less. For the 1976 Guatemala earthquake, the teleseismically recorded one-day aftershocks define a fault length of just 180 km. However the locally recorded microaftershocks scatter over the full 250 km length of the fault. A similar situation is encountered for the 1972 Sitka, Alaska strike-slip earthquake ($M_s = 7.6$); the teleseismically located aftershocks fall into one cluster 90 km from the epicenter (SCHELL and RUFF, 1988) whereas the microaftershocks define a 180 km bilateral fault zone (PAGE, 1973). Given the remote oceanic location of the 1981 Macquarie Ridge event, neither a detailed aftershock study nor fault displacement mapping are possible. However, there is a seismic station on Campbell Island operated by the New Zealand D.S.I.R. Geophysics Division, and many small one-day aftershocks were recorded by this single station. The *P* and *S* wave picks from these 16 events were provided by Dr. Warwick Smith. The most important observation is that the *S*-*P* times for all these events only range from 48.2 to 50.7 seconds, and the mainshock *S*-*P* time is 49.1 s. Hence, the distance to these aftershocks varies by less than 20 km about the mainshock distance. Fortunately, the azimuth from Campbell Island to the mainshock is not perpendicular to the fault strike. Thus, with the assumption that the aftershocks are on the fault plane, we see that the microaftershocks cluster in the epicentral region (see Figure 3).

The aftershocks imply a fault length of only 100 km, anomalously small for a large strike-slip event. This tentative conclusion motivates a more thorough analysis of the rupture process. Both surface waves and *P* waves are analyzed for information on the rupture length. We use seismograms recorded by stations of the WWSSN (in operation since 1963), GDSN (in operation since 1980), IDA (in operation since 1977), and Caltech.

3.1.2. Rayleigh and Love waves. GDSN stations are azimuthally well distributed about the 1981 earthquake (Figure 5). Figures 6 and 7 show three-component filtered traces from the GDSN digital recordings. These filtered traces emphasize the multiple orbit surface waves around a period of 250 s, the "mantle waves". The long-period spectral amplitude and phase from these high quality digital records can be inverted to obtain the seismic moment and faulting geometry. The procedure is fully documented in KANAMORI and GIVEN (1981), and has been successfully applied to many earthquakes over a large magnitude range. Due to the shallow depth of this earthquake (less than twenty kilometers), the vertical dip-slip component is unresolved, as discussed in KANAMORI and GIVEN (1981), and some type of constrained moment tensor inversion is necessary. We shall use two different modes of the constrained moment tensor inversion: (1) linear inversion for a moment tensor constrained to have no isotropic nor vertical dip-slip components, (2) nonlinear fault inversion where we constrain one nodal plane. For the linear inversion, the moment tensor is resolved into two double couples. If we assume that the larger double couple represents the actual faulting, then the fractional size of the second double couple is a measure of the reliability of the solution. The second double couple is relatively small in all cases for the 1981 event; this supports the notion that this earthquake has a single dominant focal mechanism. We have inverted various subsets of our data to test the reliability of the results. The constrained moment tensor inversion always yields a strike-slip mechanism, with right-lateral motion along the NE-SW plane. Figure 8 summarizes the solutions for

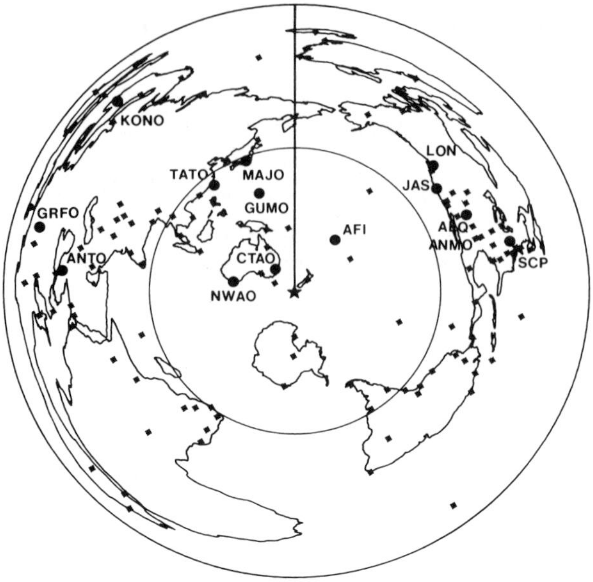

Figure 5
Azimuthal equidistant world map centered on the 1981 epicenter (star). The solid straight line passes through the North Pole. The GDSN stations used in this study are plotted as large dots, the other symbols show WWSSN station locations. Although the GDSN stations are in the two northern quadrants, multiple orbit surface waves provide excellent azimuthal coverage.

different combinations of data. Note that the strike of the NE plane is in excellent agreement with the *P* wave first-motions, though the first-motions clearly indicate a dip of $\approx 70°$ rather than $90°$ (see Figure 9). Figure 10 shows the spectral amplitude and phase for Rayleigh and Love waves with the predicted values from the constrained moment tensor inversion. Inclusion of phase data in the inversion seems to result in a lower moment estimate. To improve the solution, we can constrain one nodal plane and invert for the slip vector angle and the seismic moment. We constrain the NE trending plane to have a strike of $30°$ (clockwise from north), and vary the dip as shown in Figure 9. For the various data subsets, we always obtain right lateral strike-slip with a small oblique component. Our best estimate for the seismic moment with a point source assumption is 4.5×10^{27} dyn cm, and the best fault geometry is shown in Figure 4. The excellent correspondence between the surface wave focal mechanism and the *P* wave first motions shows that the faulting geometry remained constant for the duration of the earthquake.

DZIEWONSKI and WOODHOUSE (1983) also inverted GDSN data to obtain a moment tensor solution for the 1981 event. The major double couple of their solution has the following fault parameters: a fault strike of 31°E, fault dip of 66°, and a rake of 186°. The seismic moment is 2.7×10^{27} dyn cm, and the size of the

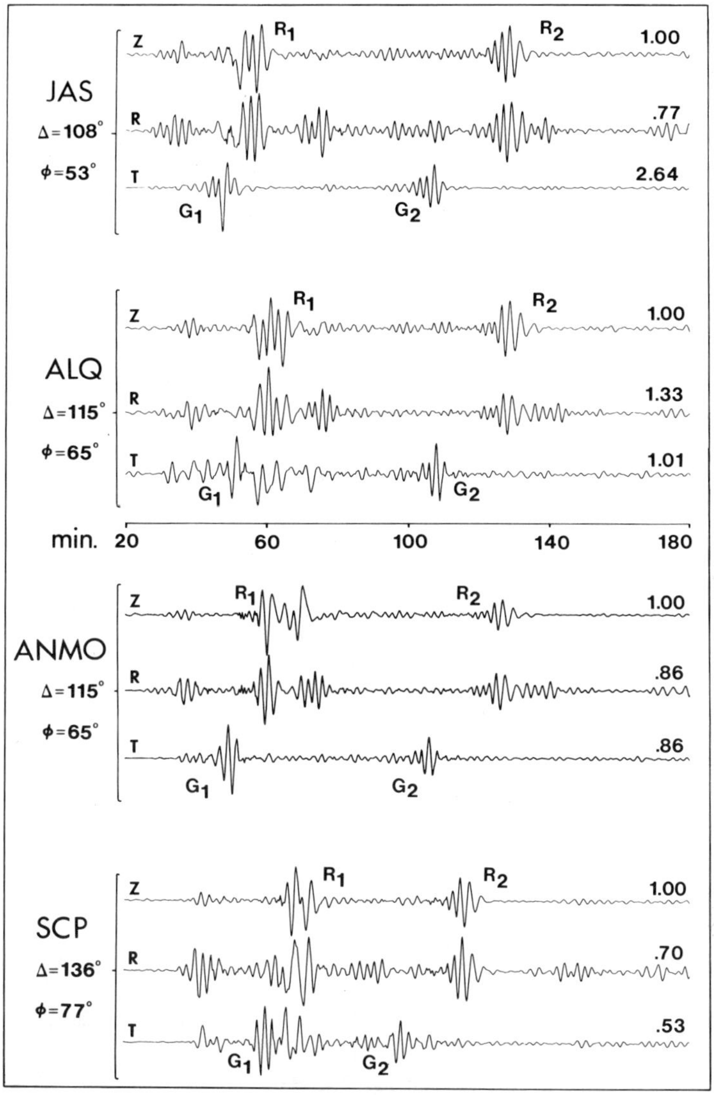

Figure 6
Filtered traces from GDSN stations, station codes are at left with distances and azimuths. Nearly three hours of three-component long period channels are filtered to enhance the "mantle wave" periods. The horizontal components are rotated into radial and transverse components, thereby isolating the Rayleigh waves on the *Z* and *R* traces and the Love waves on the *T* trace. Multiple orbit Rayleigh and Love waves are well recorded. Time scale refers to minutes after origin time. The numbers for each trace give the maximum peak-to-trough amplitude relative to the *Z* trace for each station, amplitudes are arbitrarily scaled between stations.

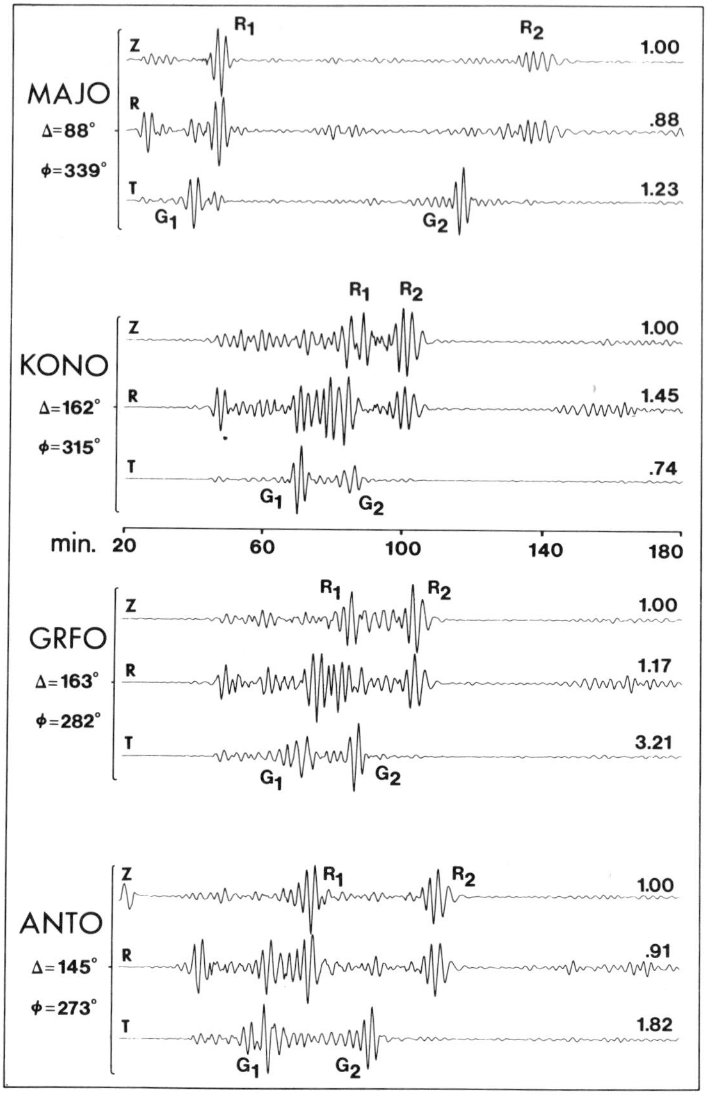

Figure 7
Same as Figure 6.

second double couple is about 1/10 the size of the major double couple. The fault geometry of the Harvard CMT solution is very similar to our solution, but their seismic moment is about half of our value. There are several significant differences between the Harvard CMT (DZIEWONSKI *et al.*, 1981) and the Kanamori and Given techniques. The Harvard inversion includes shorter period information, and inverts for the spatial and temporal centroid of the moment density function. The

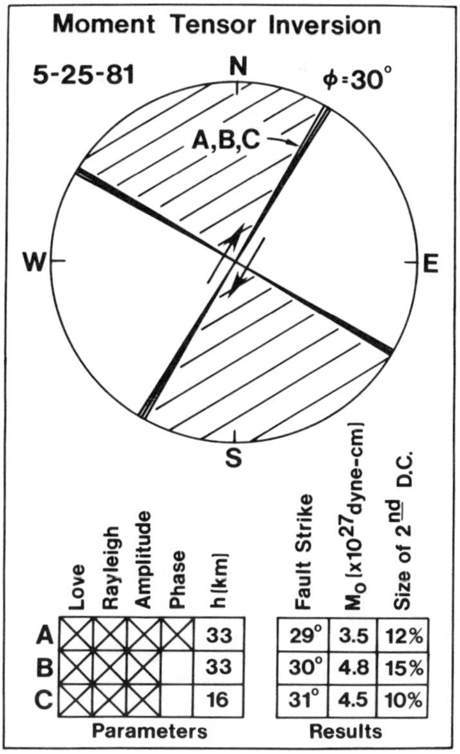

Figure 8

Summary of moment tensor inversion for three different cases (A, B, and C). The major double couple is a strike-slip focal mechanism for all cases. Box on left shows the data set and assumed depth, while the box on right shows the resultant fault strike, seismic moment, and relative size of the second double couple. Right lateral strike-slip along a fault strike of N30°E is clearly a consistent result.

CMT inversion can potentially yield more information about the earthquake, but it is also possible that lack of coherence at the shorter periods might produce a smaller moment estimate than the Kanamori and Given technique.

To test the robustness of our results, we have used the Harvard CMT spatial and temporal centroid values and recomputed the solution with the Kanamori and Given technique. We have also added several more Rayleigh wave phases recorded by IDA stations, and performed the inversion at periods ranging from 150 s to more than 300 s, with and without finite faulting; we find the faulting geometry to be extremely stable. Our overall best estimate for the seismic moment that matches both Rayleigh and Love wave lobe amplitudes for finite faulting is 5×10^{27} dyn cm. The moment magnitude (KANAMORI, 1977) for the 1981 event is then $M_W = 7.7$. We conclude that both the faulting geometry and seismic moment are well resolved.

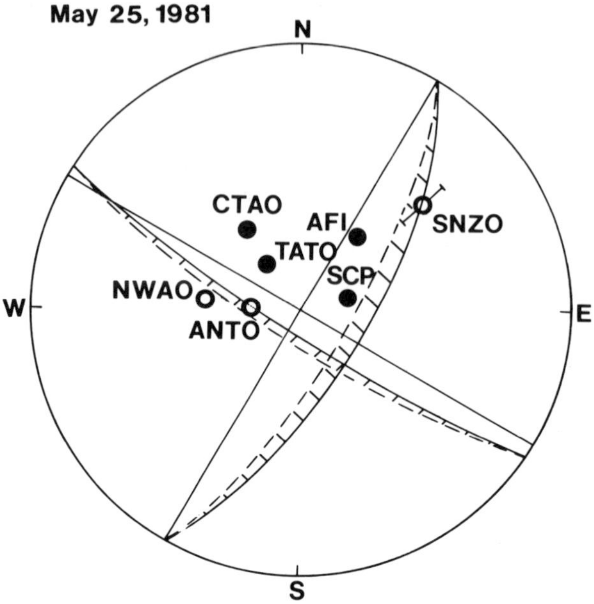

Figure 9
Summary of fault plane inversion results. The pure strike-slip mechanism with a N30°E strike is shown. *P* wave first motions from GDSN stations are plotted. The dip of the N30°E nodal plane is varied within the hachered region.

3.1.3. Surface wave directivity. We looked for evidence of fault finiteness in the long-period surface waves. Unilateral rupture propagation with uniform moment release produces a consistent azimuthal asymmetry in the long-period surface wave amplitudes (BEN-MENAHEM, 1961). The moment tensor inversion results presented in Figure 10 assume a point source; the directivity effect consists of a systematic azimuthal variation in the amplitude residuals. We can obtain useful constraints on the fault length by using the amplitudes from several different periods. It is desirable to work with the ratio of spectral amplitudes between even and odd order arrivals at a single station to cancel common effects. It is also possible to use variations in the phase, or in group delays, to obtain the temporal and spatial information (see KANAMORI and GIVEN, 1981). The Harvard CMT technique routinely uses this type of information.

The strongest directivity effects for unilateral rupture are observed at azimuths along the fault strike. Fortunately, lobes of the Love wave radiation pattern coincide with fault strike for strike-slip earthquakes. For the 1981 event, there are three GDSN stations in the western U.S.A. along the fault strike azimuth. Since the G1 phase was not reliably recorded by all stations, we perform the analysis using

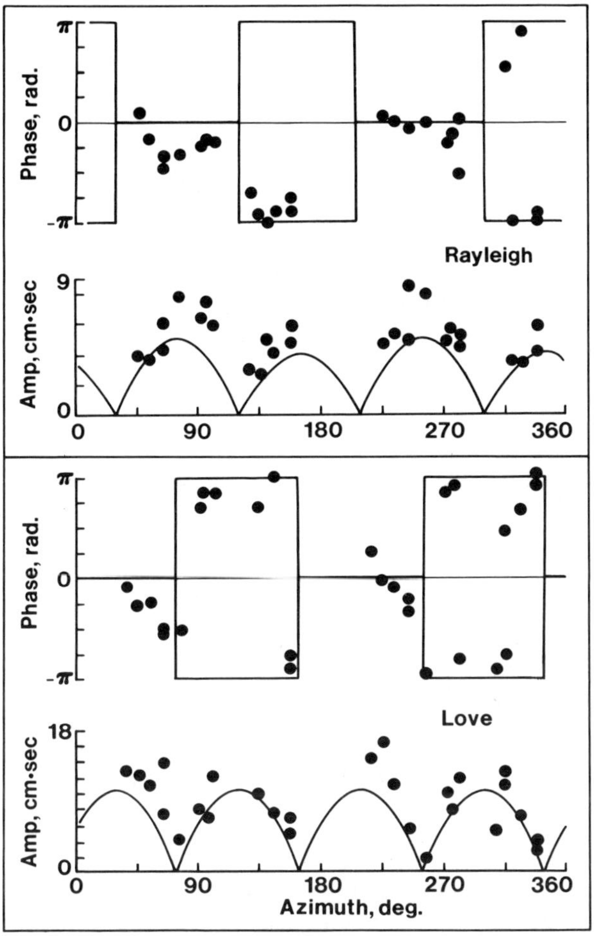

Figure 10
Detailed moment tensor results for case A of Figure 8. The inversion uses both Rayleigh and Love wave spectral amplitudes and phases at a period of 256 s. All individual data points are plotted as solid dots, with the predicted radiation pattern and phase curves also plotted. A source phase shift of 45 s has been applied to the observed phase.

G2 and G3. Figure 11 shows the observed spectral ratios for the three stations. There is little coherence between the curves for each station, the only reliable feature is the overall range of the different curves, i.e., the data band. Figure 11 also shows theoretical curves representing various rupture modes and the data band. The curve for 100 km fault length falls in the middle of the data band. The curves for a fault length of 150 km or longer clearly deviate from the data band. We conclude that a rupture mode of unilateral propagation to the southwest for about 100 km is consistent with the surface wave data.

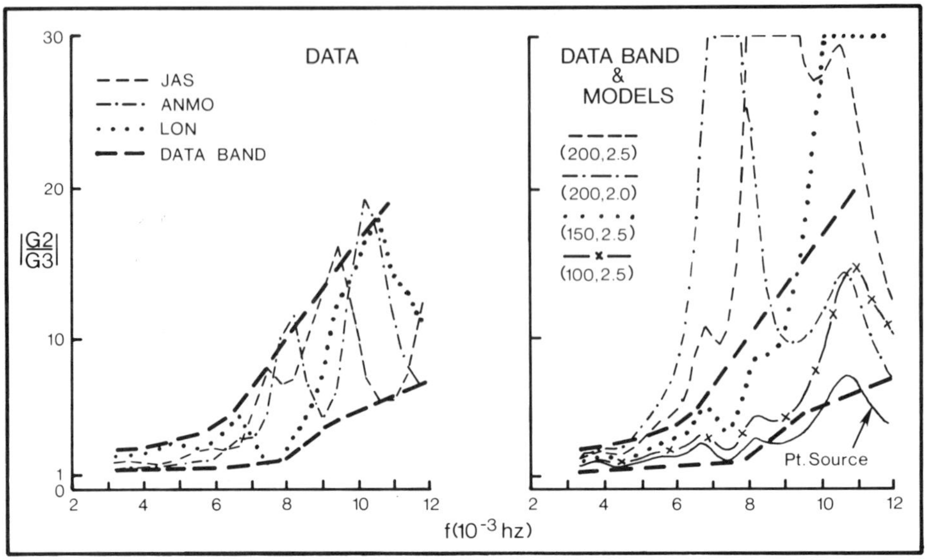

Figure 11
Test for Love wave directivity. The azimuth to western U.S.A. stations coincides with fault strike and Love wave radiation maximum. Spectral ratios between G2 and G3 waves recorded by GDSN stations are plotted in the lefthand graph. There is scatter between the results from individual stations, but together they define a data band. A comparison of this data band with curves calculated for different faulting parameters is shown in the righthand graph. The numbers in parentheses in the legend are the fault length and rupture velocity, in km and km/s respectively. The comparison implies a short fault length for the 1981 event.

3.1.4. P waves and the rupture process. Source time functions deconvolved from teleseismic *P* waves display the basic temporal history of moment release. Figure 12 shows the source time functions deconvolved from *P* waves recorded by GDSN stations and the 30–90 instrument at Pasadena, CA U.S.A. (see RUFF and KANAMORI, 1983, for discussion of the deconvolution technique). The primary feature of both the observed seismograms and the source time functions is the distinct double-event character. The second event follows rupture initiation by about 30 s. This second event seems more consistent from station to station than the first event, which splits into two subevents at some stations. The sharp rise in moment release at the onset of the second event produces a distinct break in the WWSSN long-period *P* waves (see Figure 13). The WWSSN *P* wave records provide nearly complete azimuthal coverage about the epicenter. The good azimuthal coverage and the distinct break of the second event onset allow a directivity analysis to determine the location of the second event onset relative to the epicenter. This procedure is analogous to relative earthquake location (see BECK and RUFF, 1985, for other applications). The implicit assumptions for this procedure are ideally suited for strike-slip earthquakes, where we can regard the fault surface as a thin

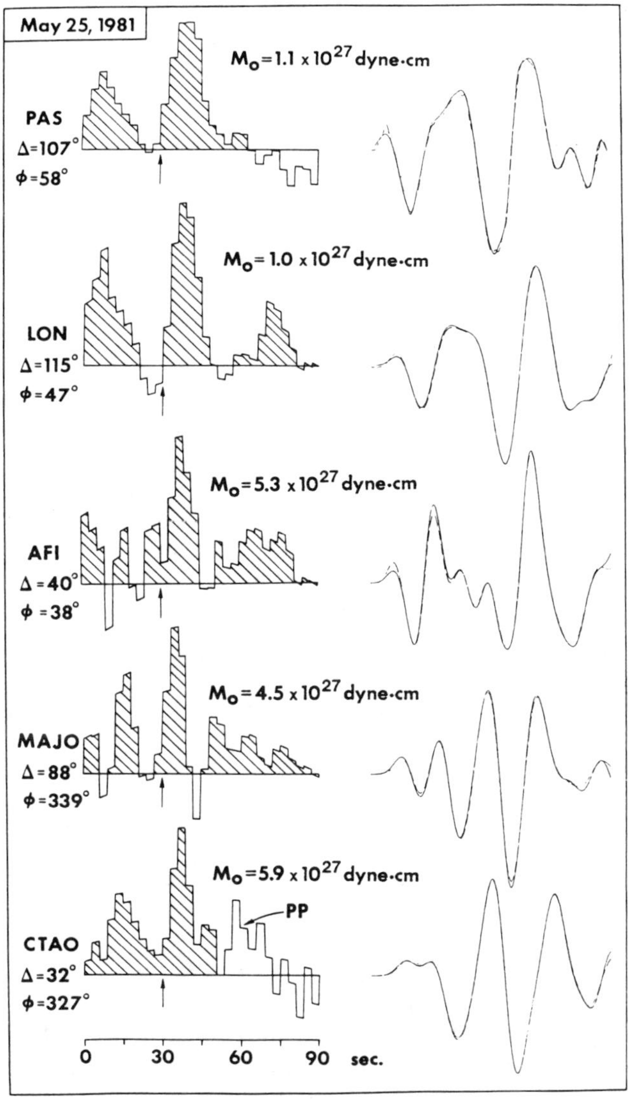

Figure 12

Source time functions and P waves for the 1981 event. Station code, distance, and event-station azimuth are listed for each station. The deconvolved source function is at left, while the observed and synthetic P wave seismograms are plotted together as solid and dashed traces, respectively (same convention in all subsequent source time function plots). The seismic moment for each time function is also shown. The arrows point to the second event onset at 30 s, and the PP arrival at CTAO is noted. The strong second event is easily seen in both the P wave seismograms and source time functions.

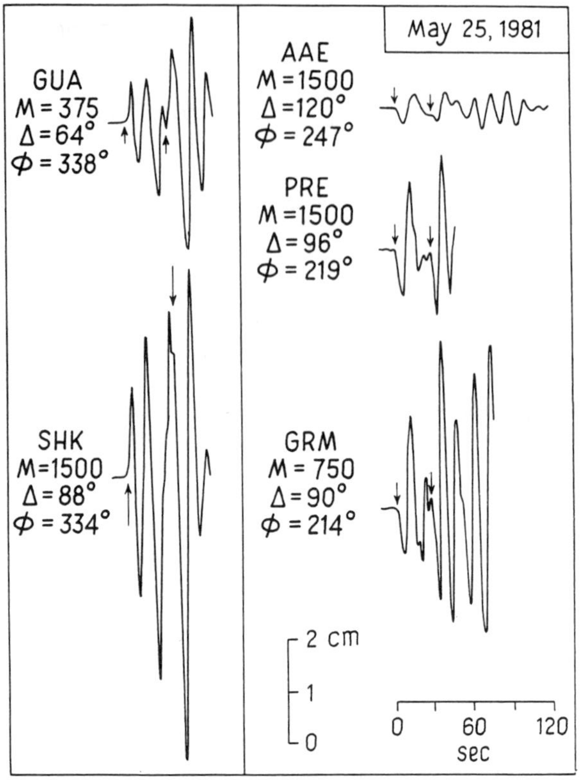

Figure 13
WWSSN long period vertical component *P* waves for the 25 MAY 1981 earthquake. Station code, epicentral distance, and azimuth are listed for each *P* wave. Amplitude scale refers to trace amplitude on the original records at listed magnifications. The arrows show the initial *P* wave arrival and the second event onset.

horizontal strip (see RUFF, 1983, for derivation of the ribbon fault model). The time difference between the first and second event onsets shows an azimuthal variation of only a few seconds. Close examination reveals that time differences at southern azimuths are perhaps 4 s less than those from northern azimuths. This implies that the second event initiated at about 50 km SW of the epicenter.

Short period records potentially offer better time resolution of source pulses. However, we must convince ourselves that a short-period break reflects a significant event in the long-period moment release, and that we can associate the same breaks between different stations. For the 1981 event, we are lucky in both respects. Figure 14 shows two WWSSN short-period records at NE and SW azimuths. There is a consistent break that seems to correspond to the second event initiation. From these two short-period records, we can see a 4.5 s directivity time shift. This time difference, combined with a *P* wave slowness appropriate for these stations, yields

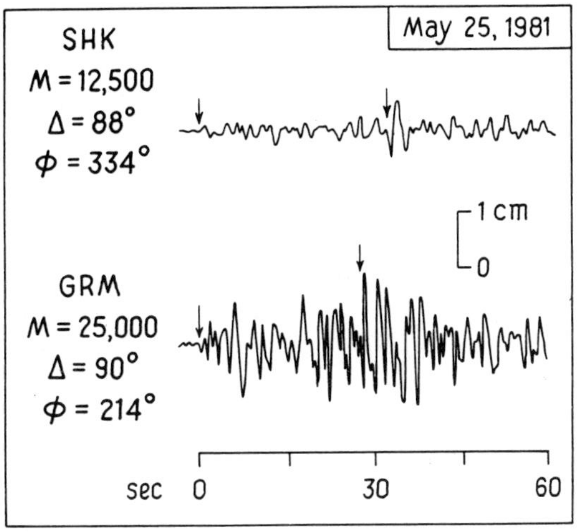

Figure 14
WWSSN short period vertical component *P* waves for the 25 May 1981 earthquake. Station code, epicentral distance, and azimuth are listed for both seismograms. Amplitude scale refers to original trace amplitude for the listed magnifications. The first arrow shows the initial *P* wave arrival. The second arrow shows a distinct break that we associate with the second event onset in the long period records.

a spatial separation of ≈ 50 km to the southwest. Thus the large second event initiates about 50 km SW of the epicenter in the vicinity of the aftershock cluster (see Figure 3). Note that the apparent rupture velocity between the epicenter and the second event onset is 1.7 km/s. The duration of the second event is about 16 s; even if we assumed a much faster rupture velocity of 4 km/s to the south, the total fault length is no more than 100 km.

Nearly all of the moment release occurs in the two events found in the *P* waves. Although there might be additional moment release after the second event, it is only a small fraction of the total. The seismic moment in the deconvolved time functions from nondiffracted stations scatters about the surface wave moment estimate. Hence, there is no evidence for an additional long-period component of moment release in the time functions (for an example of this behavior, see BECK and RUFF, 1984). Therefore, we conclude that the rupture process of the 1981 event is characterized by unilateral propagation to the southwest with most of the moment released in two events with a spatial extent of 100 km or less. There seems to be a concentration of moment release at the second event onset, 50 km SW of the epicenter, that coincides with the cluster of the largest aftershocks.

The source time functions in Figure 12 are deconvolved for a point source at 10 km depth. This implies a fault width of 20 km if the moment release is uniformly distributed down from the ocean bottom. It is in fact quite difficult to resolve the depth distribution of large shallow events. The long-period surface waves clearly cannot resolve the difference between, for example, 16 and 33 km depth. The *P* waves offer better resolution than the long-period surface waves, but there are difficulties. CHRISTENSEN and RUFF (1985) demonstrated both theoretically and empirically a characteristic behavior of source time functions that are deconvolved at the incorrect depth. Namely, the source time function becomes more "complicated" and develops extraneous multiple pulses when the assumed depth is greater than the true depth. Thus, we can deconvolve source functions for a range of depth assumptions, and the best depth estimate corresponds to the simplest source function as determined by visual inspection or some statistical measure. Figure 15 shows the focal depth test for the 1981 event for station MSO. A depth below 20 km

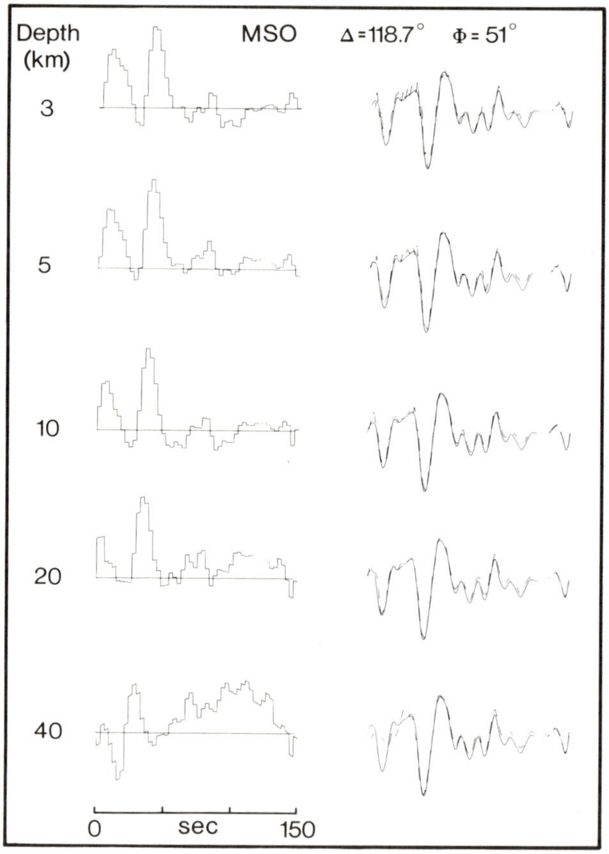

Figure 15

Focal depth test for the 25 MAY 1981 earthquake. The *P* wave recorded at MSO is deconvolved for different depth assumptions, ranging from 3 to 40 km. Plotting convention is the same as in Figure 12. Water layer multiple bounces with a 2 km ocean depth are included in the Green's function. The listed depth refers to distance below the ocean-earth interface, and a *P* wave velocity of 6.6 km/s is assumed. The simplicity criterion for depth selection indicates a best point source depth of 5 to 10 km.

is not acceptable. Although it is difficult to choose between the 5 to 10 km cases, we found that a point source depth of 10 km produces adequate results for all seismograms. Thus, the fault width is 20 km or less.

3.1.5. P waves: unexplained features. We end our *P* wave analysis by pointing out that there are still some unexplained aspects of the rupture process that are exhibited in the *P* waves. We stated above that there is no significant moment release, in terms of the total seismic moment, after the second event. Indeed, there are no large features in the source functions beyond the second event at stations in the NE quadrant, e.g., MSO. However, this is not true at other azimuths. Figure 16 shows *P* waves recorded by long period WWSSN instruments. Unlike the vertical component *P* waves in Figure 13 that eventually go off-scale, these *P* waves are on-scale for three or more minutes. *P* waves at the NW and SW azimuths show additional pulses after the second main event. Several seismograms display a nearly constant amplitude level for almost three minutes. These quasi-harmonic pulses are quite coherent between stations in any one azimuthal sector, but clearly change as a function of azimuth. Nodal seismograms will typically display this behavior, but not all of these stations are nodal. Another possibility is that these oscillations are multiple reflections within the ocean layer. These water bounces usually do not

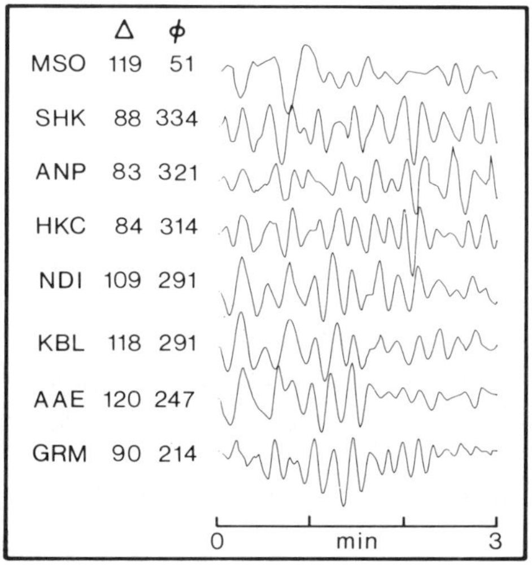

Figure 16

P waves of the 25 MAY 1981 earthquake recorded by WWSSN long period instruments. Station code, distance, and azimuth (in degrees) are listed in front of the traces. The seismograms are ordered by azimuth, counter-clockwise from NE to SW. Due to off-scale *Z* components, the *N* component records are used for stations SHK, ANP, HKC, and GRM (there is long period drift in the GRM record). Amplitudes are scaled to the same range for each trace. This comparison illustrates the azimuthal dependence of the quasi-harmonic oscillations after the second event, yet the oscillations are coherent in any one azimuthal sector.

affect long period records unless the water depth is quite deep and the focal mechanism is favorable, and even then a very shallow focal depth is required (see WARD, 1979; BECK and RUFF, 1987). At first glance, water bounces seem to present an attractive mechanism to explain the azimuthal variation: waves to the NE stations might see a water depth of less than 2 km, while waves to the NW and SW might bounce in the deeper water of the Puysegur trench. However, a preliminary quantitative exploration of this hypothesis does not satisfactorily explain the differences between the seismograms. A comparison of the P wave Green's functions indicates that the NW and SW stations might be more sensitive than the NE stations to small changes in the focal mechanism. While we are reasonably confident of the main conclusions regarding the 1981 event, i.e., seismic moment, focal mechanism, fault dimension, and double event character, there are unexplained complications in the P waves after the second main event. Earthquakes do not easily yield all their secrets.

3.1.6. Seismic displacement and static stress drop. We have determined the seismic moment and fault dimension for the 1981 event. Other seismological quantities can be inferred from these fundamental parameters. From the basic definition of seismic moment (M_0), the seismic displacement averaged over fault area is given by: $D = M_0/(\mu LW)$, where the fault area is $A = LW$, with L and W the fault length and width respectively, and μ is the source region shear modulus. With a seismic moment of 5×10^{27} dyn cm, a shear modulus of 3×10^{11} dyn cm^{-2}, and a fault length of 100 km, the average displacement is $D = (1.7 \times 10^4)W^{-1}$ m km. While it is difficult to precisely determine the fault width, our satisfactory P wave modeling with a point source depth of 10 km implies that the fault width is 20 km or less. Substituting 20 km into the above formula, $D = 8.5$ m, and any decrease in W or L increases D. To calculate the repeat time of a 1981-type earthquake for this minimum estimate for D, we use a strike-slip velocity component of about 2 cm/yr and the simplest assumption for the recurrence interval of a 1981-type event: each event ruptures the same segment with the same displacement. With these assumptions, the recurrence interval is about 400 years. There is no historical record available to check this value, so it must be regarded as an order of magnitude estimate.

The standard formula for the static stress drop of an infinitely long strike-slip earthquake that ruptures from the surface to a depth W is: $\Delta\sigma = (2/\pi)\mu(D/W)$. Of course, this formula is based on physically incorrect assumptions, e.g., the displacement is uniform and drops to zero at the fault edge, but it provides a useful basis for comparing different earthquakes. With the above parameter specifications, we obtain a stress drop of about 80 bars for the 1981 earthquake. This value is at the upper bound of observed stress drops for interplate strike-slip earthquakes (see KANAMORI and ANDERSON, 1975; PURCARU and BERCKHEMER, 1982). Note that if the fault width is 10 km rather than 20 km, the stress drop increases by a factor of four to 320 bars. For comparison, the stress drop for the 1975 North Atlantic "intraplate" strike-slip earthquake is 140 bars (LYNNES and RUFF, 1985), the average stress drop for the 1972 Sitka strike-slip earthquake is 100 bars for a 10 km fault width (SCHELL and RUFF, 1988), while the average stress drop for the 1976 Guatemala earthquake is only 30 bars (KANAMORI and STEWART, 1978). Thus, it appears that the 1981 Macquarie Ridge event has a higher average stress drop than other large interplate strike-slip events. We return to this aspect later in our seismotectonic discussion.

3.2. The 12 OCT 1979 Puysegur Trench Earthquake

The 1979 event (ISC parameters: origin time, 10:25:23.1, hypocenter, 46.54°S 165.9°E $h = 33$ km, $M_s = 7.4$) is the second largest MRC earthquake since 1963. This event is located at the northern edge of the Puysegur trench where the continental shelf of South Island is encountered. Events in 1918 and 1945 of unknown mechanism occurred in this same region. The P wave first motion data are shown in Figure 17. The focal mechanism is clearly different from the 1981 event mechanism as both SW and NW stations are dilatational. The first motion data

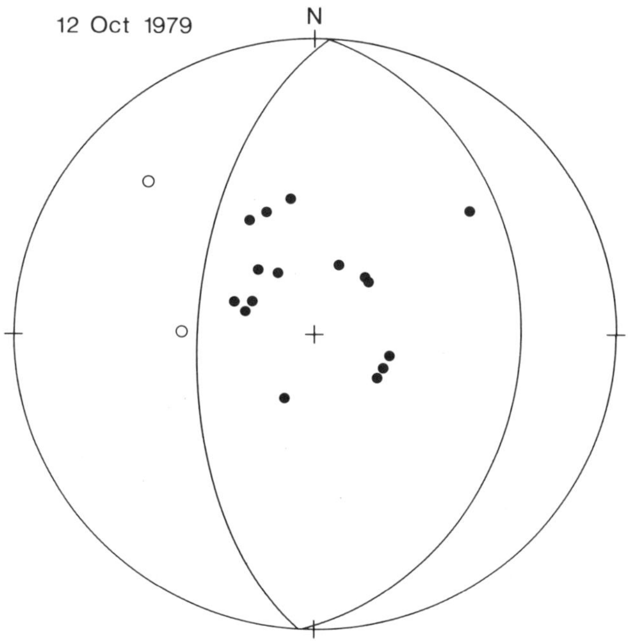

Figure 17
P wave first motions and focal mechanism solution for the 12 OCT 1979 event. The strike and dip of the steeply dipping nodal plane are determined by moment tensor inversion and first motion data. The other nodal plane is then specified by fault plane inversion.

only constrain the steeply dipping N-NE trending nodal plane. To determine the other nodal plane and the seismic moment, we use long-period Rayleigh waves. The IDA network was operating in 1979 and produced excellent long-period vertical component recordings of multiple orbit Rayleigh waves. We have applied the Kanamori and Given technique to these phases and obtained good results for the faulting geometry and the seismic moment. The constrained linear moment tensor inversion produces a thrust event with a fault strike of 3°. The first motion data constrain the west dipping plane to a dip of 58°. With these constraints on one nodal plane, the fault plane inversion determines a rake of 90.4°, i.e. pure thrust faulting, and a seismic moment of 0.6×10^{27} dyn cm. The M_W for the 1979 event is then 7.1. Figure 18 shows the Rayleigh wave spectral amplitudes and phases with the fault inversion solution.

Investigations of the seismicity at many different subduction zones find focal mechanisms similar to that for the 1979 event with the shallow dipping nodal plane defining the fault plane. The aftershock distribution for the 1979 event supports the choice of the shallow dipping plane as the fault plane (see Figure 3). The largest one-day aftershocks define a roughly equant region that is 40–60 km across; the area represents a typical fault area for an event of this magnitude. The mainshock

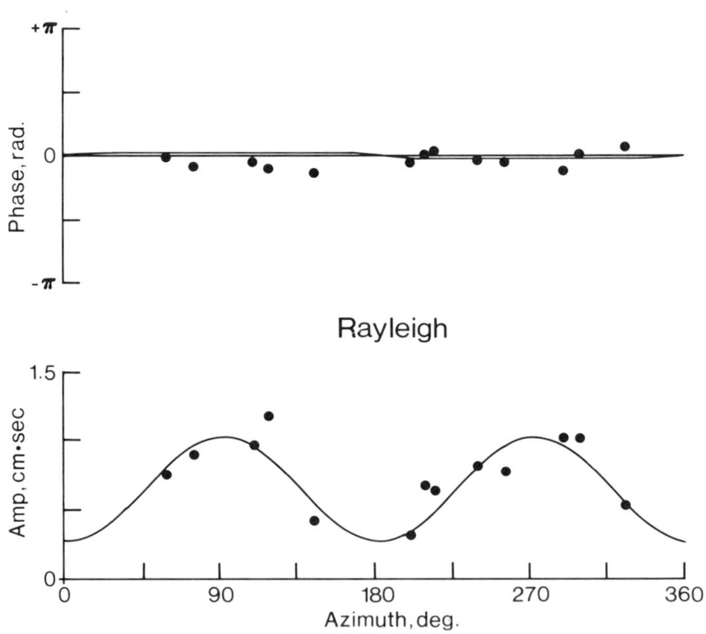

Figure 18
Rayleigh wave spectra and phase for the 12 OCT 1979 earthquake. The data (solid dots) are for a period of 256 s from R2 through R4 phases recorded by IDA stations. The fault plane inversion for depth of 33 km gives focal parameters of: $\phi = 183°$, $\delta = 58°$ (constrained), and $\lambda = 90.4°$, with a seismic moment of 6×10^{26} dyn cm. The amplitude and phase curves for this solution are plotted.

hypocenter is at the downdip edge of the rupture area, another common characteristic of subduction events.

We have also studied the long period P waves to determine the depth and moment release history for this event. Figure 19 shows the P wave at SEO deconvolved at various depths. The depth extent of rupture is quite shallow, similar to the 1981 event. The source function at 40 km clearly displays the features of overestimating the depth, i.e., a complicated double-sided source function. The source functions above 10 km are all acceptable in terms of the simplicity criterion. We have used a point source depth of 5 km for source time fuction deconvolution. Figure 20 shows the source time functions. The seismic moment of the P wave source time functions seems to be higher than the surface wave value. The 1979 event is a simple single pulse with a duration of 15 s. With a rupture velocity of 3 km/s, this duration implies a fault extent of 45 km for a semi-circular rupture, once again a typical fault dimension for an earthquake of this size. If we assume a

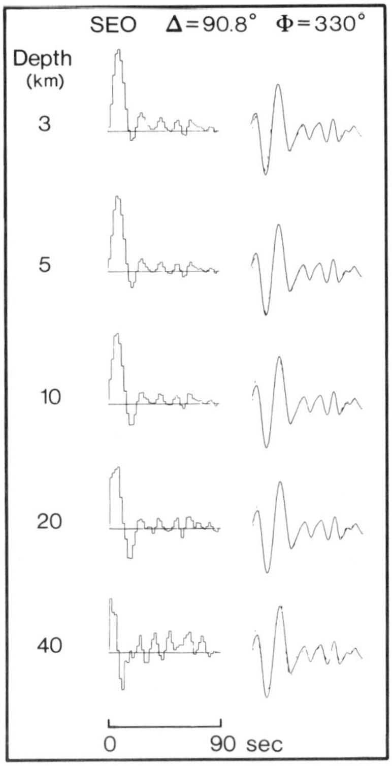

Figure 19
Focal depth test for the 12 OCT 1979 earthquake. The P wave at SEO is deconvolved for assumed depths ranging from 3 to 40 km. Water layer multiple bounces with a 1 km ocean depth are included in the Green's function. The simplicity criterion indicates a shallow depth of 3 to 10 km.

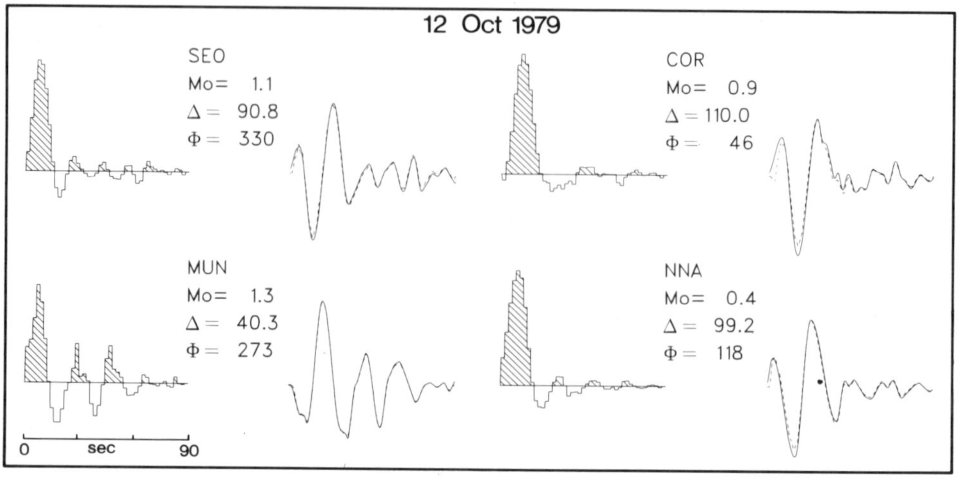

Figure 20
Source time functions for the 12 OCT 1979 event deconvolved from WWSSN long period Z component records. Seismic moment ($\times 10^{27}$ dyn cm), distance, and azimuth are listed for each station. The point source is at a depth of 5 km and water layer multiples are included. These stations at different azimuths show a consistent simple pulse of moment release with 15 s duration.

fault area of $40 \times 60\ \text{km}^2$ and a shear modulus of 3×10^{11} dyn cm, the seismic moment yields an average displacement of 0.8 m.

If we assume that the 1979 earthquake is a repeat of the 1945 event, then the recurrence interval is 34 years and the thrust velocity component from the RM2 model gives an accumulated seismic displacement of 1.2 m, within 50% of our above calculated displacement. Given the crude estimate of fault area, this approximate agreement between the seismic and cumulative displacement is consistent with the assumption that the 1945 and 1979 events ruptured the same fault area. Note that the recurrence interval between the 1918 and 1945 events is 27 years, hence it is possible that all three events ruptured the same thrust fault segment.

As previously noted, there is no indication of an anomalous stress drop in the aftershock area or rupture duration of the 1979 earthquake. The simple formula for the stress drop of a circular fault is: $\Delta\sigma = (7\pi/16)\mu(D/r)$, where r is the fault radius. To calculate the stress drop, we use the above discussed seismic moment, shear modulus, and displacement, and a fault radius for a fault area of $2400\ \text{km}^2$. These parameter choices result in a static stress drop of 12 bars, toward the lower end of observed stress drops. Although this value can clearly increase by a factor of two or more with refined fault area estimates, it will still fall within the band of typical stress drops for thrust events.

One global characteristic of subduction is underthrusting on a shallow dipping fault plane (10° to 30°). Previous seismic evidence for subduction of the Australia plate beneath South Island include the scattered intermediate depth events in the southwestern corner of South Island and the thrust-type focal mechanism for the 1960 earthquake just off the Fiordland coast of South Island (see SCHOLZ *et al.*, 1973). Our study of the 1979 event demonstrates that shallow-dipping thrusting is also occurring south of the Fiordland coast, with the oceanic lithosphere of the Australia plate thrusting beneath the continental edge. While great underthrusting earthquakes at mature subduction zones extend to a depth of 40 km, it appears that the 1979 underthrusting extends no deeper than 10 km. Nevertheless, we conclude that there is active underthrusting at the northernmost tip of the Puysegur trench.

3.3. *The 11 JUN 1970 Hjort Trench Earthquake*

The 1970 earthquake (ISC parameters: origin time, 16:46:43.7, hypocenter 58.86°S 157.6°E $h = 64$ km, $M_s = 7.2$) is the third largest event since 1963 in the Macquarie Ridge complex. It is the southernmost large event in the Hjort trench, and occurs only 300 km NW of the PAC-ANT-AUS triple junction. It is located 30 km to the west of the bathymetric deep of the Hjort trench, which locally strikes at about N20°W. There are no long-period global digital network seismograms available for this event, and we have not used moment tensor inversion. Fortunately, the focal mechanism appears to be strike-slip and is thus well constrained by *P* wave first motion data, though there are some complications. Figure 21 shows the

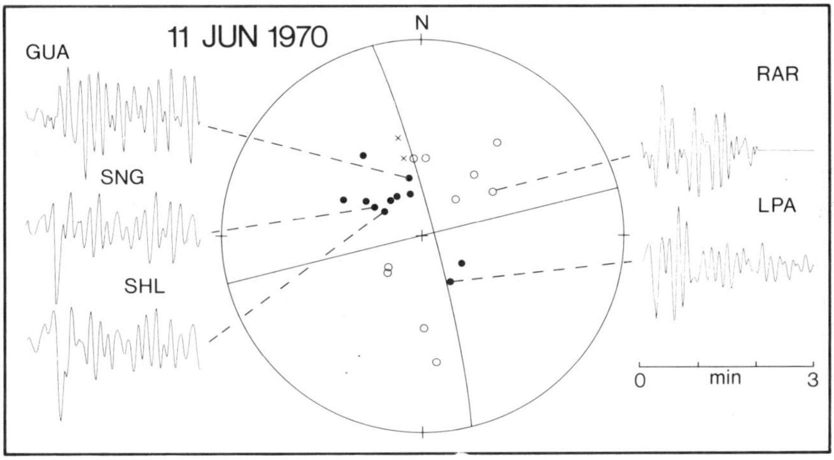

Figure 21
P wave first motions, focal mechanism, and *P* wave seismograms for the 11 JUN 1970 earthquake. The nodal plane with a strike of 346° is chosen as the fault plane. This plane has a dip in the range of 80–90°. Selected *P* waves, all recorded on WWSSN long period *Z* components, are plotted. Note the emergent character of the *P* wave which makes first motion determination difficult at many stations. The amplitude increases at about 20 s into the *P* wave. Note that LPA displays a different waveform character.

first motion data and the well constrained nodal planes. The strike of the NW trending plane is in close agreement with the local strike of the Hjort trench; hence it is reasonable to identify this plane as the fault plane. This choice for the fault plane gives right-lateral strike slip motion along this trend. There are no one-day aftershocks for the 1970 event listed in the ISC catalog.

The WWSSN long period *P* waves plotted in Figure 21 show a rather complicated character. Most of the *P* waves show an emergent arrival, making first motion determinations difficult. Careful consideration of the waveforms at noisy stations is necessary to avoid misreadings. The seismograms maintain a low amplitude for approximately 20–30 s, then a strong arrival is seen at many stations, best exemplified by the break at SNG. After this large pulse, most of the stations show quasi-harmonic oscillations for up to three minutes. These oscillations are reminiscent of the oscillations in the WWSSN *P* waves for the 1981 event. Due to the presence of these short period oscillations, we are unable to reliably deconvolve the WWSSN *P* waves. Thus, the observed seismograms are smoothed before deconvolution. It appears that reliable and consistent source functions can be obtained by eliminating the shorter periods, with an important exception that will be discussed below. This implies that the Green's functions are not correct for the higher frequencies in the observed *P* waves.

One curious feature of the ISC parameters for this event is the focal depth of 64 km, determined by reported times of *pP-P*. Looking at the seismograms in Figure 21, it seems rather unlikely that a *pP* phase could be reliably identified. We have used the RAR *P* wave to perform the focal depth test as this station is relatively insensitive to small changes in the focal mechanism. It is immediately obvious that the focal depth is above 40 km (Figure 22), thus the ISC depth is incorrect. The simplicity criterion indicates that either 3 or 5 km is the best point source depth. Hence this is a shallow event, typical of strike-slip events in the Macquarie Ridge and elsewhere.

We have deconvolved source time functions from the filtered WWSSN long-period *P* waves. Figure 23 shows the results for the focal mechanism of Figure 21 and a focal depth of 5 km. The average seismic moment from the time functions is about 1×10^{27} dyn cm ($M_W \approx 7.3$), crudely consistent with the M_s value. The long period part of the moment release shows a single pulse with an emergent ramp-type onset and a rather abrupt truncation. Sharp truncations of the moment release have been observed for several large earthquakes (e.g., BECK and RUFF, 1985; SCHWARTZ and RUFF, 1985). This truncation is responsible for the major pulse in the seismograms at about 20 s.

The LPA seismogram shows a completely different character. Note that the *P* wave onset at LPA is relatively larger than at the other stations. The arrival time has been checked, and it would appear that this is a real feature. Moreover, the *P* wave at NNA displays the same character as LPA. It is not surprising then that the source time function at LPA shows a completely different character than the other

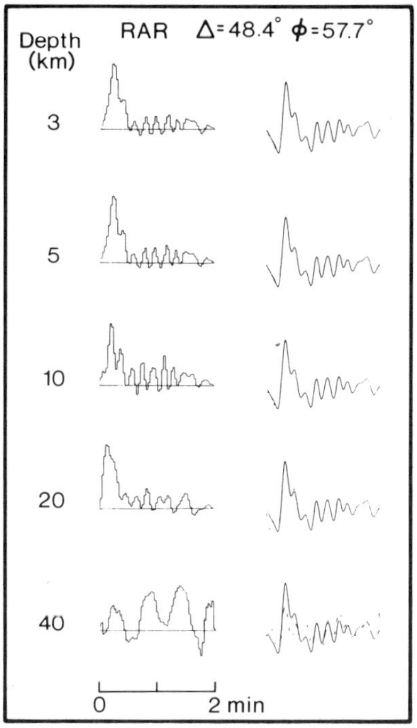

Figure 22
Focal depth test for the 11 JUN 1970 earthquake. The long period Z component at RAR is deconvolved at assumed depths ranging between 3 and 40 km. Water bounces from a 4 km ocean thickness are included, and the fault plane dip is 90°. The observed P wave is filtered to largely remove periods shorter than 6 s. The simplicity criterion indicates a shallow focal depth, perhaps 5 km.

stations. There is an obvious explanation for this discrepancy: we do not have mutually consistent Green's functions for all the stations. Given that the character of the other seismograms is widely observed at different azimuths, we prefer to interpret LPA as "anomalous". Since NNA seems to share this anomalous character, we cannot simply dismiss it as a seismograph problem or a nodal phenomenon. One possible cause of this anomalous behavior is some change in the focal mechanism that effects the SE quadrant more than the other quadrants of the focal sphere. At this time however, we are not able to suggest, much less prove, a focal mechanism change that reconciles all the observations. We can only issue a caution that moment release might have occurred with faulting geometry different from the first motion mechanism. Thus, the rupture process of the 1970 earthquake presents complications at all periods: short periods have the quasi-harmonic oscillations and the long periods are not mutually consistent at all azimuths. Since some fraction of the moment release occurred with the geometry of the first motion mechanism, the RM2 predicted motion is correct insofar as the sense of strike-slip motion is concerned. This lends some support to the notion that the PAC-AUS plate boundary follows the Hjort trench.

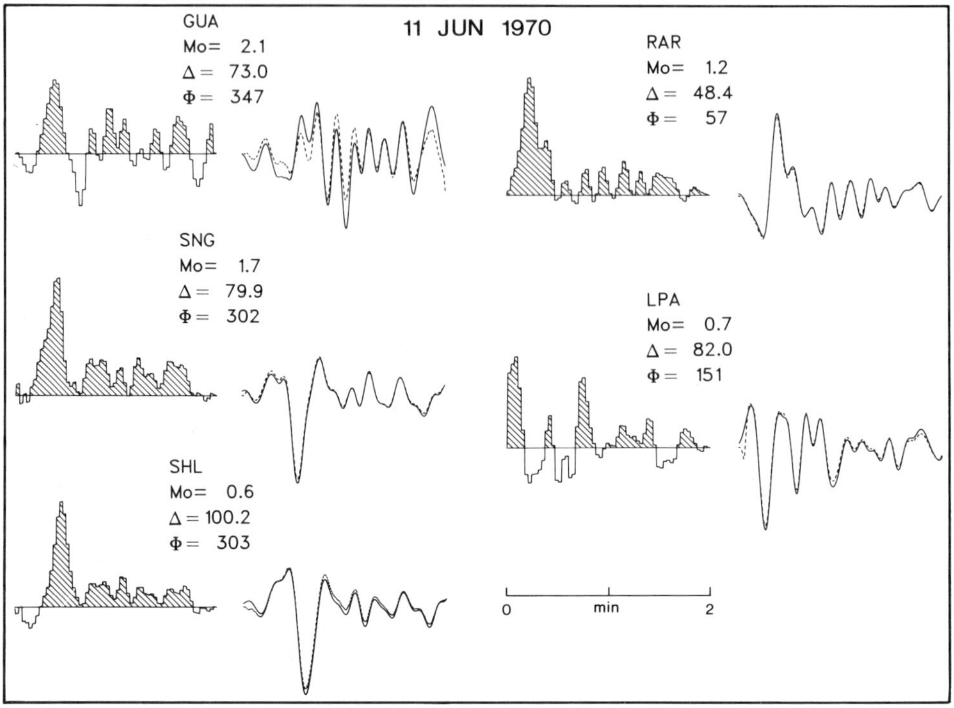

Figure 23
Source time functions for the 11 JUN 1970 earthquake deconvolved from filtered long period P wave seismograms. The effect of the smoothing filter can be seen by comparing the filtered traces to the observed traces in Figure 21. The source functions are mostly coherent in the long period band with a single large pulse of moment release, LPA is the obvious exception. Seismic moment has units of 10^{27} dyn cm.

3.4. Focal Mechanisms of Smaller Post-1963 Events

We report the focal mechanisms of smaller earthquakes that have been determined by other investigators. These determinations are based on two different procedures: first motion focal mechanisms and long-period moment tensor inversion. All mechanisms for the post-1963 earthquakes are listed in Table 1.

3.4.1. First motion focal mechanisms. We have found three sources of first motion mechanisms for Macquarie Ridge earthquakes, SYKES (1967), BANGHAR and SYKES (1969), and JOHNSON and MOLNAR (1972), for a total of five earthquakes. In general, the steeply dipping planes of these mechanisms are well constrained, while shallow dipping planes are somewhat uncertain. Hence, the strikes of the shallow dipping "fault planes" for the thrust events are not well determined, but the strike slip mechanisms and the slip vector orientation for the thrust events are generally reliable. Three of these focal mechanisms are for events

Table 1

Summary of focal mechanisms for Macquarie Ridge Complex earthquakes.

#	Date	Epicenter		Depth	Mag	Focal Mech				Ref
		Lat(°S)	Lon(°E)			P	ϕ	δ	λ	
1	12 SEP 1964	49.0	164.5	S	6.9	A	173	58	(68)	S
2	8 NOV 1964	49.0	164.2	S	5.6	A	146	78	(90)	BS
3	2 AUG 1965	55.9	157.7	S	6.6	F	184	84	210	BS
4	25 MAY 1966	52.8	160.2	S	6.5	?	(168)	(36)	(90)	BS
5	20 SEP 1967	49.6	163.9	S	6.1	A	191	72	(90)	JM
6	11 JUN 1970	58.9	157.6	5	7.2	F	346	85	180	*
7	12 OCT 1979	46.5	165.9	5	7.4	F	3	32	90	*
8	7 FEB 1980	54.2	158.8	S	6.5	F	20	90	186	*
9	25 MAY 1981	48.6	164.7	10	7.6	F	30	70	173	*
10	27 JUN 1982	55.4	160.2	S	6.1	F	214	85	134	H
11	7 JUL 1982	51.1	160.6	S	7.0	F	63	78	193	H
12	23 MAY 1984	52.0	161.1	S	5.9	F	211	34	122	H
13	2 JUL 1984	54.9	158.9	S	5.0	F	19	90	180	H
14	3 JAN 1985	54.4	155.4	S	5.1	?	240	84	160	H
15	31 JAN 1985	46.1	165.1	S	6.1	A	197	66	95	H

Epicentral parameters are ISC values through 1982, NEIS for 1984 and 1985. Magnitude is USCGS m_b through 1967, M_s afterwards. Shallow hypocentral depth is indicated by 'S'. Numerical depth values are in km, and are determined by P wave analysis. The focal mechanism parameters are strike, dip, and rake, in degrees, for the standard convention (see KANAMORI and GIVEN, 1981). The 'P' column refers to which nodal plane is described, 'F' is the fault plane and 'A' is the auxiliary plane in our interpretation. For first motion mechanisms, we describe the best-constrained plane. Unreliable values are given in parentheses. The focal mechanism references are: S, SYKES, 1967; BS, BANGHAR and SYKES, 1969; JM; JOHNSON and MOLNAR, 1972; H, the Harvard group, see ISC catalog; *, this study.

that occur in the cluster of seismicity between 49°S and 50°S (we will refer to this seismicity cluster as the "1981 cluster"), and all three events show thrust mechanisms. We assume that the east-dipping nodal planes are the fault planes. The azimuths of the poles to the auxiliary planes for these three events are N56°E, N83°E, and N101°E. Since the intermediate orientation is from the largest event, we use a slip vector orientation of N83°E as representative of the thrust motion in the 1981 cluster. The 25 MAY 1966 earthquake has compressional first motions in the center of the focal sphere and is therefore a thrust event, but the nodal planes are free to rotate to any strike. The 2 AUG 1965 event is a right lateral strike-slip event along the northerly plane and is well constrained by first motions. We will discuss this event in a later section. There is one MRC earthquake studied by BANGHAR and SYKES (1969) that is not listed in Table 1, the 12 MAY 1963 event. Their first motion data suggest a vertical dip-slip mechanism, although it is poorly constrained. The epicenter is 150 km to the east of the MRC and the ISS depth is listed at 70 km. These parameters suggest that this earthquake may be an intraplate event in a subducted slab. Only a few good quality seismograms are available for this event due to its small magnitude. A comprehensive P wave study is not possible, but preliminary analysis of the P waves at close stations indicate that this earthquake is in fact shallow. A shallow depth removes the speculation of a slab event.

3.4.2. Moment tensor inversion focal mechanisms. Several reliable focal mechanisms in the MRC have been determined as part of systematic global studies. As an early experiment in the systematic application of moment tensor inversion, KANAMORI and GIVEN (1981) inverted the long-period Rayleigh waves recorded by IDA stations to determine the constrained moment tensor for large events that occurred in 1980; their event list includes one event in the Macquarie Ridge complex. The 7 FEB 1980 event (ISC parameters: origin time, 10:49:16.3, hypocenter 54.2°S 158.8°S $h = 10$ km, $M_s = 6.5$) is located just 40 km north of Macquarie Island along the crest of the northeast striking bathymetric high. The constrained moment tensor solution in Kanamori and Given is a strike-slip mechanism with right lateral slip along a plane that trends N20°E; this trend coincides with the local strike of the Macquarie Ridge. To check this solution, we have used P waves recorded by GDSN stations. We find that only a minor modification of the Kanamori and Given focal mechanism is necessary. The first motion data and a few

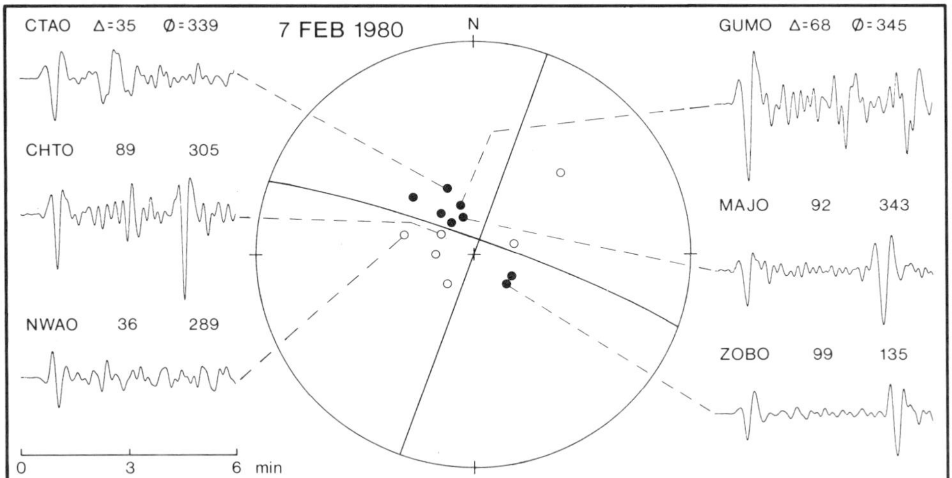

Figure 24
Focal mechanism, P wave first motions, and GDSN P waves for the 7 FEB 1980 earthquake. Station code, distance and azimuth (in degrees) are listed for these Z components. The moment tensor inversion gives a major double couple of right lateral strike-slip along a trend of N20°E. The first motion data require only a minor modification, a rake of 186° rather than 180°. The long period GDSN P waves show single simple pulses. The traces start 30 s before the P wave arrival, and the PP phase follows the P wave.

P wave seismograms are plotted in Figure 24 along with the modified focal mechanism. The 1980 event appears to be a simple shallow earthquake.

The Harvard group systematically inverts for the moment tensors of events that are well recorded by the GDSN, their solutions are now routinely published in the NEIS and ISC listings. Their catalog of solutions is fairly complete for events with magnitude larger than $5\frac{1}{2}$ from 1981 to the present, and they have extended the catalog back to 1977. A total of six earthquakes, excluding the 25 MAY 1981 event, in the Macquarie Ridge complex are in their post-1981 catalog. We have not checked any of these solutions, but given the excellent correspondence for the faulting geometry of the large 1981 event, our opinion is that the nodal planes are well determined. These six events are distributed along the Macquarie Ridge complex and include both strike-slip and thrust earthquakes. The focal mechanisms for these earthquakes are listed in Table 1. We will discuss these focal mechanisms in conjunction with all the MRC focal mechanisms in the following section.

3.5. Discussion of Post-1963 Seismicity

The focal mechanisms and locations for the entire list of post-1963 earthquakes are plotted in Figure 25 along with the simplified bathymetry of the MRC. The earthquakes are identified by numbers that correspond to the listing in Table 1. One curious feature of the mechanisms in Figure 25 is the interspersed occurrence of both strike-slip and thrust earthquakes. If we focus our attention on the 1981 cluster, we see that the smaller thrust events are located adjacent to the 1981 fault zone. The slip vector strike of N83°E for thrusting is markedly different than the N30°E strike for the 1981 event slip vector. Any estimate of the relative plate motion in this region based solely on the smaller thrust events would be incorrect. Thus, we must be cautious when using any single earthquake along the Macquarie Ridge as an indicator of the local plate motion. We can view the slip vectors from strike-slip and thrust events in the same region as extremal bounds on the direction of plate motion.

There are two events to the north of the 1981 cluster. Event #7 is the 12 OCT 1979 earthquake that we have discussed in detail, event #15 is a smaller event that also shows shallow-dipping underthrusting to the east. This latter event is too small to represent a repeat of the 1979 event and its location is to the west of the 1979 event. There are two thrust mechanisms in the McDougall trough region ($52\frac{1}{2}$°S) to the south of the 1981 cluster. Unlike the 1979 mechanism, these focal mechanisms imply underthrusting to the west. While mechanism #4 is not reliable, mechanism #12 is a Harvard CMT solution. A switch in the "polarity" of underthrusting is thus associated with the alternation of the troughs and rises between the Puysegur and McDougall regions. Although we would also anticipate west-dipping underthrusting along the small trough to the east of Macquarie Island ($54\frac{1}{2}$°S), there are only two strike-slip mechanisms in this region. Note that the Harvard CMT solution for

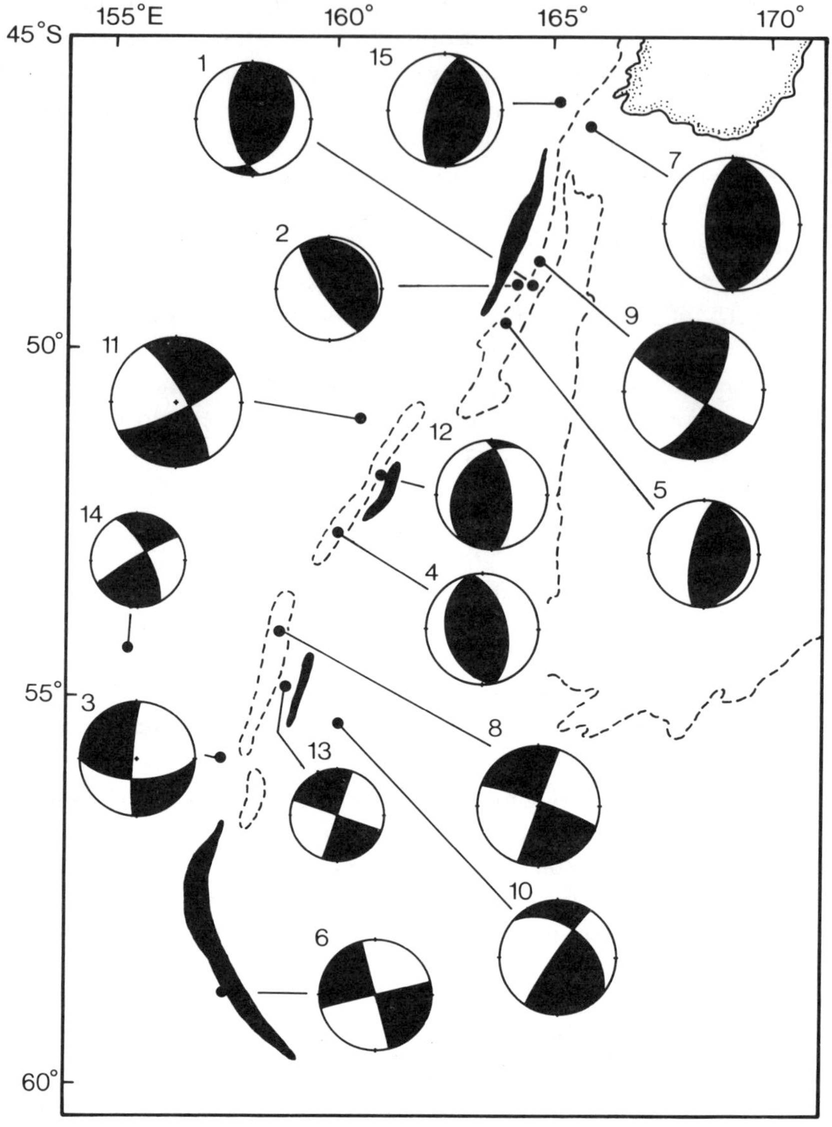

Figure 25

Focal mechanisms for Macquarie Ridge earthquakes and the primary physiographic features. All reliable mechanisms for post-1963 events are shown, the numbers refer to the event number in Table 1, and the dots show the epicentral locations from Table 1. The focal sphere size is proportional to event magnitude, and the compressional quadrants are shaded in these lower-hemisphere equal-area projections. The dashed lines show the 1000 fathom depth contours, while the troughs along the Macquarie Ridge are shaded (delineated by the 2600 fathom depth contours, from MAMMERICKS *et al.*, 1975). Part of South Island is at the upper right. The shallow shelf to the east is the Campbell Plateau. The two large troughs are the Puysegur trench to the north and the Hjort trench in the south; two smaller troughs between 50°S and 55°S are to the east of the ridge. Oblique convergence is accommodated by a dual-rupture mode of strike-slip and thrust earthquakes.

event #13 has the same fault strike as the 7 FEB 1980 earthquake discussed above (event #8). One implication of these events is that the seismically defined plate boundary is along, or slightly to the east of, the bathymetric rise between 52°S and 55°S. To the south, the strike-slip events of 2 AUG 1965 (#3) and 11 JUN 1970 (#6) both plot slightly to the west of the MRC.

There are three events in Figure 25 that plot off the main bathymetric trend of the MRC, events #10, #11, and #14. The two smaller events in the Macquarie Island region (#10, #14) are clearly displaced to either side of the MRC and represent intraplate seismicity. The mechanisms show right lateral strike-slip along northeast trending planes, hence they exhibit the same basic deformation as the slip along the PAC-AUS boundary. The large off-MRC event at 51°S (#11) deserves more emphasis. This event is located in the seismicity gap between the 1981 cluster and the McDougall trough. This seismicity gap corresponds to an offset in the bathymetric rise, and the alternation of the troughs. Perhaps the 1982 event occurs along an ancient zone of weakness that is associated with the MRC offset. Note that the right lateral motion for this event occurs along the nodal plane striking at N63°E. This fault strike is significantly different from the N30°E fault strike of the large 1981 event. Thus the slip vector for the 1982 event points in a direction intermediate to the strike-slip and thrust slip vectors in the 1981 cluster. The slip vector for the 1982 event may reflect the "true" PAC-AUS relative motion.

The seismicity gaps are interesting features. The 1981 cluster seems to be bounded by zones of reduced seismicity to the north and south. With active subduction and the occurrence of three large earthquakes in the northernmost Puysegur trench, the seismicity gap in the central portion of the Puysegur trench is particularly puzzling. Is this region a seismic gap, or is it possible that some of the surrounding pre-1963 earthquakes are actually located in this quiet region? Many of the historical earthquakes are poorly located and large shifts in their location are possible. Note that two of the large events in the northern MRC (1924 and 1926) are located off the main bathymetric expression of the Macquarie Ridge complex. Are these events mislocated, or is there significant intraplate seismic activity to the west? To complement the study of the recent large earthquakes, we have checked the locations of the pre-1963 large earthquakes and have even found focal mechanism information for two earthquakes.

4. Pre-1963 Large Earthquakes

Of the 15 large events in the Macquarie Ridge complex, 10 occurred before the establishment of the WWSSN. It is difficult to study these events as there is no centralized data archive of high-quality seismograms. Since many of the early seismographs were not well-calibrated, wave amplitude or waveshape studies are difficult. Even the polarity of many of the seismographs is not known. Station timing errors can also introduce significant travel time residuals. Despite all these pitfalls, we have systematically tested the epicentral locations of all large earthquakes in the MRC, and are able to place constraints on the faulting geometry for two events.

4.1. Epicenter Relocation

While most of the seismicity plots along the main bathymetric features of the MRC, two of the large earthquakes in Figure 1, the 24 JUL 1924 and 3 OCT 1926 events, plot west of the Puysegur trench. A major discontinuity in the oceanic lithosphere lies between the Puysegur trench and the intersection of the Balleny fracture zone with the southern edge of the Tasman plateau. This discontinuity is associated with the change in spreading direction at 50 Ma (before present) from the older Tasman sea opening to the present-day Australia-Antarctica opening (WEISSEL *et al.*, 1977). A large earthquake occurred 1 MAY 1951 along the Balleny fracture zone, and thus might be associated with the Puysegur-Balleny trend. The oceanic lithosphere between the MRC and the Balleny fracture zone is a rather narrow southward extension of the AUS plate that is bounded to the west and east by the ANT and PAC plates, respectively. Thus it is possible that the region between the MRC and the Balleny fracture zone is a mini-plate that is decoupled from the AUS plate along the Puysegur-Balleny trend. Although a few events have occurred along this trend since 1963, they are small events and the focal mechanisms have not been determined. Our analysis is thus limited to relocations of the above-mentioned three large events.

The ISS locations for the 24 JUL 1924 and 3 OCT 1926 events are quite different from the locations plotted in Figure 1. Epicenters for the pre-1950 large events are from GUTENBERG and RICHTER (1949); both the 1924 and 1926 events are shifted by about 150 km from the ISS locations. Epicentral shifts of this magnitude are not uncommon for relocations of earthquakes from the early part of this century. Gutenberg's personal notes (available through the Seismological Laboratory, Caltech, Pasadena, CA 91125) show his relocation procedure: he simply plotted travel time residuals from the ISS bulletins as a function of event-station azimuth and subjectively sketched a sinusoidal curve through the residuals. Since we are not endowed with the subjective skills of Gutenberg, we have quantified this procedure as a simple check on the locations of all large MRC events. Modern relocation techniques are quite sophisticated (see e.g., DEWEY and SPENCE, 1979) and offer relative epicentral locations accurate to 10 km or less. However, we retain a simple one-step method as we are mainly interested in much larger epicentral shifts. We have used the P arrival times as reported in the ISS bulletins. There are at least two reasons that we might expect to improve upon the ISS locations: (1) the use of modern, i.e., more accurate, travel time tables, and (2) the computational ease of testing epicentral perturbations to find the best-fit location. We have also used a subset of the data reported in the bulletins; we do not use diffracted P waves or core phases. Our selection results in fewer data, though the selected residuals are possibly more reliable.

4.1.1. Method. We use a simple relocation method that has been used by many seismologists for decades. Although this method is quite straightforward and well-known, we describe our procedure in detail for the sake of completeness.

The quantitative scheme is based on the simple first-order linearization of the travel time function for teleseismic P waves. The P wave arrival time observed at station i is ${}_{\text{obs}}T_i$. The calculated P wave time depends on three parameters: origin time, depth, and epicentral distance (t, h, and x_i, respectively). For an initial hypocentral location and origin time, we find the initial epicentral distance and calculate the travel time residual: $r_i = {}_{\text{obs}}T_i - T(t_0, h_0, {}_0x_i)$. We seek perturbations to the initial hypocentral parameters to reduce the residuals. The perturbation to epicentral location must be rewritten in terms of the perturbation in epicentral distance to the stations. If the epicenter moves a distance (dy) along a direction (θ is the angle relative to north), then the first-order relation for the travel time variation with respect to epicentral distance variation (dx_i) can be written in terms of the epicentral perturbation: $(\delta T/\delta x_i)\, dx_i = -(\delta T/\delta y)_i \cos(\phi_i - \theta)\, dy$, where ϕ_i is the event-station azimuth relative to north. The derivative $(\delta T/\delta y)_i$ is evaluated at the initial distance and is simply the ray parameter, p_i. Define the residuals for the perturbed epicentral location as dT_i, where $dT_i = {}_{\text{obs}}T_i - T(t_0 + dt, h_0 + dh, {}_0x_i + dx_i)$. Expand the travel time function as a Taylor expansion about the initial parameters, retain only the first-order linear terms, and replace $(\delta T/\delta x_i)\, dx_i$ with the above expression to obtain: $dT_i = [{}_{\text{obs}}T_i - T(t_0, h_0, {}_0x_i)] - dt - (\delta T/\delta h)_i\, dh + p_i \cos(\phi_i - \theta)\, dy$. Since we cannot resolve changes in the hypocentral depth with our data, we combine the time shift due to depth perturbation with the origin time shift. The expression in brackets is r_i, the observed travel time residual, hence we now have: $dT_i = r_i - dt + p_i \cos(\phi_i - \theta)\, dy$. The parameter perturbations, dt and dy, are linearly related to the residuals for a specified θ. The sum of the squares of residuals dT_i is minimized with respect to dt and dy by the least squares straight line through r_i plotted as a function of $p_i \cos(\phi_i - \theta)$. The intercept and slope of the least squares line through r_i are dt and $-dy$, respectively. Note that since the distance perturbation is given by the slope of the best-fit line to the residuals, the slope standard error is a measure of the distance uncertainty. We consider the epicentral relocation significant only if the standard error along the relocation direction is substantially less than the relocation distance, i.e., the slope is significantly different from zero. Since we do not know the best direction for the epicentral perturbation in advance, we try all directions, in 10° increments, and choose the direction that produces the best overall fit to the residuals. Although the relocated epicenters listed in Table 2 are quantitatively determined, our error analysis is rather simple and not completely rigorous. Thus, the listed uncertainties in latitude and longitude represent conservative subjective estimates.

Table 2

Epicentral parameters for relocated earthquakes.

Date	ISS					G & R						*			
	h	m	s	(°S)	(°E)	h	m	s	(°S)	(°E)	M	#	*dt*	(°S)	(°E)
26 JUN 1924	1	37	20.	57.0	159.0	1	37	34.	56.0	157.5	7.8	28	17.	55.0 ± 0.5	158.4 ± 1.7
24 JUL 1924	4	55	24.	48.0	159.0	4	55	17.	49.5	159.0	7.5	18	[− 16.	48.9 ± 1.0	161.5 ± 2.8]
3 OCT 1926	19	37	51.	50.5	161.0	19	38	1.	49.0	161.0	7.5	27	[34.	49.0 ± 2.2	160.2 ± 2.4]
22 FEB 1936	19	22	53.	49.0	164.4	19	22	40.	50.0	164.0	6.8	13	[7.	48.3 ± 0.6	164.0 ± 0.8]
6 SEP 1943	3	41	15.	55.1	158.5	3	41	30.	53.0	159.0	7.8	29	18.	53.9 ± 0.3	159.2 ± 0.7
1 SEP 1945	22	44	8.	46.8	165.8	22	44	10.	46.5	165.5	7.2	14	[8.	46.6 ± 0.2	165.4 ± 0.4]
1 MAY 1951	5	2	40.	50.5	149.0						7	46	[3.	50.9 ± 0.3	149.2 ± 0.4]
13 DEC 1960	7	36	13.	52.1	161.0						7.3	29	22.	50.8 ± 0.8	160.3 ± 1.5
12 SEP 1964	22	7	3.2	49.0	164.5						6.9	30	[− 2.	48.9 ± 0.5	163.5 ± 1.0]
11 JUN 1970	16	46	43.7	58.9	157.6						7.2	39	[1.	59.0 ± 0.1	157.3 ± 0.3]
12 OCT 1979	10	25	23.1	46.5	165.9						7.4	37	[0.	46.6 ± 0.1	165.2 ± 0.1]

ISS: epicentral parameters listed in ISS or ISC bulletins. G & R: epicentral parameters listed in GUTENBERG and RICHTER (1949). *: epicentral parameters from this study. Origin time is given in hour, minute, and seconds (h, m, and s respectively), latitude is in °S, longitude is in °E, M is magnitude, Gutenberg and Richter magnitude for events prior to 1960, USCGS m_b for 1964, M_s afterwards. # refers to number of observations used for relocation, dt is the origin time shift relative to ISS time. Relocation parameters in brackets are *not* significantly different from the ISS epicenter.

4.1.2. Relocation results. All relocations are relative to the ISS parameters. We recompute the station distance, azimuth, and *P* wave travel time based on the Jeffreys-Bullen tables. We apply this procedure to all large pre-1963 events along the MRC, to the 1951 event south of Tasmania, and to three post-1963 large events (see Table 2). The locations of the 1964, 1970, and 1979 events are tested to determine the bias and resolution limit of the simple relocation scheme for MRC events. Our procedure moves these events in different directions by 50 km or less. Thus, in addition to the standard error criterion, a relocation is not considered significant unless the relocation distance is greater than 50 km. Given these two criteria, we obtain significant relocations for three pre-1963 events. The events are discussed in chronological order.

The relocation of the 24 JUN 1924 event is significant, the epicenter moves 230 km northwest of the ISS location. The epicenter listed in Table 2 is along the MRC, though somewhat north of the locations of MACELWANE (1930) and Gutenberg. In the Puysegur trench region, we do not obtain significant relocations for either the 1924 or 1926 events. Though the best-fit locations are more than 100 km from the ISS locations, the standard errors include the initial epicenters. The relocation directions are to the south for the 1924 event and to the north for the 1926 event, the same directions as for Gutenberg's relocations. While the locations for these two events obtained by Gutenberg, the ISS, and our study are all to the west of the MRC, our error limits for both events extend to the MRC. Thus, we cannot reach a strong conclusion regarding the locations of these two events, but it is certainly possible that they are located west of the MRC. There is no significant relocation of the 1936 event in the 1981 seismicity cluster. Gutenberg's relocation for the 1943 event is more than 200 km north of the ISS epicenter; we obtain a significant relocation for the 1943 event that is also north of the ISS epicenter, though with a displacement of only 150 km. This relocation simply moves the 1943 event along the trend of the MRC. The relocations for the 1945 and 1951 events are not significant, the displacements are less than 50 km. Thus, the 1945 event occurred quite close to the 12 OCT 1979 event, and the 1951 event does indeed occur along the Balleny fracture zone. The relocation of the 13 DEC 1960 event is probably the most significant result of our relocation study. The relocated epicenter is displaced 160 km along a direction of N20°W from the ISS epicenter. This places the 1960 event in the low-level seismicity region south of the 1981 cluster, virtually the same location as the 7 JUL 1982 event. Thus, the occurrence of two large events in this "quiet" region implies continued slip in this region and underlines the importance of the 1982 focal mechanism. The focal mechanism of the 1960 event will be discussed in the next section.

The epicenters for all large earthquakes that we have considered are along the MRC except for the 1 MAY 1951 event and possibly the 24 JUL 1924 and 3 OCT 1926 events. We have searched the seismicity catalogs for other large events along the Puysegur-Balleny trend but have found none. There are a few very small events, but they do not define an obvious linear trend.

4.2. Focal Mechanisms

Reliable determinations of first motion focal mechanisms usually require a global seismogram collection with known instrument polarities and good azimuthal coverage. Due to special circumstances, we can constrain the focal mechanisms for two events: the 26 JUN 1924 earthquake at the northern end of the Hjort trench, and the 13 DEC 1960 earthquake south of the Puysegur trench.

A focal mechanism for the 1924 event is only possible because MACELWANE (1930) collected the seismograms for this event soon after its occurrence, he determined the instrument polarities, and published photographic reproductions of the seismograms with all pertinent information. RUFF and CAZENAVE (1985) plotted the 1924 *P* wave first motions on the focal mechanism for the nearby 2 AUG 1965 event to show that the 1924 faulting geometry differs from the strike-slip mechanism of the 1965 event (see Figure 26). The 1924 first motion data constrain one steeply dipping nodal plane to strike north. RUFF and CAZENAVE (1985) speculated that the second nodal plane has a shallow dip to the east: the 1924 earthquake then represents underthrusting on a shallow dipping fault plane. Since the combination of right lateral strike-slip and underthrusting motion occurs elsewhere along the Macquarie Ridge, this interpretation seems reasonable. The 1924 event would then be the largest thrust event in the Macquarie Ridge complex.

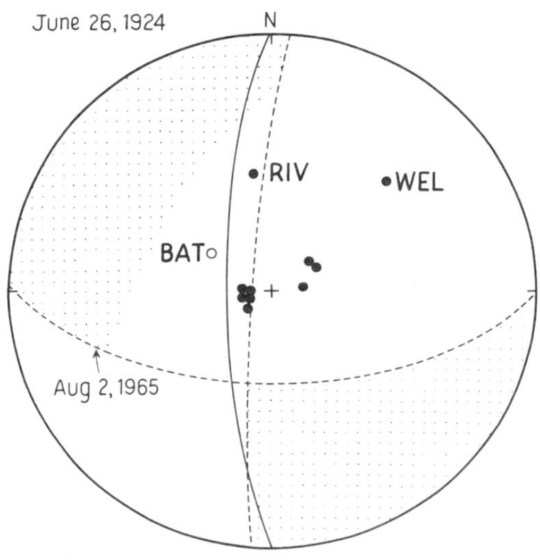

Figure 26
P wave first motions for the 26 JUN 1924 earthquake. Most arrivals are compressional and plot near the center of the focal sphere. Compressional arrivals were also recorded in Australia (RIV) and New Zealand (WEL). There is a reliable dilatational arrival at Batavia (BAT). The first motion data constrain one of the 1924 nodal planes (solid line). The dashed nodal planes are for the 2 AUG 1965 event (see Table 1), the compressional quadrants are dotted. While the 1924 event appears to be a thrust event, the second nodal plane is poorly constrained and hence is not plotted.

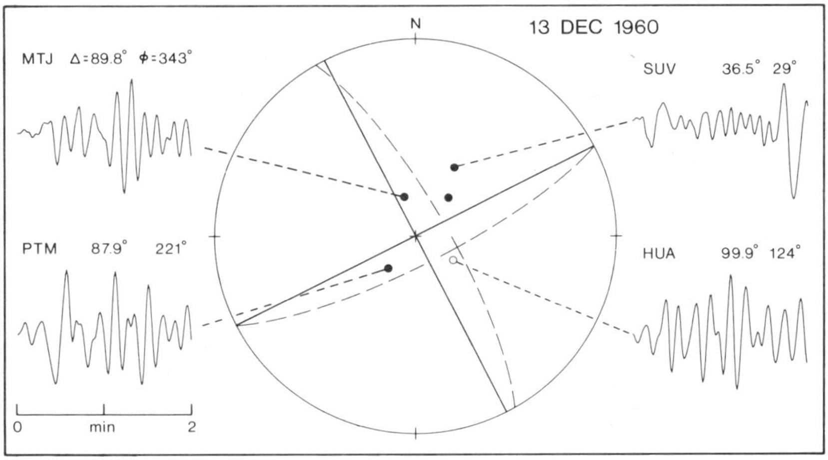

Figure 27
P wave first motions and IGY long period seismograms for the 13 DEC 1960 earthquake. The 7 JUL 1982 focal mechanism is plotted as the dashed nodal planes. The solid lines show a pure strike-slip (right lateral) focal mechanism with the same fault strike as the 1982 event.

There were several experiments with globally deployed long-period seismographs before the WWSSN. The IGY seismograph program started operating in 1957. Many excellent long period seismograms were recorded, and the instrument polarities are generally well-known. Also, the seismology group at Lamont-Doherty Geological Observatory collected and archived most of the seismograms. We have obtained copies of these long-period records for the 13 DEC 1960 earthquake and are able to read several *P* wave first motions. Figure 27 plots the 1960 first motion data, the nodal planes for the nearby 7 JUL 1982 earthquake, and a pure strike-slip mechanism with the same fault strike as the 1982 event. Note that the 1960 data do not allow the vertical fault plane to strike at N30°W, the trend for the large 1981 event to the north. Although the 1960 focal mechanism is not exactly the same as for the 1982 event, it seems that both earthquakes might share the same fault strike, and possibly the same fault plane.

5. *Discussion of Seismotectonics*

We now consider the seismotectonics of the Macquarie Ridge complex in light of the newly compiled results on several earthquakes. Two main questions are addressed: (1) does the PAC-AUS plate boundary coincide with the Macquarie Ridge complex, and if so, does the RM2 model adequately describe the PAC-AUS plate motions, and (2) is there a simple explanation for the seismicity pattern and varying focal mechanisms.

5.1. Plate Boundary Location and Plate Motion

Although there is seismicity within the Australia plate between Balleny fracture zone and Puysegur trench, our interpretation is that these events are intraplate earthquakes and the PAC-AUS plate boundary coincides with the Macquarie Ridge complex from South Island to the triple junction at the southernmost end of the Hjort trench. Relative to the seismic activity along the MRC, the intraplate events in the Balleny-Puysegur trend are infrequent in number, smaller in size, and scattered in location with no consistent linear trend. In addition, the trend of the Balleny fracture zone south of 60°S is consistent with ANT-AUS plate motion and the relatively high level of seismic activity along the fracture zone indicates plate boundary slip. Thus, the narrow slice of oceanic lithosphere between Balleny fracture zone and the MRC (the "Tasman spur") moves as a rigid extension of the Australia plate.

5.1.1. BPF: the PAC-AUS rotation pole.* If the interaction between the Australia and Pacific plates along the Macquarie Ridge complex follows from rigid plate tectonics, then the instantaneous relative motions are described by rotation about a pole. Plate motion generally occurs as a combination of right lateral strike-slip and thrust events along most of the Macquarie Ridge, hence the net slip vector at a location should be a weighted average of the strike-slip and thrust vectors. It is difficult to specify the weighting of the two different slip directions as the recurrence intervals and fault area overlaps are not well determined. The motion in the southern Puysegur trench is clearly a mixture of strike-slip and thrust motion with slip azimuths of N30°E and N83°E, respectively. The rotation pole should produce a slip vector orientation intermediate to these two directions. Therefore, the rotation pole should fall within the sector bounded by the two great circles that are normal to the strike-slip and thrust slip vector azimuths (see "pole lines" in Figure 28). Although there is a mixture of strike-slip and thrust earthquakes further south along the MRC, they are not spatially clustered as in the 1981 cluster. In the Macquarie Island region, we know the slip vector for the 1980 strike-slip earthquake. With the assumption that there is a missing thrust component, the rotation pole should lie south of the 1980 slip vector "pole line" (Figure 28). This restriction places the rotation pole south of the Minster and Jordan BFP. In a similar fashion, the 12 OCT 1979 thrust earthquake provides a further constraint, the rotation pole must lie to the east of the 1979 "pole line".

While the RM2 pole falls in the allowed sector in Figure 28, the Minster and Jordan BFP is clearly eliminated. Why does the RM2 pole better satisfy the expanded slip vector data set than the BFP pole? The BFP pole determination used only two events in the MRC, the 8 NOV 1964 event in the 1981 cluster and the 2 AUG 1965 strike-slip event at the northernmost Hjort trench that has the highest "importance" of any datum for the Pacific-India boundary (see MINSTER and

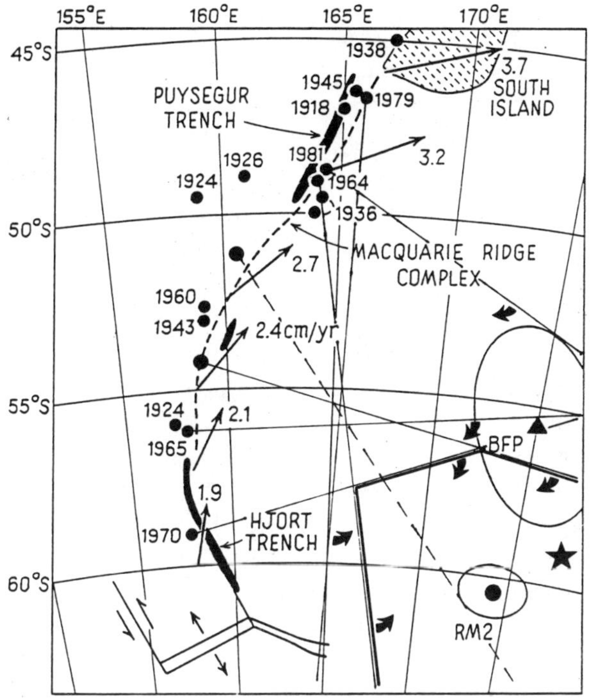

Figure 28
Slip vector constraints on the PAC-AUS rotation pole. Same as Figure 1 except that slip vector pole lines and BFP* have been added. Pole lines are perpendicular to earthquake slip vectors. The two pole lines from the 1981 cluster are from the strike-slip and thrust earthquakes, the rotation pole should lie between these lines, indicated by bold curved arrows. Other regions are characterized by either strike-slip or thrust events: the rotation pole should lie to the east of the thrust pole lines and south of the strike-slip pole lines. The best-fit rotation pole must then be located in the polygonal region bordered by the heavy line. The "true" PAC-AUS best-fit rotation pole (BFP*) is determined by the PAC-ANT and ANT-AUS best-fit rotation poles and is plotted as the star. The 7 JUL 1982 event pole line is dashed.

JORDAN, 1978, for discussion of "importance"). Indeed, notice that their BFP lies along the 1965 "pole line". We now realize that using only one event in a region of mixed focal mechanisms will bias the rotation pole, i.e., the rotation pole should not lie on the pole line from a single earthquake. The Minster and Jordan BFP is shifted to the final RM2 location by the constraint of triple junction closure. If the relative motion data for a plate boundary is biased, then a pole location based on triple junction closure could be more accurate. Of course, this assumes that the information for the other two plate boundaries is reliable. Since the PAC-ANT and ANT-AUS boundaries are both spreading centers with abundant transform faults and magnetic lineations, we would anticipate their relative motion to be well determined. Recall that the RM2 pole plotted in Figure 28 is actually the Pacific-India pole, and is determined by the RM2 poles for PAC-ANT and ANT-India.

The internal deformation of the India plate causes the ANT-India RM2 pole to be displaced from the ANT-India BFP. Since the data for the ANT-India BFP determination are from the spreading system between Antarctica and Australia, the Minster and Jordan ANT-India BFP is the best representation of ANT-AUS plate motions. Thus, we construct the "best-fit" PAC-AUS rotation pole by applying triple junction closure to the Minster and Jordan BFP's for ANT-India and PAC-ANT. This "best-fit" PAC-AUS pole is referred to as BFP* and is plotted as the star in Figure 28. Note that it plots slightly to the north of the RM2 pole and is within the allowed sector. The fact that the location of BFP* is compatible with the new slip vector constraints supports the notion that the Tasman spur (lithosphere between Balleny fracture zone and MRC) is a rigid extension of the AUS plate. Recall that the 7 JUL 1982 event occurs at 51°S between the Puysegur trench and McDougall trough and has a slip vector azimuth that is intermediate to the strike-slip and thrust slip vector azimuths in the 1981 cluster. The slip vector pole line for the 1982 event lies to the south of BFP*. Hence the Puysegur trench and McDougall trough regions might be connected by a transform fault segment and the 1982 event occurs on the extension of this fault.

Returning to the plate boundary question, there is one caveat that we must mention. Plate motion is described by the rotation pole location and the rotation rate. Since we can only locate the position of the rotation pole with earthquake slip vectors, the Tasman spur could theoretically be a separate plate from the Australia plate. The two rotation poles, PAC-AUS and PAC-Tasman, could share the same location with respect to PAC. The Tasman spur could then be decoupled from the AUS plate by a transform fault along a small circle around their rotation pole. However, even if there is *some* deformation in the Tasman spur, we repeat our contention that the high level of seismic activity in the MRC indicates that most of the plate boundary slip occurs along the MRC. Thus, the Macquarie Ridge complex is the PAC-AUS plate boundary and the earthquakes are responding to present-day kinematic conditions.

5.2. Oblique Subduction

The MRC serves as the PAC-AUS plate boundary with a rotation pole close to RM2; the relative motion along most of the MRC is then characterized by oblique subduction. This oblique subduction is accommodated by a combination of two earthquake types: right lateral strike-slip motion on nearly vertical fault planes and thrust motion on fault planes with shallow dips. The primary example of this dual-rupture mode is the activity at the southern end of the Puysegur trench where the 1981 event occurred. The smaller thrust events have occurred along the entire rupture zone of the large 1981 strike-slip earthquake and these shallow dipping thrust planes are adjacent to the vertical strike-slip fault plane. This dual-rupture mode is unusual behavior. Oblique plate motion at other subduction zones is accommodated by earthquakes with oblique slip on a shallow dipping fault plane (for example: the Alaska-Aleutian subduction zone, WU and KANAMORI, 1973; the Japan subduction zone, KANAMORI, 1971; the Ecuador-Colombia subduction zone, KANAMORI and MCNALLY, 1982). We propose that the dual-rupture mode present in the MRC indicates an immature subduction zone, and is a simple consequence of transitional tectonics where the rotation pole is migrating at a geologically significant rate.

Plate tectonic reconstructions show that the PAC-AUS rotation pole has migrated south relative to the eastern part of South Island and the MRC, i.e., the Pacific plate (see MOLNAR *et al.*, 1975). During this migration, the Puysegur trench region has been characterized by right lateral strike-slip motion, and possibly changed from a tensional to compressional environment. Thus, a long continuous vertical fault developed in this region and served as the primary locus of plate motion, just as the Alpine fault in South Island became a dominant tectonic feature. As the rotation pole migrated southward, the compressive component of motion became more significant and required some type of deformation. Horizontal shortening cannot be accommodated by slip on a vertical fault plane. New fault planes must develop if the deformation occurs by fault slip. In our interpretation, the small thrust events beneath the bathymetric rise in the 1981 cluster region represent slip on relatively new faults with a dimension of 20–50 km. The three thrust focal mechanisms show different strikes and dips of the fault planes. These planes are not yet aligned as a smoothly continuous horizontal plane along the trench. Thus, these new fault planes cannot easily accommodate the strike-slip motion, and indeed the older vertical fault is still quite capable of that. As thrusting motion continues on the mosaic of new shallow-dipping fault planes, we speculate that they eventually coalesce into a single throughgoing shallow-dipping plate interface. At this point, the strike-slip component can then be accommodated by this fault surface with oblique slip. The role of the older strike-slip fault plane depends partly on where the new shallow-dipping fault forms. If the upper part of the strike-slip fault is on the plate that is overthrust, then it will be preserved and can even remain active. On the other hand, if the strike-slip plane is left on the plate that is underthrust, then it is subducted and disappears (see Figure 29). In our view, the accommodation of oblique subduction by large strike-slip and small thrust earthquakes is a temporary situation that is due to the recent transition to compression.

5.2.1. Tectonic analogy: San Andreas fault system. There is an analogous situation that has developed along the San Andreas fault system in California. One segment of the southern San Andreas is presently not aligned with the RM2 plate motions: the "great bend" segment in southern California that crosses the transverse ranges. There is a compressive component of motion across the strike of the San Andreas in this segment between Fort Tejon and San Gorgonio Pass. Two large

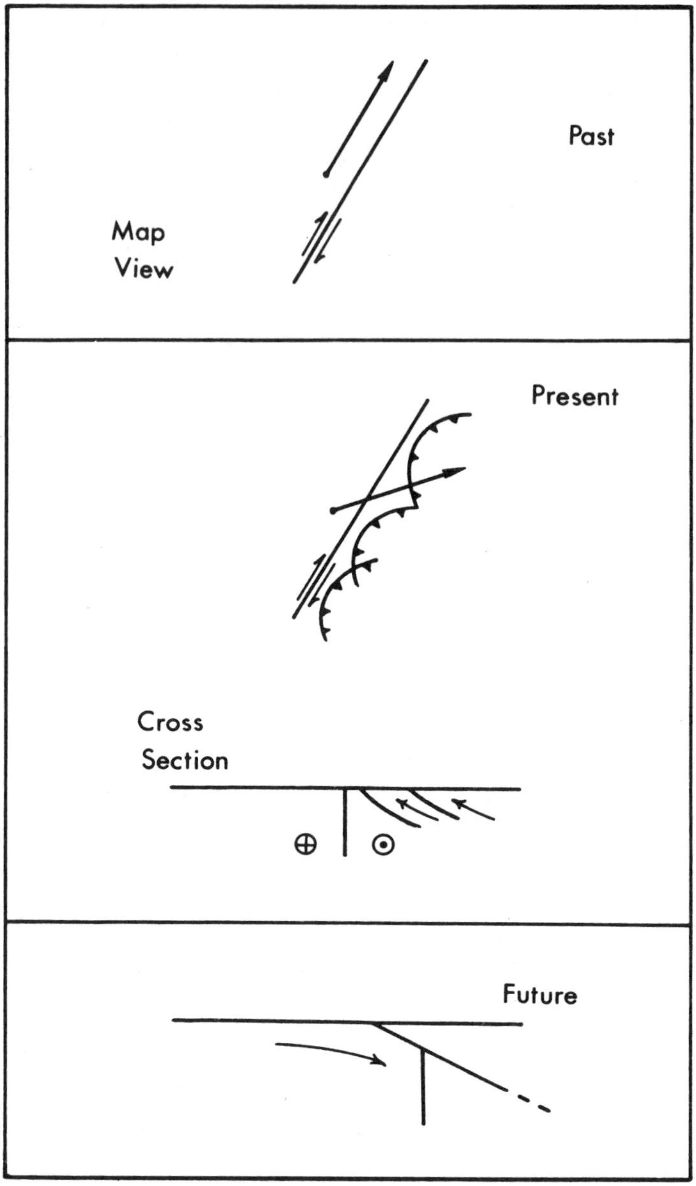

Figure 29
Transitional tectonics and dual-rupture mode. Schematic representation of the thrust interface development in the 1981 cluster region. Top diagram shows a vertical fault trace, the plate motion velocity vector (heavy arrow), and the subsequent right lateral strike-slip faulting. The middle diagram depicts the present-day situation in the 1981 cluster: the plate motion has shifted to oblique convergence and slip is also occurring on new small thrust faults. The thrust faults eventually coalesce into a throughgoing thrust interface (bottom diagram) that can accommodate oblique convergence. In the mega-crack propagation hypothesis, the transition from the middle to bottom diagram occurs in the crack-tip region that propagates along the plate boundary.

shallow-angle thrust earthquakes have occurred along this segment, the 21 JUL 1952 Kern County and 9 FEB 1971 San Fernando events, adjacent to the San Andreas fault. Yet, the largest earthquake in this region is presumably the great strike-slip earthquake that occurred in 1857. Thus, the oblique convergence is accommodated by the dual-rupture mode as observed in the MRC. Although this analogy cannot be pressed too far, it points out that the situation in the southern Puysegur trench is not an isolated occurrence and may have quite general implications for other transitional tectonic environments. We will now return to the MRC and place the seismotectonics into a larger framework.

6. Transitional Tectonics and Subduction Initiation

There is a distinct discontinuity in the level of speculation between this section and the preceding sections of this paper. Since we will now extrapolate from the present-day tectonic picture and discuss the mechanisms of plate interaction, we must necessarily stray from our "factual" seismological realm. This paper would be incomplete without some discussion of this nature. The Macquarie Ridge presents us with a rare view of subduction initiation, hence we must take advantage of the opportunity and speculate on the underlying processes. Our approach will be to state an *ad hoc* "straw man" model for subduction initiation in the Puysegur trench environment. We then concentrate on the Hjort trench region and show that this environment is "controlled" by triple junction behavior, and we finish with a speculative account of the evolutionary history of the MRC.

6.1. Puysegur Trench: Subduction Initiation

The southward migration of the PAC-AUS rotation pole should impart a gradient in deformational style from mature to immature as we proceed from north to south. There is currently a compressive component of motion across South Island and the Alpine fault. It seems that this motion has been accommodated by broad-scale crustal shortening and mountain building (WALCOTT, 1978). The plate boundary switches to an oceanic-continental contact in the Fiordland region as the Alpine fault runs offshore. The intermediate-depth seismic activity beneath Fiordland presumably indicates that a relatively mature subduction environment now exists at the northern end of the MRC. Although a volcanic arc has not yet developed, perhaps the Solander cone just off the southern coast of South Island is a harbinger of future volcanism. If the Fiordland coast region is already characterized by subduction, the "leading edge" of subduction should be located to the south in the Puysegur region.

6.1.1. Mega-crack propagation model. Let us assume an initial plate boundary configuration of a well-developed pure transform boundary (such as in the top part of Figure 29). A mature subduction zone does not simply materialize just when the plate motion changes to a small compressive component along the established strike-slip trend. Based on our observations of the Puysegur trench seismotectonics, we propose a "straw man" model for subduction initiation that is akin to crack propagation: A "crack tip" region propagates along the plate boundary, and the velocity of propagation is similar to a plate tectonic rate. The primary function of the crack tip region is to organize the mosaic of small unconnected thrust fault planes into a large scale horizontally continuous shallow-dipping thrust interface. In front of the crack tip, oblique convergence is accommodated by strike-slip motion on the vertical fault and by lithospheric deformation along the strike slip boundary, possibly including small thrust earthquakes. Behind the crack tip, oblique convergence can be accommodated by oblique slip on the thrust interface, although the old vertical strike-slip fault may still be active. The final stages of competition and coalescence of the small thrust faults could lead to a locally high level of seismicity and, by analogy to crack propagation, the tectonic stress levels might be higher in the crack tip region which might cause high stress drop earthquakes. Note that the mega-crack propagation model implies that subduction does not initiate simultaneously along the entire length of the plate boundary.

Given the above model description, the 1981 cluster region is obviously identified as the crack tip region. A relevant observation is that the 1981 cluster is just south of the deepest portion of the Puysegur trench with a considerable reduction in bathymetric contrast along the 1981 fault zone. This offers some geometric support for the 1981 cluster representing the "tip of the crack" that is propagating south. After the crack extends, *via* the thrusting activity in the 1981 cluster, the underthrusting proceeds more easily and the bathymetric trough quickly extends southward. Hence subduction initiation is an "unzipping action" that may actually lower the tectonic stress level. This model predicts that the Puysegur trench to the north now has a continuous thrust interface. One possible explanation of the low seismicity level along the deepest part of the Puysegur trench might be a transition to aseismic slip after formation of the new underthrusting fault plane. Another more ominous explanation would be that this interface now ruptures as great underthrusting earthquakes with long recurrence time.

6.1.2. Objections to the mega-crack propagation model. We can immediately place some objections to the above simple model. One complication is the seismic activity at the northern end of the Puysegur trench. The mismatch between BFP* and the 1979 event pole line implies a missing strike-slip component of motion. If the region at 47°S is analogous to the 1981 region, then we should anticipate the occurrence of a large strike-slip earthquake in this region. The two focal mechanisms from the 47°S region show thrust motion, but note that previous to the 1981 event all mechanisms in the 1981 cluster were thrust mechanisms.

If we look at the detailed bathymetry in Figure 3, we notice that the 1979 event occurred just north of the deepest portion of the Puysegur trench: perhaps the trench is "propagating" both south and north. This ruins the idea that subduction initiation has smoothly propagated south from the Fiordland coast; rather there would be nucleation sites from which the mega-crack propagates in both directions. There is however a geometric problem if the 1979 region is the tip of a mega-crack. Intermediate depth seismicity under South Island presumably indicates that a substantial length of AUS plate has been subducted at Fiordland; how can we explain the presence of a crack-tip next to a region of plate subduction?

Another complication is related to the low seismicity region south of the 1981 cluster. The proposed fault plane of the 7 JUL 1982 earthquake can be extended from the 1982 epicenter to the NE along the fault strike of N63°E. This extension intersects the 25 MAY 1981 fault trend at 50°S, just at the southern terminus of the 1981 seismicity cluster. Thus, the high level of seismicity in the 1981 cluster may not extend further south because the active plate boundary is offset by a transition region with a lower seismicity level. Thus, the 1981 cluster might be special because of the transition to the McDougall trough region, rather than for being at the southern tip of the Puysegur trench.

Further south along the MRC, we see the independent development of the McDougall trough with underthrusting to the west rather than to the east. Thus, the 1981 crack-tip may not be the unique locus of "unzipping" action. It is unknown whether the 1981 crack-tip can propagate south and switch the underthrusting polarity in the McDougall trough.

To conclude, the simple notion of a single crack-tip propagating southward certainly does not adequately explain all of the information. However, the evidence supports the basic premise that oblique subduction in a transitional environment is accommodated by the growth of new thrust faults. We cannot demonstrate that the mosaic of new thrust planes will eventually coalesce into a major throughgoing underthrusting boundary, but how else does a subduction zone form?

6.2. Hjort Trench: Triple Junction Subduction

The Hjort trench exceeds a depth of 6 km for nearly 300 km along an arcuate trend. There is no continuous major bathymetric rise associated with this trench, but the bathymetry is asymmetric about the axis: young lithosphere to the west is shallower than the older lithosphere to the east. This bathymetric difference is most pronounced at the southern end and disappears at the northernmost end of the Hjort trench. The trench shallows to the north where the 1924 and 1965 events occurred, and shallows to the south. It does not extend to the PAC-ANT-AUS triple junction. The only focal mechanism along the Hjort trench is for the 1970 earthquake, and this event is 350 km north of the triple junction. The basic level of seismicity is quite low between the 1970 event and the triple junction (see Figure 2).

The direct evidence for subduction along this trench consists of three parts: (1) the shorter length of the AUS magnetic lineations (WEISSEL *et al.*, 1977), (2) character of the short-wavelength geoid anomalies, and (3) the speculative focal mechanism for the 1924 earthquake. Given our above conclusion that the PAC-AUS boundary follows the MRC, perhaps the most compelling evidence for subduction comes from considerations of the PAC-ANT-AUS triple junction. This leads into general discussion of the evolution of the MRC.

6.2.1. Triple junction closure. Triple junction closure and stability greatly influence plate interaction at the Hjort trench due to the proximity of the Hjort trench to the PAC-ANT-AUS triple junction. WEISSEL *et al.* (1977) pointed out that the local condition of triple junction closure implies subduction along the Hjort trench. This approach leads to a prediction for the plate boundary geometry that can be tested.

Based on seismicity and bathymetric features, the PAC-ANT-AUS triple junction is located at 61.6°S 161°E. One of the fundamental tenets of plate tectonics is triple junction closure. Moreover, the condition for triple junction stability places a constraint on the plate boundary interaction. We will consider local velocity vectors based on two different sets of rotation poles: (1) RM2 poles, and (2) the "true" best-fitting poles discussed above (Table 3). Since both sets have been derived with explicit use of the triple junction closure condition, it is no surprise that the local velocity vectors calculated for a location of 61.6°S 161°E sum to zero. We show the local velocity vectors for these two different pole sets in Figure 30. The PAC-AUS vector can be regarded as the difference between two larger and roughly parallel vectors. Hence, small changes in the PAC-ANT and the AUS-ANT velocity vectors result in relatively large changes for the PAC-AUS velocity vector. The north trending motion of AUS relative to PAC implies convergence across a northwest trending boundary. To see how the convergence occurs, we must use the condition for triple junction stability.

6.2.2. Triple junction stability. The condition for triple junction stability further constrains the plate boundary interaction. The ANT-AUS plate boundary is a spreading segment with approximately symmetric perpendicular spreading. While

Table 3

Velocity vectors at PAC-ANT-AUS triple junction for two rotation pole sets.

Plate Pair	RM2		BFP*	
	Direction	Rate (cm y^{-1})	Direction	Rate (cm y^{-1})
AUS-ANT	S26°E	6.20	S28°E	6.65
ANT-PAC	N34°W	4.82	N35°W	4.78
PAC-AUS	N2°W	1.55	N10°W	2.00

Triple junction location is 61.6°S 161.0°E. Velocity vectors give motion of the second plate relative to the first plate.

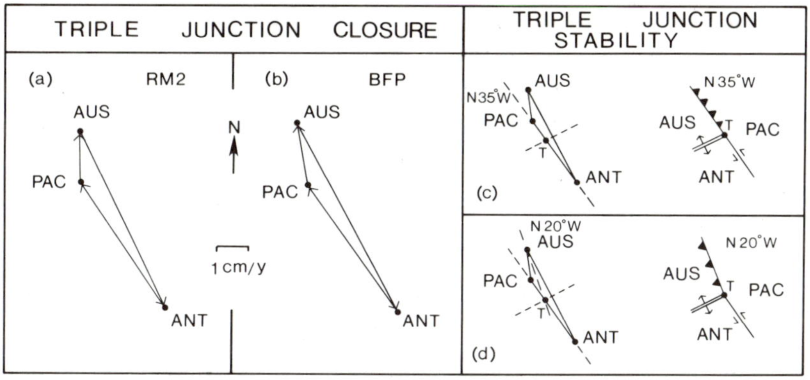

Figure 30
Triple junction closure and stability. For the PAC-ANT-AUS triple junction at $61\frac{1}{2}$°S 161°E, (a) and (b) show the velocity vector closure for two sets of rotation poles, RM2 and BFP (listed in Table 3). Velocity scale and north arrow are shown. The BFP vector diagram is repeated in (c) and (d), north direction is the same but velocity vectors are drawn at half-scale. The triple junction point, *T*, must plot in velocity space along the dashed lines: (c) shows the case for AUS subducting beneath PAC with a subduction zone strike of N35°W, plate configuration is shown to the right; (d) shows the case for PAC subducting beneath AUS with a subduction zone strike of N20°W.

the PAC-ANT boundary might be a leaky transform, we will treat it as a pure transform for now. These two plate boundary descriptions place the intersection for a stable triple point along the PAC-ANT vector (Figure 30). There are two possible stable orientations for the subduction zone along the PAC-AUS boundary: (1) AUS subducts beneath PAC and the subduction zone strike is the same as the PAC-ANT transform strike, and (2) PAC subducts beneath AUS and the subduction zone strike is intermediate to the PAC-ANT transform and PAC-AUS plate motion strikes. Thus, a simple measurement of the local strike of the PAC-AUS plate boundary provides a strong check on the polarity of subduction. There is a problem however, the Hjort trench does not extend to the triple junction. There is a 200 km gap between the southernmost Hjort trench and the triple junction, and this region is devoid of well defined linear bathymetric features. Recall that the local strike of Hjort trench in the 1970 earthquake region is N20°W, and the Hjort trench strike clearly rotates to a more westerly strike to the south. Although it cannot be measured with great precision, it appears that the strike of the southernmost Hjort trench falls in the range of N30-45°W as the trench curvature seems to increase as it shallows. A line between the triple junction and the 1970 earthquake location gives a strike of N34°W; this is in excellent agreement with the predicted value of N35°W in Figure 30c. This interpolation implies that the present day plate motions are applicable over this 350 km projection of the plate boundary. Given the present-day migration rate of 1.5 cm/yr for *T* relative to PAC (see Figure 30c), it would take more than 20 Ma for the triple junction to migrate from the 1970 event location to its present position. The PAC-ANT rotation pole has shifted in this time period, thus the above correspondence may not be diagnostic of polarity. Another possible interpolation is over a 250 km projection to the southernmost tip of the Hjort trench; due to the curvature in the southernmost Hjort trench, this measurement gives a strike of N20°W. This latter measurement seems preferable to the former as the interpolation distance is smaller, but the consequence is that now the PAC plate subducts beneath the AUS plate! If this latter conclusion is true for the 250 km plate boundary segment adjacent to the triple junction, could the polarity of subduction then switch at the Hjort trench?

Before we make fine distinctions concerning the PAC-AUS plate boundary, we should consider the effect of a leaky transform along the PAC-ANT boundary. Although it is difficult to discern the plate tectonic elements just south of the triple junction, the low-level seismicity has a trend of about N70°W over a distance of 300 km. If we approximate the PAC-ANT boundary as a leaky transform with highly oblique spreading along this trend, the triple junction point in Figure 30c moves off the PAC-ANT velocity vector to a location slightly to the west. This causes the strike of the PAC-AUS subduction zone to rotate toward a more northerly strike; we could easily select the parameters to obtain a strike of N20°W for the AUS plate subducting beneath PAC.

In conclusion, convergence is occurring along the PAC-AUS boundary north of the triple junction. Given the poor definition of both the PAC-AUS and PAC-ANT plate boundary segments, we cannot determine whether the AUS plate subducts beneath the PAC plate, or *vice versa*, or both. Thus it is difficult to decipher the nature of subduction initiation along the plate boundary segment between the triple junction and the southernmost Hjort trench. In any event, the aseismic nature of this segment implies that a mega-crack propagation model for subduction initation is not appropriate for this tectonic regime. The next section further examines the recent tectonic history of the Hjort trench by using additional information and arguments.

6.2.3. Triple junction evolution: southern Hjort trench. The triple junction stability condition can be used to obtain a view of the recent evolution of the triple junction region. As already mentioned, the average strike between the triple junction and the midpoint of the Hjort trench is N34°W, which is *consistent* with the AUS plate subducting at the Hjort trench. More evidence for the subduction of the AUS plate is based on evolution of the ANT-AUS spreading segment (see Weissel *et al.*, 1977). Referring to the triple junction plots in Figure 30, it is clear that regardless of the details of the PAC-ANT and PAC-AUS boundaries, the ANT-AUS spreading ridge segment that lies between the Balleny fracture zone and the triple junction will become shorter with time, i.e., the T point is to the west of the ANT-AUS velocity vector. Thus, the AUS plate created a few million years ago will have a greater length between the Balleny fracture zone and the plate edge than the present-day length of the spreading ridge segment. In fact, the width of the AUS plate measured perpendicular to the Balleny fracture zone decreases to a minimum value at the Hjort trench midpoint. This is not surprising given our previous measurement of a N34°W trend from the triple junction to the Hjort trench midpoint while the Balleny fracture zone follows a trend of N22°W. Part of the AUS plate is located to the east of the southern Hjort trench. Recall that the bathymetric difference across the southern Hjort trench indicates that the Hjort trench is the edge of the Pacific plate. Thus, part of the AUS plate has been subducted at the Hjort trench.

Triple junction interaction is important for the entire MRC, however the simple extrapolation used for the southern Hjort trench will not be valid further north due to significant changes in plate motions and geometry. We now discuss triple junction evolution over an extended time and distance; and thus enter a more speculative arena.

6.3. Triple Junction Evolution: Origin of the Macquarie Ridge Complex

We use three key assumptions to extrapolate the PAC-ANT-AUS triple junction interaction: (1) the AUS-ANT plate motion has remained relatively constant during the formation of the AUS plate along the MRC (i.e., the "Tasman spur"), (2) the PAC-ANT-AUS triple junction has continuously migrated southward from South Island to its present location, and (3) the present-day MRC records the geometry of the PAC-AUS plate boundary at the time of triple junction interaction. The first assumption is supported by the magnetic lineations as reported in WEISSEL *et al.* (1977). Thus, the ANT-AUS velocity vector remains constant while the other two vectors vary with time (see Figure 31). Assumptions (2) and (3) arise from considerations of the plate geometry and triple junction closure diagram. If the PAC-ANT and ANT-AUS velocity vectors trend in the same basic direction, then the triple junction will migrate south relative to the PAC plate if the PAC-ANT rate is more than about half the ANT-AUS rate at the triple junction. Also, the sense of strike-slip displacement component along the PAC-AUS boundary will be right lateral as long as the PAC-ANT rate ($r_{\text{PAC-ANT}}$, magnitude of the PAC-ANT velocity vector at the triple junction) is less than the ANT-AUS rate ($r_{\text{ANT-AUS}}$, similar definition). Thus, by imposing the rate constraint: $\frac{1}{2} < (r_{\text{PAC-ANT}}/r_{\text{ANT-AUS}}) < 1$, we ensure the southern migration of the triple junction with respect to PAC, and right lateral strike-slip along the MRC.

Whether the PAC-AUS boundary has a convergent or divergent component depends on whether the PAC-ANT vector strikes to the east or west of the ANT-AUS vector. Figure 31 shows the field of allowed PAC-ANT velocity vectors. To obtain the location of the triple junction point in velocity space, we must make assumptions concerning the nature of the PAC-ANT and PAC-AUS plate boundaries. We assume that the PAC-ANT plate boundary has been a transform or

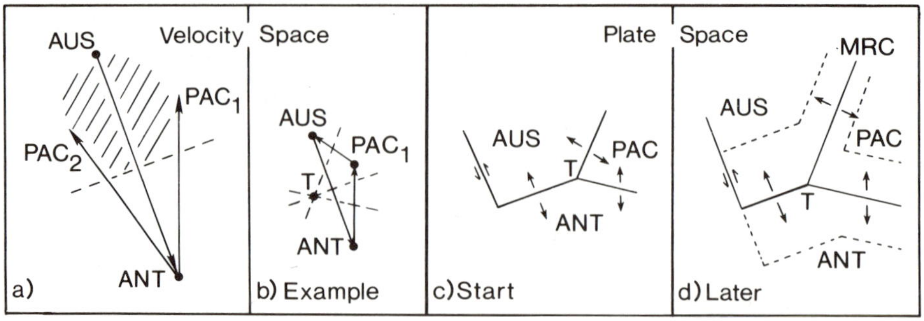

Figure 31

Triple junction evolution and Macquarie Ridge formation. Diagram on left (a) is a velocity vector diagram for the PAC-ANT-AUS triple junction. ANT-AUS plate motion is taken to be constant throughout Macquarie Ridge formation. The PAC-ANT vector can plot anywhere in the hatched region, two possible vectors are drawn (PAC$_1$ and PAC$_2$). The PAC-AUS velocity vector is then determined as the difference vector. The PAC-AUS boundary is a "soft" boundary that accommodates the mismatch between the ANT-AUS and PAC-ANT motions. If the PAC-ANT vector plots to the east of the ANT-AUS vector (e.g., PAC$_1$), then the PAC-AUS boundary will have a divergent component (b). The diagrams to the right (c and d) depict the evolution of the triple junction region for this example. Note that the PAC-AUS boundary is a leaky transform that strikes N-NE and that the ANT-AUS spreading segment shortens with time. The basic geometry of the Macquarie Ridge plate boundary (MRC) can be explained by triple junction evolution changing the PAC-AUS boundary from a divergent state to the present-day convergent state (e.g., PAC$_2$). This switch occurs as the PAC-ANT velocity vector slowly rotates to a more westerly strike, relative to ANT-AUS motion, at the migrating triple junction.

a leaky transform. Note that most of the MRC lies just west of the Campbell plateau, and it therefore seems unlikely that a major PAC-ANT spreading segment has occupied the space between the MRC and the Campbell plateau. The basic geometry of the MRC can be produced by assuming that the PAC-AUS boundary was originally in a divergent state (as suggested by MOLNAR *et al.*, 1975), and the PAC-ANT velocity vector slowly rotated to impart a convergent component. For the divergent MRC environment, we can always find a combination of leaky transforms for the PAC-AUS and PAC-ANT boundaries that give a stable triple point. Given the erosion of the ANT-AUS spreading segment through time, the triple point should be located to the west of the ANT-AUS vector. Leaky transforms along the PAC-AUS and PAC-ANT boundaries could feed off the primary ANT-AUS spreading segment. As shown in Figure 31, the PAC-AUS leaky transform can strike to the northeast, i.e., the trend for the present-day northern and central MRC. Recall that the local strike of the Hjort trench in the 1970 region is about N20°W; this could mark the location of the triple junction at the crossover time when the PAC-ANT velocity vector rotated to the west of the ANT-AUS vector at the triple junction. After the crossover time, the plate boundary segment immediately north of the triple junction would be a convergent boundary. This implies that only the southern Hjort trench region was formed originally as a convergent plate boundary, while the other parts have evolved from a leaky transform to a convergent region. This lends circumstantial support to the notion that the small thrust faults in the northern MRC represent new breaks in the lithosphere.

6.4. Macquarie Ridge Complex: A "Soft" Plate Boundary

The MRC plate boundary has evolved at the junction of two major spreading systems. The PAC-ANT spreading system reorganized when Australia rifted from Antarctica. In particular, the Tasman Sea spreading stopped, and the PAC-ANT system linked up with the AUS-ANT system. These two spreading systems are not perfectly "matched" in terms of their spreading direction and rate. The PAC-AUS boundary along the MRC has served to accommodate this mismatch.

To take a more global view, the major spreading systems are currently connected as a single sinuous strand. However, many of these spreading systems originally developed somewhat independent of each other. If the major spreading systems retain their rotation poles, i.e., their fracture zones fall along consistent small circles, then there must be some conflict in the local plate motions at the junction between two of these spreading systems. A third plate boundary is clearly required at this junction; this third boundary is a "soft" boundary in that it accommodates the mismatch in the motion of the other two spreading center boundaries. The motion across the "soft" boundary changes as necessary for triple junction closure and stability (see Figure 32). In general, the velocity vectors for the

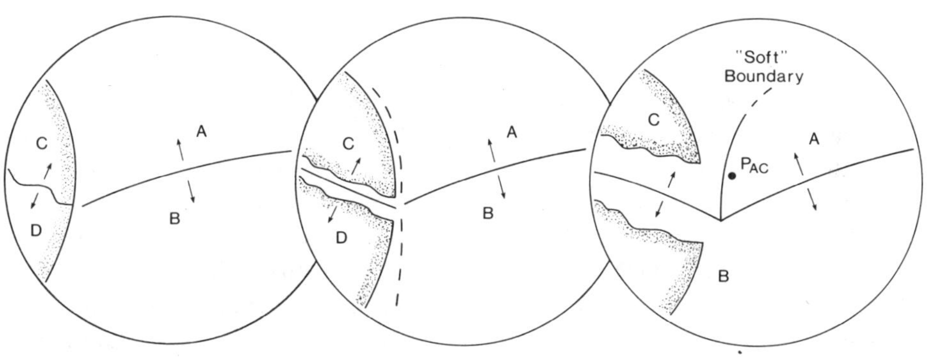

Figure 32

Schematic development of a "soft" plate boundary to accommodate mismatch between two major spreading systems. The diagrams sketch the global-scale tectonic features at three stages. The initial situation is depicted at left, one oceanic spreading system separates plates A and B, while the other system will fragment the continent into plates C and D. At a later time in the middle diagram, the continent has fragmented but the connection between the two systems is immature, perhaps there are still four plates with two auxiliary plate boundaries (dashed). In the evolved three-plate configuration at right, plates B and D are now fused, and the two major spreading systems continue their activity with the "soft" boundary absorbing the mismatch. The rotation pole (P_{AC}) for the "soft" boundary is located close to the triple junction. Subduction initiation can easily occur along the "soft" plate boundary.

two spreading boundaries at the triple junction will change with time. Thus, the "soft" velocity vector must be capable of changing to keep up with the evolution. If the velocity vectors (at the triple junction) for the two major boundaries are almost "matched", i.e., nearly parallel with comparable rates, then the magnitude of the velocity vector for the "soft" boundary is small and the direction of motion may change rather quickly. The "soft" boundary can easily serve this function if its rotation pole follows close to the triple junction. In this case, small changes in the pole location can readily change the direction of the velocity vector at the triple junction, yet these pole changes will not greatly affect the tectonic environments far from the triple junction along the "soft" boundary, e.g., for the case of the PAC-AUS "soft" boundary, we do not want to switch from subduction to spreading at the Kermadec trench in a few million years. Thus, the migration of the PAC-AUS rotation pole is simply a consequence of the pole following the triple junction as it migrates south.

Subduction initiation at the triple junction has occurred along the southernmost Hjort trench simply because the slow migration of the triple junction and the PAC-ANT pole places the "soft" PAC-AUS boundary in a convergent state at the triple junction. Subduction initiation to the north at the Puysegur trench is due not only to the migration of the PAC-ANT pole relative to the PAC-AUS pole, but also to the northward transport of the Puysegur trench region by the spreading systems.

6.4.1. Predictions for other "soft" plate boundaries. The above described circumstances that require a "soft" boundary are quite general. Hence we expect to see other "soft" plate boundaries as offshoots along the major spreading systems. Given a triple junction with at least two spreading segments, our prediction is that the rotation pole for the "soft" boundary should be located close to the triple junction. Indeed there are several candidates for "soft" boundaries: the Eurasia-Africa plate boundary may be the "soft" boundary between the north and central Atlantic spreading segments; the Nazca-Cocos plate boundary may be a "soft" boundary for the Pacific spreading ridge; and the India-Arabia plate boundary may be the "soft" boundary for spreading systems in the Indian Ocean and Gulf of Aden. The implication is that the connection and evolution of the major spreading systems can generate secondary spreading segments or initiate subduction as the need arises. Placed in this context, subduction initiation does not present a major obstacle to plate boundary evolution. Perhaps an efficient mechanism such as mega-crack propagation makes subduction initiation a common tectonic process. A newly initiated subduction zone can develop into a major subduction zone as it continues to move away from the triple junction. Perhaps there is a threshold at which the new subduction zone changes from a "soft" to "hard" boundary.

7. Conclusions

It is difficult to simply list the conclusions of this paper due to the gradient in scientific style from our detailed investigation of the 1981 earthquake to our global-scale speculations. Hence, we group our primary conclusions into three categories: strong, moderate, and weak.

Strong conclusions: The PAC-AUS plate boundary coincides with the MRC. Our study of the largest earthquakes in the MRC finds a combination of right lateral strike-slip and thrust earthquakes along the MRC, and the overall sense of motion is consistent with the BFP* rotation pole located on the basis of triple junction closure with the well determined BFP's for the PAC-ANT and AUS-ANT plate boundaries. Oblique convergence in the southern Puysegur trench is accommodated by a dual-rupture mode: a large strike-slip earthquake on vertical fault plane and smaller thrust earthquakes on shallow-dipping fault planes that are not continuous.

Moderate conclusions: A subduction initiation model applicable to the Puysegur trench region uses the basic idea of a propagating crack. The crack tip region separates a disjointed mosaic of small thrust faults in front of the crack tip from a through-going single thrust interface behind the crack tip. The 1981 cluster is identified as the crack tip region. The tectonics in the Hjort trench region are dominated by triple junction behavior and the PAC-AUS boundary to the north of the triple junction presently evolves as a convergent boundary. The subduction initiation process with newly created lithosphere is apparently aseismic.

Weak conclusions: The MRC is a "soft" plate boundary that accommodates the mismatch between the PAC-ANT and ANT-AUS spreading ridge systems. Soft boundaries are necessary to connect the major spreading systems. Apparently, two spreading systems can easily initiate a new subduction zone.

Many of our interpretations are based on the basic principles of plate tectonics. It is well-known that: triple junctions will evolve; rotation poles migrate in a three-plate system; and that subduction initiation must somehow occur. The novelty of our development is not in the description of a specific history for the MRC, but rather in the speculation on the controlling factors and processes. For example, we have ventured beyond the kinematic fact that subduction initiation must occur and have proposed the mega-crack propagation model to initiate subduction. In addition, the concept of a "soft" third boundary to help connect two major spreading systems implies that new subduction zones can be readily created by the "hard" boundaries of major spreading systems. The interpretations and conclusions of this paper are nonunique; different investigators may use the same information presented above and arrive at different conclusions. However, the importance of the MRC tectonic setting will remain. The Macquarie Ridge complex displays spatial and temporal gradients of plate boundary interaction, and thus represents a natural laboratory for the study of plate boundary evolution and subduction initiation.

Acknowledgments

Several people have contributed materials for this project: H. Eissler, L. Stewart, and W. Smith, in particular. Les Thurston drafted several figures. We benefited from discussions with Hiroo Kanamori and Anny Cazenave. Special thanks to Melissa Schell for her help in the final stages of figure and manuscript preparation. This project has been supported by grants from the National Science Foundation: initially to Prof. Kanamori at Caltech and then later to L. J. R. at the University of Michigan (EAR8351515 and EAR8407786). Also, a grant from the Shell Companies Foundation has helped the completion of this project.

Seismicity Update: Another magnitude 7 event has occurred in the MRC: 3 SEP 1987, 6:40:13.9, 58.9°S 158.5°E, $M_s = 7.3$. This location is quite close to the 1970 Hjort trench event. The best double couple from the Harvard CMT solution has a fault strike of 155°, dip of 69°, and rake of 188°—right lateral strike-slip along a trend of N25°W, close to the N14°W fault strike of the 1970 event. The Harvard CMT solution for a large ($M_s = 6.8$) aftershock shows a substantial thrust component—another example of the curious dual rupture mode along the MRC.

References

BANGHAR, A. R. and SYKES, L. R. (1969), *Focal mechanisms of earthquakes in the Indian Ocean and adjacent regions*, J. Geophys. Res. *74*, 632–649.

BECK, S. L. and RUFF, L. J. (1984), *The rupture process of the great 1979 Colombia earthquake: Evidence for the asperity model*, J. Geophys. Res. *89*, 9281–9291.

BECK, S. L. and RUFF, L. J. (1985), *The rupture process of the 1976 Mindanao earthquake*, J. Geophys. Res. *90*, 6773–6782.

BECK, S. L. and RUFF, L. J. (1987), *Rupture process of the great 1963 Kurile Islands earthquake sequence: Asperity interaction and multiple event rupture*, J. Geophys. Res. *92*, 14,123–14,138.

BEN-MENAHEM, A. (1961), *Radiation of seismic surface waves from finite moving sources*, Bull. Seism. Soc. Am. *51*, 401–435.

CHRISTENSEN, D. H. and RUFF, L. J. (1985), *Analysis of the trade-off between hypocentral depth and source time function*, Bull. Seism. Soc. Am. *75*, 1637–1656.

DEWEY, J. W. and SPENCE, W. (1979), *Seismic gaps and source zones of recent large earthquakes in coastal Peru*, PAGEOPH *117*, 1148–1171.

DZIEWONSKI, A. M., CHOU, T., and WOODHOUSE, J. H. (1981), *Determination of earthquake source parameters from waveform data for studies of global and regional seismicity*, J. Geophys. Res. *86*, 2825–2852.

DZIEWONSKI, A. M. and WOODHOUSE, J. H. (1983), *An experiment in systematic study of global seismicity: CMT solutions for 201 moderate and large earthquakes of 1981*, J. Geophys. Res. *88*, 3247–3271.

GUTENBERG, B. and RICHTER, C. F., *Seismicity of the Earth* (Princeton University Press, 1949).

JOHNSON, T. and MOLNAR, P. (1972), *Focal mechanisms and plate tectonics of the southwest Pacific*, J. Geophys. Res. *77*, 5000–5033.

KANAMORI, H. (1971), *Focal mechanism of the Tokachi-Oki earthquake of May 16, 1968: Contortion of the lithosphere at a junction of two trenches*, Tectonophysics *12*, 1–13.

KANAMORI, H. (1977), *The energy release in great earthquakes*, J. Geophys. Res. *82*, 2981–2987.

KANAMORI, H. and ANDERSON, D. L. (1975), *Theoretical basis of some empirical relations in seismology*, Bull. Seism. Soc. Am. *65*, 1073–1095.

KANAMORI, H. and GIVEN, J. W. (1981), *Use of long-period surface waves for rapid determination of earthquake source parameters*, Phys. Earth Planet. Int. *27*, 8–31.

KANAMORI, H. and MCNALLY, K. (1982), *Variable rupture mode of the subduction zone along the Ecuador-Colombia coast*, Bull. Seism. Soc. Am. *72*, 1241–1253.

KANAMORI, H. and STEWART, G. S. (1978), *Seismological aspects of the Guatemala earthquake of February 4, 1976*, J. Geophys. Res. *83*, 3427–3434.

LYNNES, C. S. and RUFF, L. J. (1985), *Source process and tectonic implications of the great 1975 North Atlantic earthquake*, Geophys. J. R. Astr. Soc. *82*, 497–510.

MACELWANE, J. B. (1930), *The south Pacific earthquake of June 26, 1924*, Beitr. Geophys. *28*, 165–227.

MAMMERICKS, J., SMITH, S. M., TAYLOR, I. L., and CHASE, T. E. (1975), *Topography of the South Pacific* (*Scripps Institute of Oceanography, La Jolla, CA*).

MINSTER, J. B. and JORDAN, T. E. (1978), *Present-day plate motions*, J. Geophys. Res. *83*, 5331–5354.

MOLNAR, P., ATWATER, T., MAMMERICKS, J. and SMITH, S. M. (1975), *Magnetic anomalies, bathymetry, and the tectonic evolution of the south Pacific since the late Cretaceous*, Geophys. J. R. Astr. Soc. *40*, 383–420.

PAGE, R. (1973), *The Sitka, Alaska earthquake of 1972: An expected visitor*, Earthquake Information Bull. *5*, 4–9.

PLAFKER, G. (1976), *Tectonic aspects of the Guatemala earthquake of 4 February 1976*, Science *193*, 1201–1208.

PURCARU, G. and BERCKHEMER, H. (1982), *Quantitative relations of seismic source parameters and a classification of earthquakes*, Tectonophysics *84*, 57–128.

RUFF, L. J. (1983), *Fault asperities inferred from seismic body waves*, In *Earthquakes: Observation, Theory, and Interpretation* (ed. Kanamori, H. and Boschi, E.) (North-Holland, Amsterdam) pp. 251–276.

RUFF, L. and CAZENAVE, A. (1985), *SEASAT geoid anomalies and the Macquarie Ridge complex*, Phys. Earth Planet. Int. *38*, 59–69.

RUFF, L. and KANAMORI, H. (1983), *The rupture process and asperity distribution of three great earthquakes from long-period diffracted P-waves*, Phys. Earth Planet. Int. *31*, 202–230.

SCHELL, M. M. and RUFF, L. J. (1988), *Rupture of a seismic gap in southeastern Alaska: The 1972 Sitka earthquake* (M_s 7.6), Phys. Earth Planet. Int., in press.

SCHOLZ, C. H., RYNN, T. M., WEED, R. W., and FROHLICH, C. (1973), *Detailed seismicity of the Alpine fault zone and Fiordland region, New Zealand*, Geol. Soc. Am. Bull. *84*, 3297–3316.

SCHWARTZ, S. Y. and RUFF, L. J. (1985), *The 1968 Tokachi-Oki and the 1969 Kurile Islands earthquakes: Variability in the rupture process*, J. Geophys. Res. *90*, 8613–8626.

STEIN, S. and OKAL, E. A. (1978), *Seismicity and tectonics of the Ninetyeast Ridge area: Evidence for internal deformation of the Indian plate*, J. Geophys. Res. *83*, 2233–2245.

SYKES, L. R. (1967), *Mechanisms of earthquakes and nature of faulting on the midocean ridges*, J. Geophys. Res. *72*, 2131–2153.

WALCOTT, R. I. (1978), *Present tectonics and late Cenozoic evolution of New Zealand*, Geophys. J. R. Astr. Soc. *52*, 137–164.

WARD, S. (1979), *Ringing P waves and submarine faulting*, J. Geophys. Res. *84*, 3057–3062.

WIENS, D. A. DEMETS, C., GORDON, R., STEIN, S., ARGUS, D., ENGELN, J., LUNDGREN, P., QUIBLE, D., STEIN, C., WEINSTEIN, S., and WOODS, D. (1985), *A diffuse plate boundary model for Indian Ocean tectonics*, Geophys. Res. Lett. *12*, 429–432.

WEISSEL, J. K., HAYES, D. E., and HERRON, E. M. (1977), *Plate tectonics synthesis: The displacements between Australia, New Zealand, and Antarctica since the late Cretaceous*, Marine Geol. *25*, 231–277.

WU, F. T. and KANAMORI, H. (1973), *Source mechanism of February 4, 1965, Rat Island earthquake*, J. Geophys. Res. *78*, 6082–6092.

(Received August 1, 1986, revised December 1, 1987, accepted January 1, 1988)

PAGEOPH, Vol. 129, Nos. 1/2 (1989)

0033-4553/89/020131-71\$1.50 + 0.20/0

Estimation of Strong Ground Motions from Hypothetical Earthquakes on the Cascadia Subduction Zone, Pacific Northwest

THOMAS H. HEATON[1] and STEPHEN H. HARTZELL[1]

Abstract—Strong ground motions are estimated for the Pacific Northwest assuming that large shallow earthquakes, similar to those experienced in southern Chile, southwestern Japan, and Colombia, may also occur on the Cascadia subduction zone. Fifty-six strong motion recordings for twenty-five subduction earthquakes of $M_s \geq 7.0$ are used to estimate the response spectra that may result from earthquakes $M_w < 8\frac{1}{4}$. Large variations in observed ground motion levels are noted for a given site distance and earthquake magnitude. When compared with motions that have been observed in the western United States, large subduction zone earthquakes produce relatively large ground motions at surprisingly large distances. An earthquake similar to the 22 May 1960 Chilean earthquake (M_w 9.5) is the largest event that is considered to be plausible for the Cascadia subduction zone. This event has a moment which is two orders of magnitude larger than the largest earthquake for which we have strong motion records. The empirical Green's function technique is used to synthesize strong ground motions for such giant earthquakes. Observed teleseismic P-waveforms from giant earthquakes are also modeled using the empirical Green's function technique in order to constrain model parameters. The teleseismic modeling in the period range of 1.0 to 50 sec strongly suggests that fewer Green's functions should be randomly summed than is required to match the long-period moments of giant earthquakes. It appears that a large portion of the moment associated with giant earthquakes occurs at very long periods that are outside the frequency band of interest for strong ground motions. Nevertheless, the occurrence of a giant earthquake in the Pacific Northwest may produce quite strong shaking over a very large region.

Key words: Earthquake, strong motion, subduction, Cascadia, Washington, Oregon.

Introduction

This is the last in a sequence of papers that lead to estimates of the type of ground motions that might be caused by earthquakes on the Cascadia subduction zone. HEATON and KANAMORI (1984) compared physical characteristics of many world-wide subduction zones and concluded that the Cascadia subduction zone is similar to other subduction zones with strong seismic coupling. HEATON and HARTZELL (1986) extended those studies and concluded that the Cascadia subduction zone is most similar to the subduction zones of southern Chile, southwestern Japan, and Colombia; each of which have experienced sequences of very large

[1]U.S. Geological Survey, 525 S. Wilson Ave., Pasadena, CA 91106, U.S.A.

shallow subduction earthquakes. HEATON and HARTZELL (1986) also presented several hypothetical earthquake sequences that may be plausible for the Cascadia subduction zone. HARTZELL and HEATON (1985) compared the nature of the time history of energy release for large subduction earthquakes by studying broad-band teleseismic P-wavetrains from sixty of the largest events in the last fifty years. This survey gave insight to the variation in source parameters and shorter-period radiated energy (2 to 50 sec) for earthquakes from different subduction zones. More informed judgements could then be made concerning the use of available strong motion data from world-wide subduction zones to predict the ground motion at a specific site. In this paper, we present estimates for the strong ground motions that might be expected if large subduction earthquakes do occur on the Cascadia subduction zone.

We assume that gap-filling earthquake sequences that are similar to those already observed in southern Chile, southwestern Japan, and Colombia may also occur in the northwestern United States. The largest earthquakes in these sequences range in size from M_w 8 to M_w $9\frac{1}{2}$. Strong ground motion records are available for shallow subduction earthquakes as large as M_w 8.2, but strong ground motions have not yet been recorded for larger events. In this study, we assume that ground motions from M_w 8 earthquakes on the Cascadia subduction zone are not systematically different from the motions that have been recorded during M_w 8 earthquakes on other subduction zones. Although the comparison of teleseismic P-waves from large shallow subduction earthquakes on differing subduction zones (HARTZELL and HEATON, 1985) indicates that there is a great variability between earthquakes in the time histories of energy release, no obvious systematic pattern could be recognized that would suggest inherent differences in the nature of energy release from earthquakes at subduction zones that are similar to the Cascadia subduction zone. For hypothetical earthquakes of M_w less than $8\frac{1}{4}$, our approach is to simply collect and categorize existing ground motion records according to earthquake size and site distance. We then create a suite of ground motions that have been recorded under conditions similar to those existing at a site for which ground motion estimates are desired. Motions within the suite are scaled to account for variations in earthquake size and site distance. This procedure is similar to that described by GUZMAN and JENNINGS (1976) and also HEATON *et al.* (1986).

As discussed by HEATON and HARTZELL (1986), there are several similarities between the Cascadia and southern Chile subduction zones. The 1960 Chilean earthquake (M_w 9.5) is thought to be the largest earthquake of this century (KANAMORI, 1977); it is also the largest event which we consider to be feasible in the northwestern United States. Estimation of strong ground motions for the 1960 earthquake is necessarily problematic since its seismic moment is approximately one hundred times that of the largest earthquake for which strong ground motions have been recorded. In order to obtain some estimate of the nature of shaking from such giant earthquakes, we employ the empirical Green's function technique

(HARTZELL, 1978, 1985; KANAMORI, 1979; IRIKURA, 1983; HOUSTON and KANAMORI, 1986). That is, we model a giant earthquake as a summation of smaller earthquakes for which we have ground motion records. Unfortunately, there are many poorly justified assumptions that are required in the application of this technique. We use teleseismic P-wave recordings, of both the earthquakes that we wish to model and the earthquakes for which we have strong motion recordings, to help constrain model assumptions.

Strong Motion Estimates for Earthquakes of $M_w < 8\frac{1}{4}$

Fifty-six recordings of strong ground motion from twenty-five shallow subduction earthquakes of $M_w \geq 7.0$ were collected for this study. Figures A1–A20 in Appendix A show the locations of recording sites relative to aftershock zones, the time histories of one component of horizontal ground acceleration and velocity for each recording, and the time histories of the mainshock moment release as inferred from teleseismic body waves by HARTZELL and HEATON (1985). These figures only briefly summarize this important data set; more detailed descriptions of the nature of each earthquake and the geologic setting of each recording site would undoubtedly allow us to rank the significance of individual records in our study of the Pacific Northwest.

There are several notable features in the figures in Appendix A. In most cases, the recordings sites are at vast distances from the earthquakes when compared with the strong motions data set recorded in the western United States. For instance, we do not know of any western U.S. records having site distances of over 150 km and peak accelerations of over 0.05G. However, this situation is fairly common in the figures in Appendix A. HEATON *et al.* (1986) conclude that the distribution of peak values of ground motion observed in Japan is quite different from that observed in the western U.S. Although the very largest ground motions have been observed at relatively close distances in the western U.S., impressively large motions have been recorded at distances of 100 km and more in Japan. HEATON *et al.* (1986) conclude that a major reason for this difference is simply due to differences in station distributions and earthquake sizes. That is, there have been many near-source recordings of moderate-sized U.S. earthquakes, but earthquakes of $M > 7$ have been rare. In contrast, there are numerous recordings of large Japanese earthquakes, but very near-source records are rare.

Every record in Appendix A is plotted on a common time scale which is quite compact so that these long-duration records fit within the confines of the figures. Although these ground motion records may, at first, appear to be very high frequency, a closer inspection shows that many of the ground accelerations records are dominated by frequencies of less than several hertz.

Comparison of the strong ground motions with the teleseismic time functions of

HARTZELL and HEATON (1985) gives somewhat inconsistent results. In some cases, there is fairly good correspondence between envelopes of the strong motion recordings and the envelope of the teleseismic time function. For example, the 1968 Hyuganada earthquake and the 1978 Miyagi-Oki earthquake both have relatively simple and short duration accelerograms and teleseismic time functions. Furthermore, the 1968 Tokachi-Oki earthquake, the 1979 St. Elias earthquake, and the 1983 Akita-Oki earthquake all have more complex teleseismic time functions whose envelopes are comparable to the envelopes of their strong motion recordings. These examples indicate that teleseismic records can be used to infer the overall duration and complexity of large earthquakes. Unfortunately, there are also examples where there is clearly no correlation between the teleseismic time function and the recorded strong ground motion; the 1968 Iwate-Oki earthquake, the 1973 Nemuro-Oki earthquake, and the 1966 Peruvian earthquake all have teleseismic time functions that are considerably longer than the observed strong ground motions. Without more detailed study of these earthquakes and local site effects, we can only speculate on the causes of the discrepancies between the strong motion records and the teleseismic time functions. HARTZELL and HEATON (1985) show that their time functions are very complex when they are derived from stations that are near theoretical body wave nodes. Problems of this type are undoubtedly present in some of the teleseismic time functions shown in Appendix A. However, despite the poor comparison of the teleseismic time function and the strong ground motion for the 1973 Nemuro-Oki earthquake, the teleseismic time function of HARTZELL and HEATON (1985) compares well with the teleseismic time function that KIKUCHI and FUKAO (1987) derived for this earthquake using an entirely independent set of teleseismic P-waveforms. We believe that these comparisons demonstrate that the teleseismic time functions provide useful estimates of the source duration and complexity as observed in the near-source region. However, there are also inconsistencies in these comparisons and one should be careful not to overinterpret the teleseismic time functions of HARTZELL and HEATON (1986).

In Figure 1, we show the response spectra of all of the horizontal components of ground motion for the records presented in Appendix A. The spectra are grouped according to earthquake size and the distance of the recording sites from the earthquakes. Of course, earthquakes are described by a complex 3-dimensional rupture surface and the definition of distance is necessarily ambiguous. For simplicity, we use the epicentral distance. However, because of the large fault dimensions of the 1968 Tokachi-Oki earthquake, we choose to measure distance from a point that is more central to the rupture surface than the epicenter is. This point is shown in Appendix A.

One of the most striking features of Figure 1 is the large degree of scatter in the spectra for ground motions observed at similar distances and from similar sized earthquakes. This scatter is quite troublesome when one is confronted with the problem of estimating the ground motions that a particular site may experience.

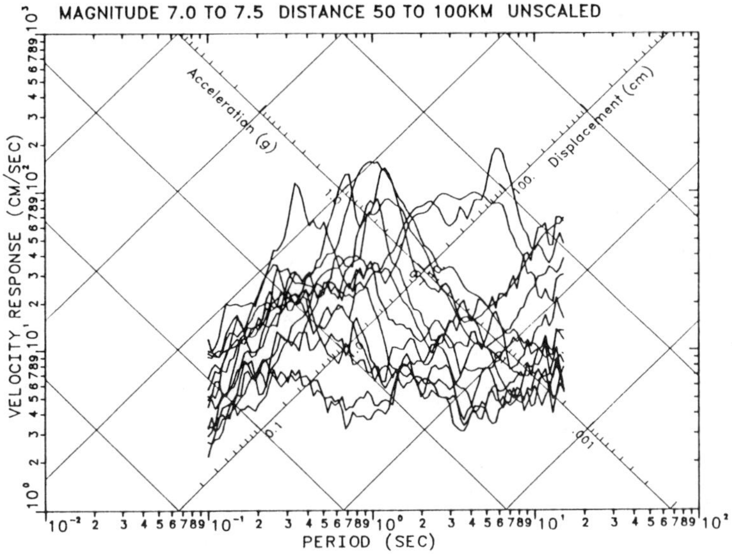

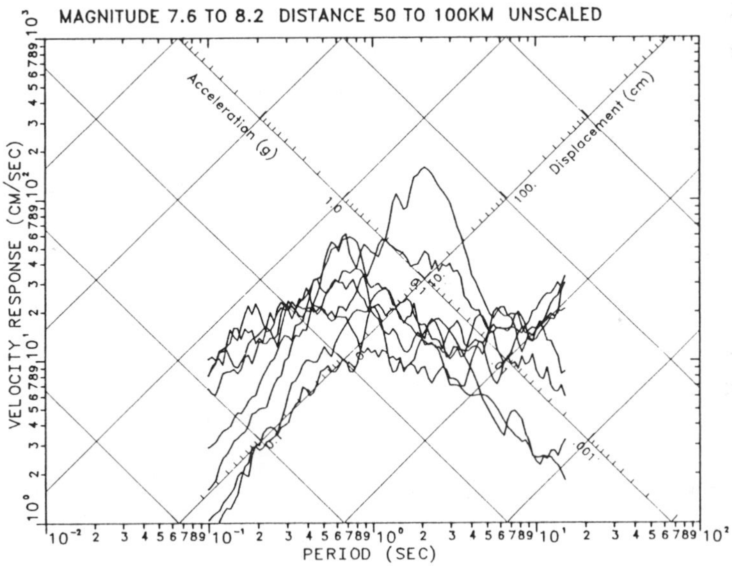

Fig. 1(a).

Figure 1

Pseudo-velocity response spectra (5% damped) of horizontal components of ground motion from large shallow subduction earthquakes. Time histories of one of the components are shown in Appendix A. The spectra from earthquakes of $7.0 \leq M_w \leq 7.5$ are shown above and spectra from earthquakes of $7.6 \leq M_w \leq 8.2$ are shown below. The spectra are further segregrated by site distance: a) 50 to 100 km, b) 101 to 140 km, c) 151 to 200 km, d) 201 to 250 km, and c) 251 to 300 km.

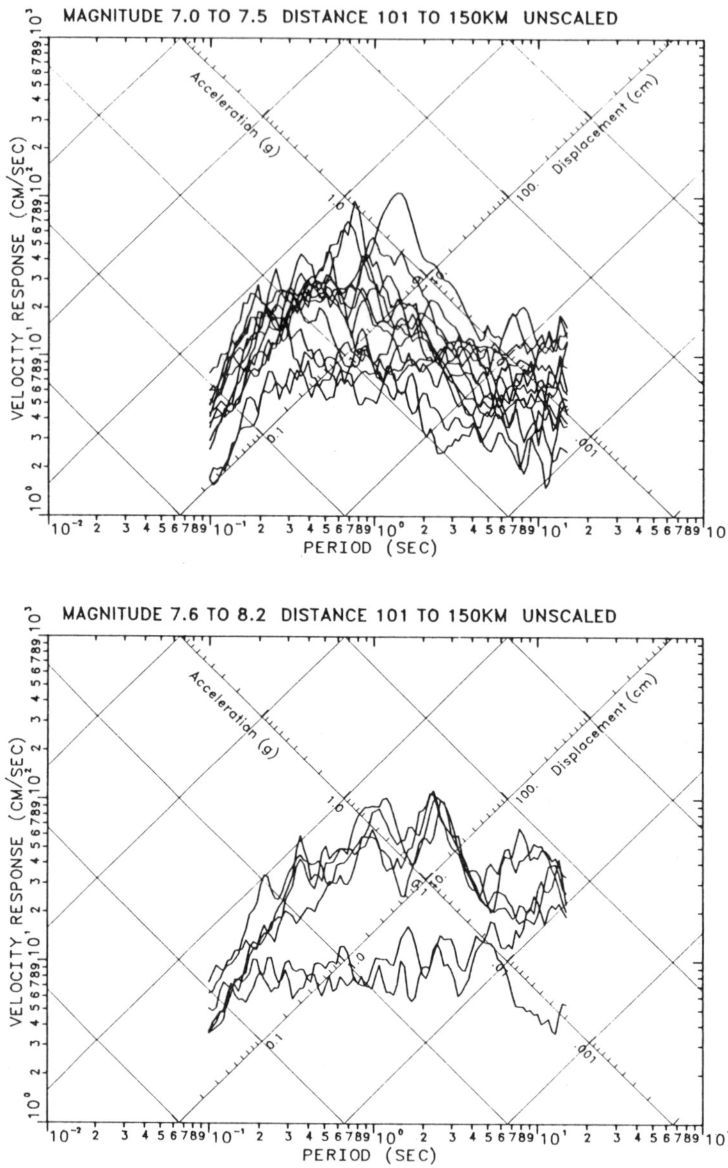
MAGNITUDE 7.0 TO 7.5 DISTANCE 101 TO 150KM UNSCALED
VELOCITY RESPONSE (CM/SEC)
PERIOD (SEC)
Acceleration (g)
Displacement (cm)
MAGNITUDE 7.6 TO 8.2 DISTANCE 101 TO 150KM UNSCALED
VELOCITY RESPONSE (CM/SEC)
PERIOD (SEC)
Acceleration (g)
Displacement (cm)

Fig. 1(b).

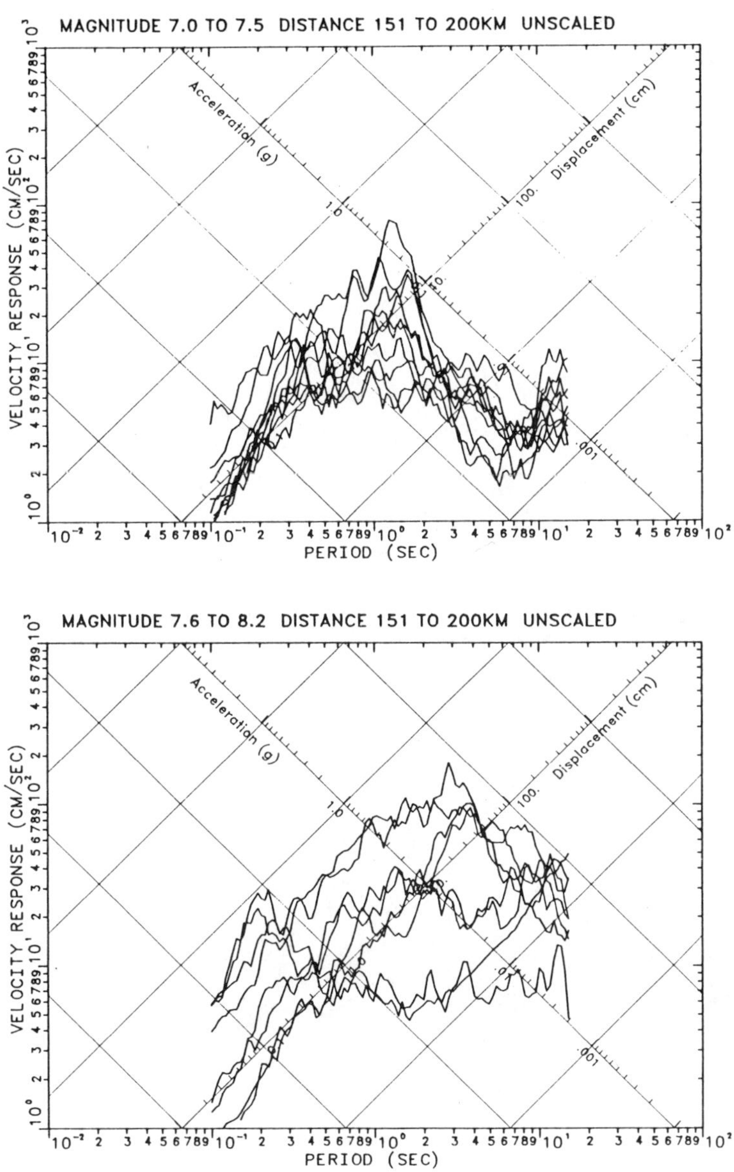
MAGNITUDE 7.0 TO 7.5 DISTANCE 151 TO 200KM UNSCALED
VELOCITY RESPONSE (CM/SEC)
PERIOD (SEC)
Acceleration (g)
Displacement (cm)
MAGNITUDE 7.6 TO 8.2 DISTANCE 151 TO 200KM UNSCALED
VELOCITY RESPONSE (CM/SEC)
PERIOD (SEC)
Acceleration (g)
Displacement (cm)

Fig. 1(c).

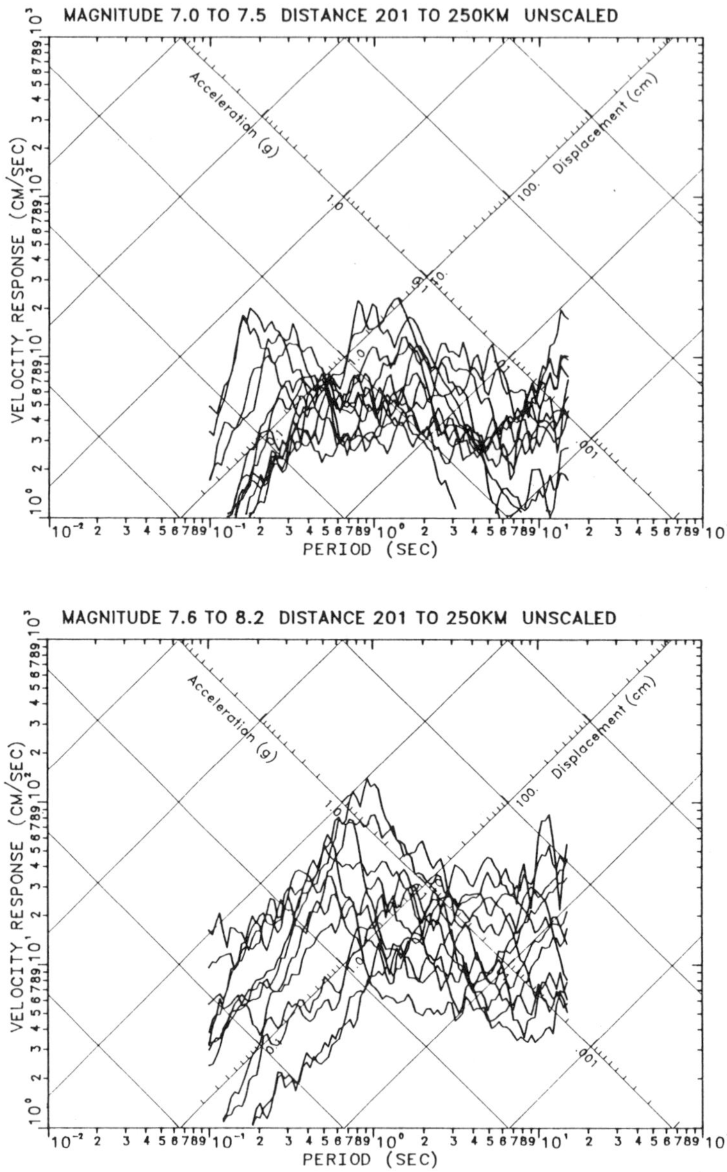

MAGNITUDE 7.0 TO 7.5 DISTANCE 201 TO 250KM UNSCALED
VELOCITY RESPONSE (CM/SEC)
Acceleration (g)
Displacement (cm)
PERIOD (SEC)
MAGNITUDE 7.6 TO 8.2 DISTANCE 201 TO 250KM UNSCALED
VELOCITY RESPONSE (CM/SEC)
Acceleration (g)
Displacement (cm)
PERIOD (SEC)

Fig. 1(d).

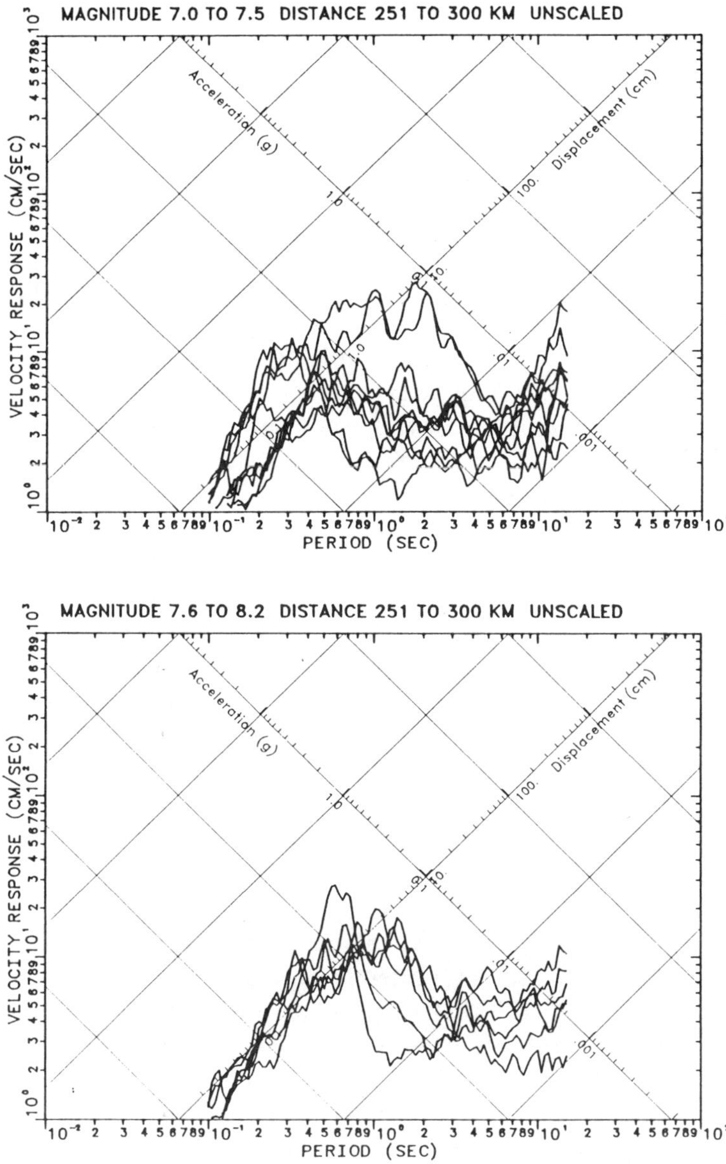
MAGNITUDE 7.0 TO 7.5 DISTANCE 251 TO 300 KM UNSCALED
VELOCITY RESPONSE (CM/SEC)
PERIOD (SEC)
Acceleration (g)
Displacement (cm)
MAGNITUDE 7.6 TO 8.2 DISTANCE 251 TO 300 KM UNSCALED
VELOCITY RESPONSE (CM/SEC)
PERIOD (SEC)
Acceleration (g)
Displacement (cm)

Fig. 1(e).

Even if the earthquake magnitude and distance is known (and it usually is not), the resulting ground motions are uncertain by more than a factor of ten. In many respects, the uncertainties arising from this scatter are larger than those in estimating the earthquake magnitude or distance. This problem is not restricted to this data set; strong ground motions recorded in the western United States are plagued by scatter of similar magnitude (HEATON *et al.*, 1986).

Is there some other way to regroup these records so that more certain estimates of ground motion can be obtained at a particular site? TRIFUNAC (1976) and JOYNER and BOORE (1982) present convincing evidence that lower frequency (less than 2 Hz) ground motions are systematically larger, by about a factor of two, at soil sites than at rock sites for ground motions recorded in the western United States. Higher frequency ground motions, however, show no significant correlation with site conditions. KAWASHIMA *et al.* (1984) report similar conclusions for ground motions from earthquakes of less than *M* 7 in Japan. They also report that soft soil sites experience low-frequency (less than 1 Hz) ground motions of about four times that of rock sites. LIU and HEATON (1984) suggest that the larger ground motions observed at soft sites result from the excitation of surface waves within basins on which soft sites usually sit.

In Figure 2, we compare response spectra that are obtained from a variety of earthquakes that are recorded at the same site. It is clear that the characteristics of the motion observed at a particular site are very similar from one earthquake to another and can dominate the source effects. Although site effects can sometimes be anticipated using simple models of plane body waves in horizontally layered media, there are undoubtedly important effects from complex three-dimensional geologic structures such as basins (LIU and HEATON, 1984). One promising technique for recognizing the effect of the recording site on strong gound motions is to record and analyze the weak motions from smaller earthquakes. Similar transmission effects should be seen in both strong and weak ground motions.

Of course, the details of the earthquake rupture process also affect the characteristics of strong ground motion. Directivity, fault finiteness, and slip distribution must control the overall characteristics of strong ground motion. Although significant peaks in response spectra are probably caused by reverberations within some geometrically regular velocity structure, the partition of energy in different frequency bands may be strongly affected by the nature of the rupture process. CROUSE *et al.* (1988) investigate response spectra from shallow earthquakes in seven different subduction zones and they conclude that, for the most part, systematic differences are not seen from one zone to another. However, they do present evidence that motions recorded in Peru/northern Chile and New Britain/ Bougainville regions were significantly smaller than other zones in the period range from 1 to 3 seconds. Although source characteristics may be important, they are very difficult to anticipate for design purposes. Transmission effects are also clearly important and may be quite complex. However, to some degree they can be anticipated through the study of weak ground motions from smaller earthquakes.

In Figure 3, we compare the smoothed response spectra that are the averages of the spectra in each of the distance and magnitude ranges presented in Figure 1. The average spectra were obtained by taking the average of the logarithmic values. Tables 1 and 2 give spectral velocities and standard deviations for these averaged spectra at selected periods. Although the large scatter in the data makes even these smoothed, average spectra appear to be somewhat chaotic functions of distance and magnitude, some general trends can be seen. At periods of less than 0.3 seconds, spectral velocities from magnitude 7.6 to 8.2 earthquakes are not significantly greater than those from magnitude 7.0 to 7.5 earthquakes. However, at periods greater than 1.0 seconds, the larger earthquakes are clearly associated with larger spectral velocities. Curiously, spectral velocities in the closest distance range, 50 to 100 km, do not seem significantly larger than those at distances from 101 to 150 km. However, the scatter in this data set is so large, that this observation may not be significant.

KAWASHIMA *et al.* (1984) and CROUSE *et al.* (1988) have both applied multiple regression analyses to response spectra from earthquakes at convergent plate boundaries (principally from Japan). In addition to considering ground motions from the large earthquakes included in this study, they also considered ground motions from earthquakes as small as M 5.0. Since their regression analyses presume that the motions can be characterized by smooth monotonic functions of distance and magnitude, the estimation of ground motions using their formulae does not lead to such apparent paradoxes as having the motions increase with distance as they sometimes do in Figure 3. Of course, a price is paid for this aesthetically pleasing smoothness. Ground motion predictions, particularly for magnitudes and distances near the end members of the data set, depend upon the specific functional forms chosen and upon data taken at other magnitudes and distances.

A comparison of ground motions predicted by the formulae of KAWASHIMA *et al.* (1984) and CROUSE *et al.* (1988) with the logarithmic averages of the data in this report is found in Tables 1 and 2. Ground motion predictions for shallow crustal earthquakes in the western United States are also given for comparison (JOYNER and BOORE, 1981; JOYNER and BOORE, 1982). In Figure 4, we compare the predicted ground motions of KAWASHIMA *et al.* (1984) with those for the western United States for earthquakes of M_w 7.9. The ground motions predicted for subduction earthquakes are dramatically larger than those for the western United States, at least at distances greater than 50 km. Unfortunately, the western United States curves are uncertain since there have been no strong motion recordings from earthquakes of this size in the western United States and these curves are extrapolations outside of the data. Curiously, peak ground accelerations and velocities for earthquakes of magnitude less than 7 are not dramatically different for Japanese

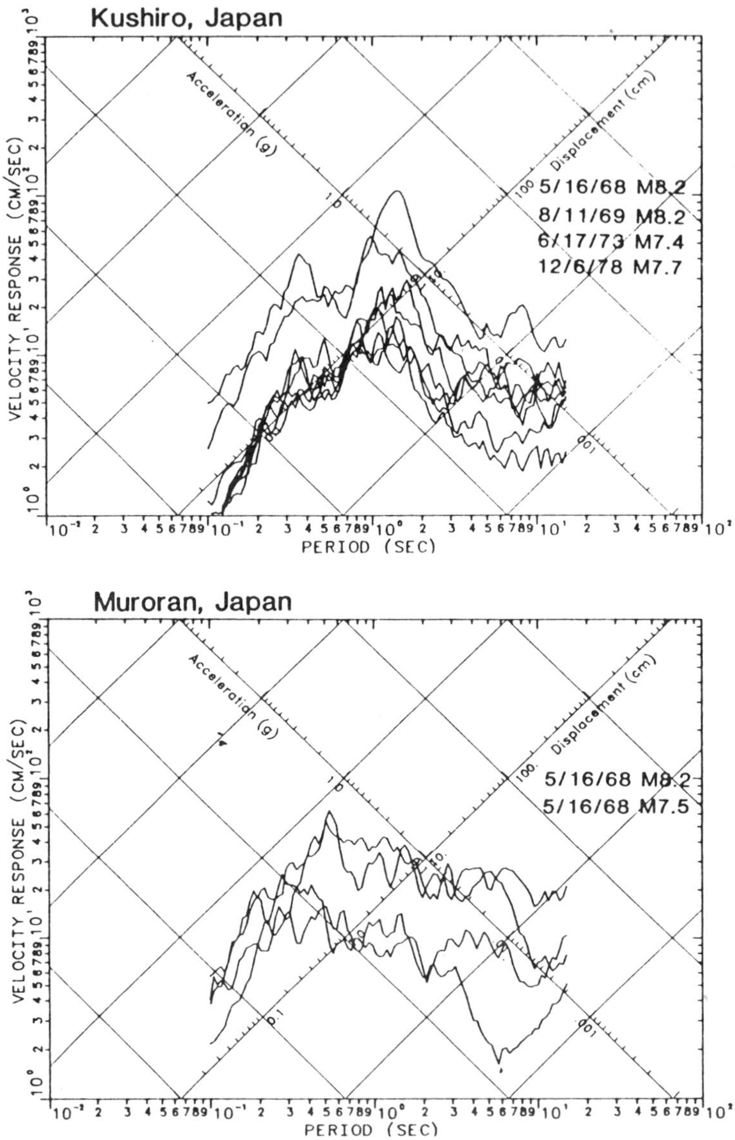
Kushiro, Japan
VELOCITY RESPONSE (CM/SEC)
PERIOD (SEC)
Acceleration (g)
Displacement (cm)
5/16/68 M8.2
8/11/69 M8.2
6/17/73 M7.4
12/6/78 M7.7
Muroran, Japan
VELOCITY RESPONSE (CM/SEC)
PERIOD (SEC)
Acceleration (g)
Displacement (cm)
5/16/68 M8.2
5/16/68 M7.5

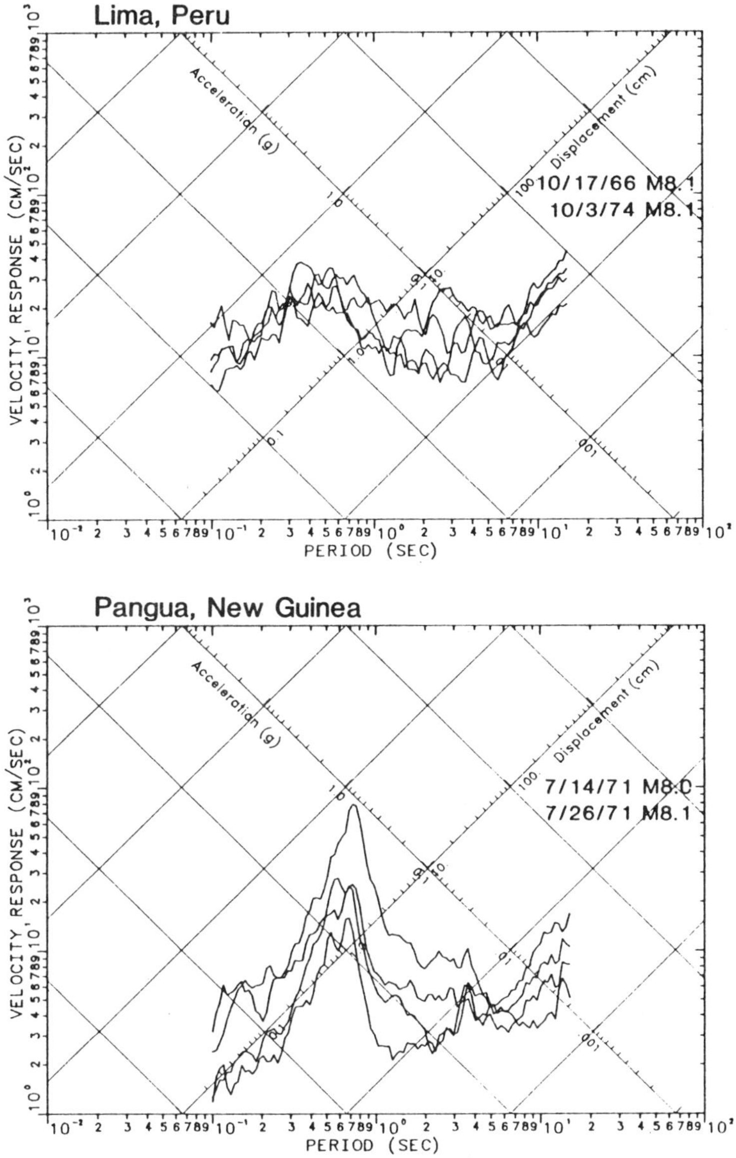

Figure 2
Pseudo-velocity response spectra (5% damped) of horizontal components of ground motion grouped by recording site.

Table 1

Magnitude 7.0 to 7.5.

	Observed, This Study				
	50–100 km	101–150 km	151–200 km	201–250 km	251–300 km
$\ddot{U}_{max}$ (cm/sec^2)	230. (392. / 135.)	161. (247. / 105.)	66. (104. / 42.)	43. (91. / 20.)	39. (66. / 23.)
$\dot{U}_{max}$ (cm/sec)	19.0 (44.0 / 8.2)	12.2 (21.6 / 6.9)	6.5 (10.9 / 3.9)	3.6 (5.8 / 2.2)	3.1 (5.3 / 1.8)
U_{max} (cm)	5.2 (13.1 / 2.0)	2.9 (5.2 / 1.6)	1.7 (2.5 / 1.2)	0.9 (1.7 / 0.5)	0.7 (1.3 / 0.4)
PSV (0.1) cm/sec	5.1 (8.6 / 3.0)	3.9 (6.0 / 2.5)	1.3 (2.1 / 0.7)	0.9 (2.0 / 0.4)	0.9 (1.3 / 0.6)
PSV (0.3) cm/sec	17.9 (32.7 / 9.8)	17.1 (26.5 / 11.0)	8.0 (12.7 / 5.0)	4.9 (8.7 / 2.7)	5.0 (8.3 / 3.0)
PSV (1.0) cm/sec	23.0 (72.1 / 7.3)	16.4 (35.7 / 7.6)	16.1 (27.4 / 9.5)	5.9 (10.8 / 3.3)	5.1 (11.9 / 2.2)
PSV (3.0) cm/sec	14.6 (35.6 / 6.0)	8.7 (15.3 / 5.0)	6.2 (8.4 / 4.5)	4.4 (7.6 / 2.5)	4.4 (8.1 / 2.4)
PSV (10.0) cm/sec	13.6 (30.0 / 6.2)	6.1 (10.2 / 3.7)	4.0 (5.3 / 3.0)	3.4 (6.2 / 1.9)	4.3 (6.8 / 2.7)

	Crouse, Depth = 30 km, *M* 7.25				
	75 km	125 km	175 km	225 km	275 km
PSV (0.1) cm/sec	2.4	1.6	1.3	1.0	0.9
PSV (0.3) cm/sec	11.7	7.7	5.6	4.4	3.5
PSV (1.0) cm/sec	18.9	12.6	9.3	7.3	6.0
PSV (3.0) cm/sec	11.5	7.6	5.6	4.4	3.6

	Kawashima, *M* 7.25				
$\ddot{U}_{max}$ (cm/sec^2)	149.2	92.8	66.0	50.6	40.7
$\dot{U}_{max}$ (cm/sec)	12.5	7.7	5.5	4.2	3.4
U_{max} (cm)	2.3	1.4	1.0	0.7	0.6
PSV (0.1) cm/sec	4.4	2.8	2.0	1.5	1.3
PSV (0.3) cm/sec	16.8	10.6	7.6	5.9	4.8
PSV (1.0) cm/sec	31.3	19.8	14.2	11.0	8.9
PSV (3.0) cm/sec	12.7	8.0	5.8	4.4	3.6

	Joyner and Boore, *M* 7.25				
$\ddot{U}_{max}$ (cm/sec^2)	51.0	22.9	12.2	7.1	4.3
$\dot{U}_{max}$ (cm/sec)	9.5	4.2	2.2	1.3	0.78
PSV (0.1) cm/sec	1.1	0.3	0.1	0.03	0.01
PSV (0.3) cm/sec	4.4	1.4	0.5	0.2	0.09
PSV (1.0) cm/sec	10.4	3.8	1.6	0.7	0.4
PSV (3.0) cm/sec	17.5	10.5	7.4	5.8	4.7

( ) = ± one standard deviation.

Table 2

Magnitude 7.6 to 8.2.

	Observed, This Study				
	50–100 km	101–150 km	151–200 km	201–250 km	251–300 km
$\ddot{U}_{max}$ (cm/sec^2)	144. (233. / 89.)	199. (289. / 137.)	108. (202. / 57.)	104. (262. / 41.)	47. (57. / 39.)
$\dot{U}_{max}$ (cm/sec)	18.5 (28.3 / 12.0)	18.5 (42.8 / 8.0)	17.2 (36.8 / 8.1)	14.7 (28.3 / 7.6)	5.0 (5.7 / 4.3)
U_{max} (cm)	6.7 (10.2 / 4.4)	7.8 (18.0 / 3.4)	8.9 (24.1 / 3.3)	5.3 (13.1 / 2.1)	1.0 (1.4 / 0.7)
PSV (0.1 cm/sec	3.4 (8.8 / 1.3)	4.7 (6.3 / 3.5)	2.3 (5.4 / 1.0)	2.3 (7.3 / 0.7)	1.0 (1.2 / 0.8)
PSV (0.3) cm/sec	13.0 (23.9 / 7.1)	18.6 (33.0 / 10.5)	9.5 (17.3 / 5.2)	9.7 (24.3 / 3.9)	5.4 (6.9 / 4.2)
PSV (1.0) cm/sec	25.0 (38.2 / 16.3)	35.9 (95.4 / 13.5)	20.7 (48.9 / 8.8)	21.8 (50.4 / 9.4)	8.1 (15.5 / 4.3)
PSV (3.0) cm/sec	16.4 (38.0 / 7.1)	34.3 (77.8 / 15.1)	36.5 (109.7 / 12.1)	14.6 (25.6 / 8.3)	3.9 (4.8 / 3.1)
PSV (10.0) cm/sec	9.2 (20.7 / 4.1)	22.4 (48.4 / 10.3)	27.6 (44.9 / 16.9)	14.9 (33.5 / 6.6)	4.7 (6.8 / 3.2)
	Crouse, Depth = 30 km, *M* 7.9				
	75 km	125 km	175 km	225 km	275 km
PSV (0.1) cm/sec	2.9	2.1	1.6	1.4	1.1
PSV (0.3) cm/sec	16.0	11.0	8.2	6.6	5.4
PSV (1.0) cm/sec	36.1	25.2	19.2	15.3	12.7
PSV (3.0) cm/sec	26.3	18.3	13.8	11.1	9.2
	Kawashima, *M* 7.9				
$\ddot{U}_{max}$ (cm/sec^2)	238.4	148.3	105.5	80.9	65.0
$\dot{U}_{max}$ (cm/sec)	23.7	14.7	10.5	8.0	6.4
U_{max} (cm)	5.4	3.3	2.3	1.8	1.4
PSV (0.1) cm/sec	6.6	4.2	3.0	2.3	1.9
PSV (0.3) cm/sec	28.1	17.8	12.8	9.9	8.0
PSV (1.0) cm/sec	71.2	45.0	32.4	25.0	20.2
PSV (3.0) cm/sec	30.5	19.3	13.9	10.7	8.7
	Joyner and Boore, *M* 7.9				
$\ddot{U}_{max}$ (cm/sec^2)	74.0	33.2	17.7	10.3	6.3
$\dot{U}_{max}$ (cm/sec)	19.7	8.8	4.7	2.7	1.6
PSV (0.1) cm/sec	1.1	0.3	0.1	0.04	0.01
PSV (0.3) cm/sec	4.7	1.5	0.5	0.2	0.1
PSV (1.0) cm/sec	13.2	4.8	2.0	1.0	0.5
PSV (3.0) cm/sec	23.0	13.7	9.8	7.6	6.2

( ) = ± one standard deviation.

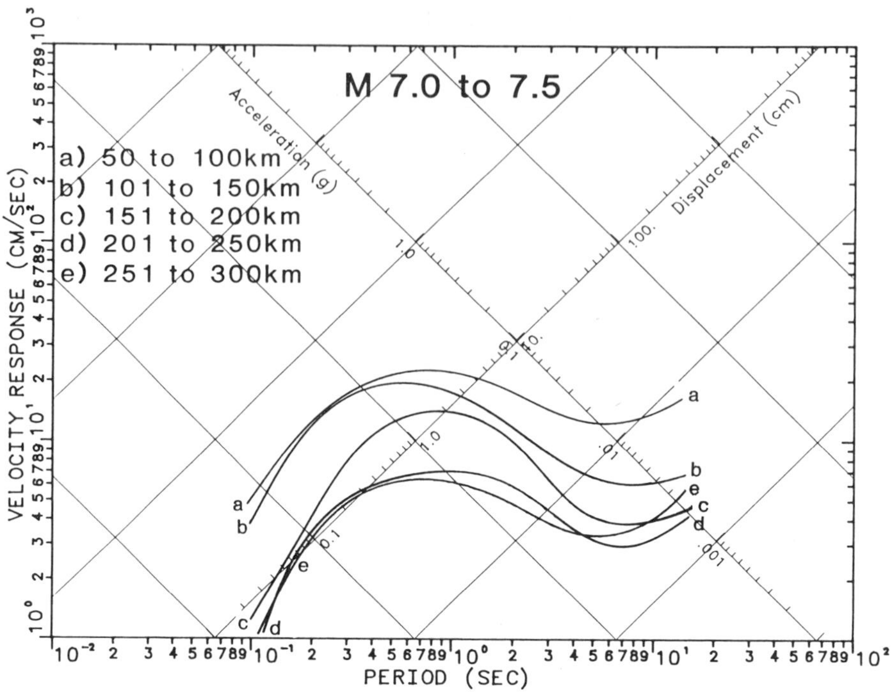
M 7.0 to 7.5
a) 50 to 100km
b) 101 to 150km
c) 151 to 200km
d) 201 to 250km
e) 251 to 300km
Acceleration (g)
Displacement (cm)
VELOCITY RESPONSE (CM/SEC)
PERIOD (SEC)

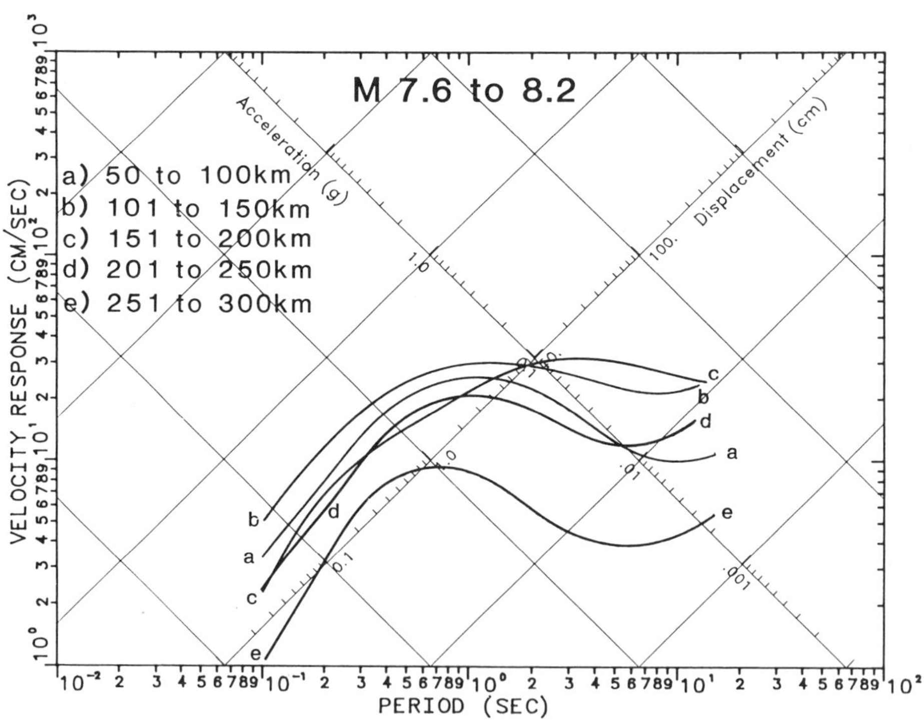
M 7.6 to 8.2
a) 50 to 100km
b) 101 to 150km
c) 151 to 200km
d) 201 to 250km
e) 251 to 300km
Acceleration (g)
Displacement (cm)
VELOCITY RESPONSE (CM/SEC)
PERIOD (SEC)

and western United States earthquakes (HEATON *et al.*, 1986). It may be that the distance attenuation is a function of earthquake size, with large subduction earthquakes having a relatively small distance attenuation.

Choosing a ground motion for design at a particular site is obviously a difficult and somewhat philosophical problem given the large scatter in the observed ground motions. However, once a design motion is chosen for a site, it may be very sobering to plot that design level against the actual data plotted in Figure 1.

Estimating Strong Ground Motions for Earthquakes of $M_w > 8\frac{1}{4}$

The Cascadia subduction zone has not experienced any large, shallow subduction earthquakes over its greater than 1000 km length for at least 150 years despite an apparent convergence rate of about 4 cm/yr. Although it is conceivable that convergence is occurring continuously through aseismic creep, the Cascadia subduction zone has many similarities to the subduction zones in southern Chile, southwestern Japan, Colombia, and Mexico (HEATON and HARTZELL, 1986). Each of these zones have experienced very large historic shallow subduction earthquake sequences and can be considered to be strongly coupled. If the Cascadia subduction zone is also strongly coupled, then earthquakes far larger than any of the events for which we presently have strong motion records can be postulated for this zone. In particular, the largest earthquakes experienced on these other subduction zones were: The 22 May 1960 Chilean earthquake (M_w 9.5), the 1707 Hoei earthquake of southwestern Japan ($M_w > 8\frac{1}{2}$), the 31 January 1906 Colombian earthquake (M_w 8.8), and the 3 June 1932 Jalisco earthquake of Mexico (M_s 8.2). HEATON and HARTZELL (1986) present circumstantial evidence that suggests (but does not prove) that large earthquakes along the Cascadia subduction zone may have an average repeat time of 400 to 500 years. Given the length of the apparent seismic gap and a suggested repeat time of more than 400 years, it seems difficult to eliminate the possibility of earthquakes comparable to the 22 May 1960 Chilean earthquake, the largest earthquake recorded this century (KANAMORI, 1977).

What might the ground motions look like from a giant earthquake such as the 1960 Chilean earthquake? This earthquake is in a different class from the earthquakes for which strong ground motions have been recorded; its seismic moment is at least one hundred times larger than that of the largest earthquake for which we have data. There are several approaches that could be utilized to estimate the nature

Figure 3
Smoothed (with period) logarithmic averages of pseudo velocity response spectra (5% damped) that are shown in Figure 1. Numerical values for these curves and standard deviations of the data about the average are found in Tables 1 and 2.

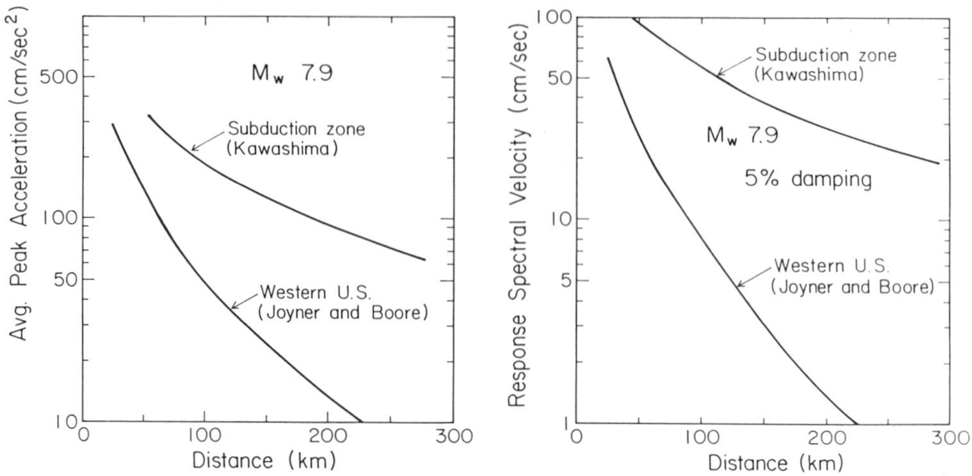

Figure 4
Comparison of predictions of average peak ground motions obtained by regression analysis of data from the southwestern United States and from Japan. Soil sites and 5% damping are assumed for the response spectra.

of motions from giant earthquakes. One approach would be to simply use the regression analyses of data from smaller earthquakes and extrapolate to very large magnitudes. However, there is little confidence that the functional form that was chosen to fit the data is appropriate to extrapolate outside the region for which there is data. A second approach would be a slight variation to the first. Once again data from smaller earthquakes could be extrapolated into the range of very large magnitudes, but the functional forms would be based on spectral scaling laws and similarity conditions. However, there is little reason to expect that an earthquake having a fault length of as much as 1000 km and rupture that may extend to the uppermost mantle will be similar to an earthquake with a fault length of 100 km and fault width that is confined to the offshore accretionary wedge.

Fortunately, teleseismic data are available from some of the most significant events of this century (HARTZELL and HEATON, 1985; HOUSTON and KANAMORI, 1986). In particular, teleseismic P-waveforms contain information about energy radiated by these earthquakes at periods as short as 1 second. Unfortunately, strong ground motions and teleseismic P-waveforms result from different complex combinations of source and wave propagation features. To infer strong ground motions directly from teleseismic P-waveforms would require a great leap of faith. In principal, we could use the teleseismic P-waveforms to deduce the detailed rupture history of a large event. We could then derive strong motion recordings from that rupture history. Of course, this presupposes that we accurately know the propagation effects of earth structure and that we have sufficient teleseismic data to infer the details of the rupture process at very short wavelengths. Although this approach can be dismissed as being impractical, it has fundamental similarities to the empirical Green's function technique that we employ in this study.

The Empirical Green's Function Technique

The basic idea behind the empirical Green's function technique is the notion that large earthquakes can be considered to be a linear combination of smaller ones. That is, to model a large earthquake, we merely superpose N smaller ones, where N is the integral ratio of the moment of the larger event to the smaller one. Since waves from the smaller and larger events travel through the same seismic velocity structure in the same manner, this technique allows us to remove this unknown from the modeling process. Furthermore, since the source of the smaller event has rupture properties that are probably similar to that of the larger event, this technique also allows us to model effects due to statistical irregularity of the rupture process. Descriptions and applications of the technique are given by HARTZELL (1978, 1985), KANAMORI (1979), IRIKURA (1983) HOUSTON and KANAMORI (1986), and others. A more quantitative discussion of this technique is also found in Appendix B of this paper.

Unfortunately, things are not quite as simple as they might appear at first glance. There are difficult questions involved in deciding how to sum these records together. For instance, if an earthquake rupture is a smooth, coherent process, then smaller ruptures must be summed in such a way that an irregular rupture process does not result. Just how to fit these records together is a very fundamental question that is explored further in Appendix B. The nature of the irregularity of the rupture process may be investigated by asserting that earthquake ruptures are self-similar with respect to size. However, this is an important issue and self-similarity should not be accepted as a matter of faith. Fortunately, this same empirical Green's function technique can be used to model teleseismic P-waveform data of historic giant earthquakes. This teleseismic data helps constrain the assumptions in the summation process. In this study, we are most concerned with estimating strong ground motions in the spectral band from 10 seconds to 10 hertz. We are not so concerned about ground motions in the spectral band comparable to the duration of giant earthquakes which may be more than 100 seconds. Thus we apply the empirical Green's function technique in such a way that we match the characteristics of short-period teleseismic waveforms. Because of effects introduced by the randomness that we assume in the timing of the Green's functions, summing enough records to match the moment ratio of the smaller and larger events may result in a significant overestimate of the short-period motions. This problem is discussed in Appendix B. To alleviate this problem, we sum just enough events to match the characteristics of the short-period teleseismic waveforms, even though this may not be consistent with the moment ratio of the two events.

In order to apply this technique, we must have an adequate set of records to use as empirical Green's functions. In this study we primarily use the records from two earthquakes for which we have the best combination of strong motion and teleseismic data; these are the 16 May 1968 Tokachi-Oki earthquake (M_w 8.2) and the 12 May 1978 Miyagi-Oki earthquake (M_w 7.5). The Miyagi-Oki event appears to be a rather simple single source, whereas the Tokachi-Oki event appears

Table 3
Empirical Green's Function.

Earthquake	Date	M_w	Green's Function Code	Station	Record Code	Distance (km)
Hyuganada	04/01/68	7.4	A	Hyaga	S–213	80
			B	Kochi	S–211	162
Tokachi-Oki	05/16/68	8.2	C	Muroran	S–234	201
			D	Aomori	S–235	175
			E	Hachinohe	S–252	130
			F	Miyako	S–236	180
Miyagi-Oki	06/12/78	7.5	H	Hachinohe	S–1202	262
			I	Akita	S–1200	240
			J	Ofunata	S–1210	101
			K	Shiogama	S–1201	95
			L	Onahama	S–1191	175
			M	Yamashita	M–217	370
St. Elias	02/28/79	7.8	Q	Icy Bay		77
			R	Yakutat		170

to be relatively complex. The actual strong motion records chosen to be Green's functions are given in Table 3. An inspection of these records (found in Appendix A) shows that there is a wide range in the character of the motions from station to station. As was discussed earlier, much of this character seems to be a site effect. In our summation procedure, we sum the records from several different sites and thus our synthetic records represent motions that are averaged in a poorly defined way over several sites. If ground motions are desired at a specific site, it may be better to sum records taken only from stations that appear to have simple site responses (hard rock sites?) and then apply a site correction that is appropriate. Of course, this is easier said than done and we have not attempted to eliminate Green's functions with complex site responses in this study.

Definition of the Source Geometry

The source characteristics of six of the largest earthquakes for which we have short-period teleseismic data are given in Table 4. Source parameters are also given for the four events in Table 3 chosen as empirical Green's functions. We will describe the details of the model geometry for only the largest of these, the 22 May 1960 Chilean earthquake (M_w 9.5). Unfortunately, there is still very little known about the spatial and temporal distribution of seismic energy release for this, or any other giant earthquake. There is little question that the rupture length was very long, on the order of 1000 km. For simplicity, we assume that the characteristics of the rupture do not change significantly along the fault length, although we present no real data to justify this assumption. The question of how the rupture varies down the dip is both problematic and of considerable concern since such variation can significantly affect

Table 4
Earthquake Parameters.

Earthquake	Date	M_w	$M_0 \times 10^{28}$ dyne-cm	Length (km)	Width (km)
Hyuganada	04/01/68	7.4	0.17	—	—
Miyagi-Oki	06/12/78	7.5	0.22	100	75
St. Elias	02/28/79	7.8	0.65	—	—
Tokachi-Oki	05/16/68	8.2	2.8	180	140
Kuriles	10/13/63	8.5	7.5	250	150
Rat Island	02/04/65	8.7	14.0	500	150
Aleutians	03/09/57	9.1	15.0	800	150
Kamchatka	11/04/52	9.0	35.	650	200
Alaska	03/28/64	9.2	75.	700	300
Chile	05/22/60	9.5	270.	950	200

the spatial distribution of the strong shaking. For example, we would like to know whether or not large amounts of seismic energy are radiated from directly beneath a site for which we simulate ground motions. Since it is likely that the rheology of fault zone materials changes with depth, it also seems likely that the rupture characteristics change with depth. It may be that significant slip occurs slowly at depth with little radiation of energy in the frequency band of interest. Because there is so little constraints on the assumption of fault width, we present the results both from models in which short-period radiation is confined mainly to the offshore regions and also from models in which seismic radiation is allowed to extend significant distances inland.

The geometries of several models of the Chilean earthquake are shown in Figures 5 and 6 and a summary of parameters for models presented in this study is found in Table 5. In Figure 5, we show several different ways to simulate a Chilean earthquake using the records from Miyagi-Oki-size earthquakes (4/1/68 Hyuganada *M* 7.4, 6/12/78 Miyagi-Oki *M* 7.5, and 2/28/79 St. Elias *M* 7.8). First consider the geometry of the model M–1140–200; the M stands for Miyagi-Oki, 1140 is the total number of records that are summed, and 200 is the fault width. The long-period moment ratio of the 1960 Chilean earthquake and the Miyagi-Oki earthquake is 1140. In this model, each subfault is assumed to be of dimensions 50 km long by 40 km wide. The letters in the subfaults tell which records were used as Green's functions for that subfault (see Table 3). The Green's functions were chosen such that the difference in distance between that required by the model geometry and that at which they were observed is a minimum. The Green's functions are also scaled for distance using the distance attenuation relationship of CROUSE *et al.* (1988). The subfaults with plus signs have Green's functions that are a simple average of their closest neighbors. In the synthesis of teleseismic *P*-wave records, the Green's function is assumed to be the same for each subfault. The rupture is assumed to propagate radially at a velocity of 3.2 km/sec from the hypocenter shown. In model M–1140–200, each subfault ruptures 12 times within a period of 60 seconds after the onset of rupture. The actual rupture times are random within 12, 5-second, evenly-spaced time windows.

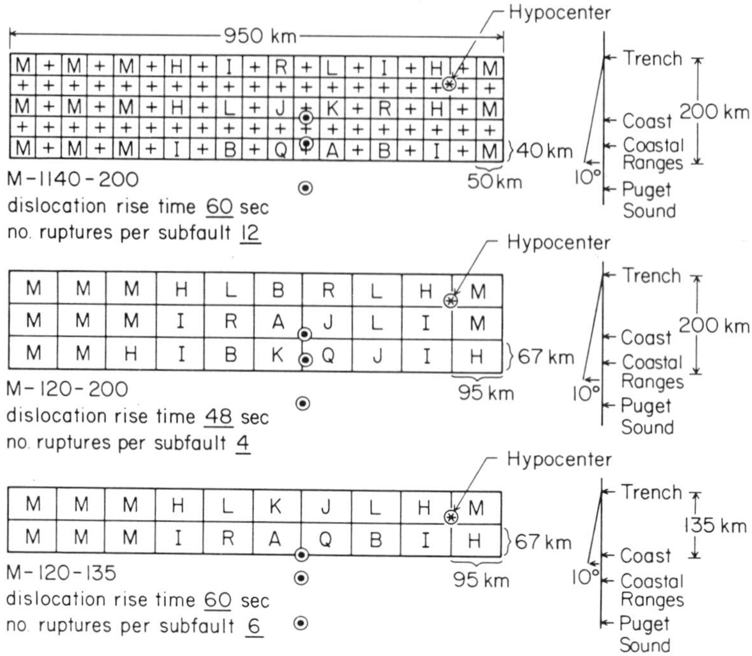

Figure 5
Geometry of fault models used to simulate an earthquake in the Pacific Northwest similar to the 22 May 1960 Chilean earthquake (M_w 9.5). Each box represents a subfault whose response is simulated by summing the ground motion from M_w 7.5 earthquakes. Letters in the boxes refer to Table 3 and they designate which records are used as Green's functions. Boxes with a + use a Green's function that is a linear interpolation of the Green's functions in adjacent boxes. Cross-sections, showing the relative position of the fault and the observation points are shown to the right.

The fault is assumed to dip 10 degree landward from a surface trace that is about 120 km offshore. The earthquake is observed at three different locations near the center of the rupture. One site is assumed to be on the coast, another is located within the coastal ranges about 50 km inland, and the final site is located in the Puget Sound region. When the rupture is assumed to have a width of 200 km, it extends beneath both the coastal and coastal ranges sites.

The second model, M–120–200, is very similar to the first except the total number of Green's functions that are summed is reduced to only 120. As discussed later, fewer Green's functions are required by the short-period teleseismic data than is required by the moment ratio. In this second model, each subfault is assumed to be 95 km long and 67 km wide. Each subfault ruptures 4 times within a period of 48 seconds after the onset of rupture. In the third model, M–120–135, the rupture surface is assumed to be significantly narrower (135 km) and is confined to mainly the offshore regions. Although it seems likely that the Chilean earthquake ruptured over a larger fault width than this, it may be that short-period waves emanated only from the shallower portions of the fault. The total number of Green's functions summed is

Table 5

Model Parameters.

Model Code	Length (km)	Width (km)	Dislocation Rise Time (sec)	*	$m*$	$n*$
1960 Chile M_w 9.5						
M–1140–200	950	200	60	19	5	12
M–120–200	950	200	48	10	3	4
M–120–135	950	135	60	6	2	6
T–96–200	900	200	64	6	2	8
T–24–200	900	200	50	6	2	2
T–6–100	900	100	50	6	1	2
1964 Alaska M_w 9.2						
M–364–300	700	300	32	13	7	4
M–88–300	700	300	34	11	4	2
1957 Aleutian M_w 9.1						
M–66–150	800	150	12	11	3	2
M–22–150	800	150	6	11	2	1
1952 Kamchatka M_w 9.0						
M–165–200	650	200	15	11	5	3
M–36–200	650	200	14	6	3	2
1965 Rat Island M_w 8.7						
M–55–150	500	150	5	11	5	1
M–18–150	500	150	4	6	3	1
1963 Kuriles M_w 8.5						
M–36–150	250	150	7	4	3	3
M–12–150	250	150	2	4	3	1
1968 Tokachi-Oki M_w 8.2						
M–9–140	180	120	4	3	3	1

* is the number of subfaults along the fault length; m is the number of subfaults along the fault width; n is the number of times each subfault ruptures.

120, the same as in the previous model. However, the number of times that each subfault ruptures is increased to 6 during a total dislocation rise time of 60 seconds.

In Figure 6, we show an alternate set of models for the Chilean earthquake in which records from the 1968 Tokachi-Oki earthquake (M_w 8.2) are used as empirical Green's functions. Most features of these models are very similar to those presented in Figure 5. However, since the Tokachi-Oki earthquake is a much larger event, only 96 of them are required to match the moment of the Chilean earthquake. Once again, the letters in each subfault refer to specific records listed in Table 3. Models T–96–200 and T–24–200 both have the same overall subfault geometries and are 900 km long by 200 km wide; the only difference being that each subfault ruptures 8 times over a total dislocation time of 64 seconds in T–96–200, whereas each subfault ruptures only twice over a dislocation time of 50 seconds in model T–24–200. In the last model, T–6–100, the Chilean earthquake is modeled as a single line of Tokachi-Oki earthquakes where short-period radiation is confined to

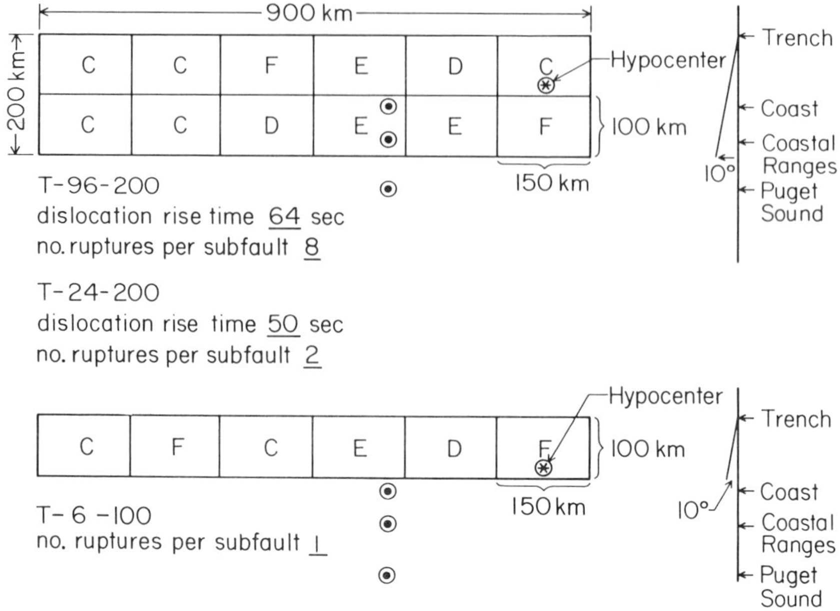

Figure 6
Same as Figure 5, except that 16 May 1968 Tokachi-Oki earthquake (M_w 8.2) records are used as Green's functions.

the offshore region. In this case only 6 Tokachi-Oki earthquakes are summed with each subfault experiencing only one rupture. Of the Tokachi-Oki Green's function simulations, this model fits the teleseismic P-wave amplitudes the best.

Results for Teleseismic P-waveforms

Before we discuss the strong ground motions that we obtained from our modeling procedure, we will present the salient features of our models of the teleseismic waveforms of giant earthquakes. These teleseismic models provide important constraints on the model parameters of our strong motion models. Simulations of teleseismic recordings of the 22 May 1960 Chilean earthquake are shown in Figure 7 using Miyagi-Oki earthquake records (M_w 7.5) as Green's functions and in Figure 8 using Tokachi-Oki (M_w 8.2) records as Green's functions. We model the Pasadena, California long-period Benioff seismograms (1–90) and the Tinemaha, California short-period Benioff seismograms (1–0.7) since both of the earthquakes that we use as Green's functions and also the giant earthquakes that we wish to model were well recorded on these seismometers. The peak amplitude of each record is given where the units are microns for the long-period Benioff records and the units are in centimeters (not corrected for magnification) for the short- period Benioff records. Since the Japanese earthquakes are observed at a range of about 75 degrees and the Chilean

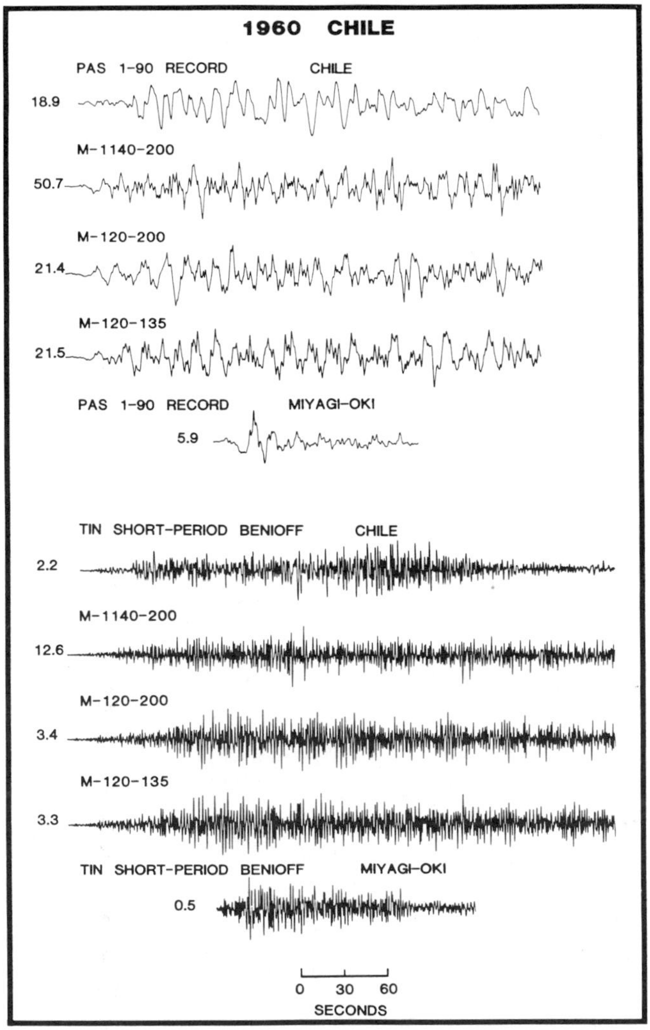

Figure 7
Comparison of observed long- and short-period vertical Benioff teleseismic P-wave seismograms with records synthesized using the 1978 Miyagi-Oki earthquake seismograms as Green's functions. Details of the model parameters are found in Figure 5 and the text. The Miyagi-Oki seismograms that were used as Green's functions are also shown. Peak amplitudes of the long-period Benioff records have been corrected for instrument magnification and are given in microns. Peak amplitudes for the short-period Benioff records are given in centimeters on the original seismograms.

earthquake is observed at about 85 degrees, the Japanese records were corrected for spherical spreading in the summation procedure, about a factor of 0.87 in this case.

Although the moment of the Chilean earthquake that is derived from long-period waves is about 1000 times that of an M_w 7.5 earthquake, it seems clear that summing the records from that many Miyagi-Oki earthquakes in a random way inevitably overestimates the observed waveforms (Figure 7). In a similar manner,

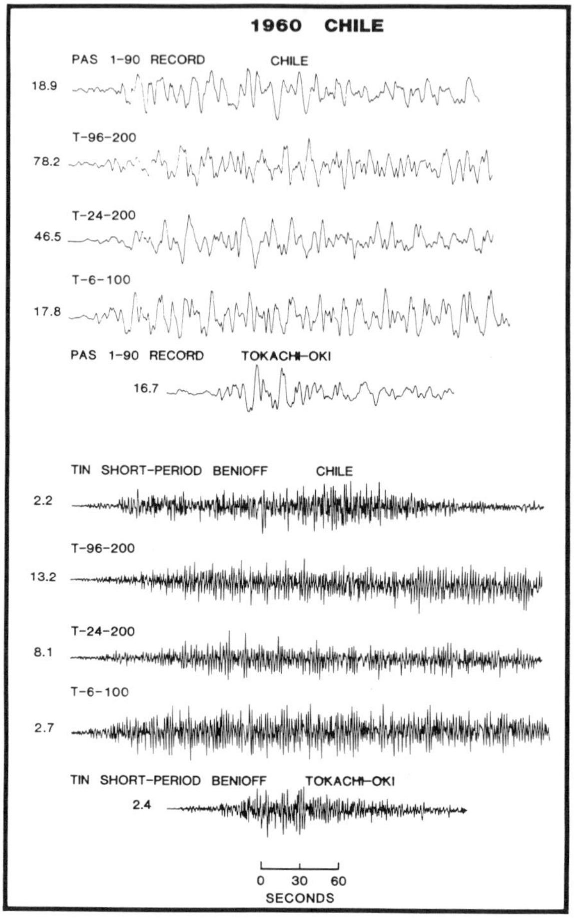

Figure 8
Same as Figure 7, except that seismograms from the 1968 Tokachi-Oki earthquake are used as Green's functions. Model parameters are found in Figure 6 and the text.

summation of the records from 96 Tokachi-Oki earthquakes results in synthetic records that are too large (Figure 8). In fact, the records from the Tokachi-Oki earthquake have peak amplitudes comparable to those observed for the Chilean earthquake. Thus it is important that our modeling procedure increase the duration of the signal without significantly increasing the peak amplitudes. In Figure 8 we see that it is necessary to sum only about 6 Tokachi-Oki earthquakes in order to match the overall amplitude and duration of the Chilean earthquake records. It should be noted that the deconvolutions of the Pasadena Benioff 1–90 records by HARTZELL and HEATON (1985) for both the Miyagi-Oki and Tokachi-Oki earthquakes yielded moments equal to the accepted long-period estimates. These results indicated that the Pasadena Benioff records are representative records for these two events. Below we present teleseismic simulations of other great earthquakes using Miyagi-Oki as a Green's function which support the results obtained in the simulation of the Chilean earthquake, and argue against these results being due to nodal records.

A comparison of synthetic short-period Benioff records with that observed during the 1964 Alaskan earthquake (M_w 9.2) is shown in Figure 9. The teleseismic P-waves for this event are among the largest we observed for any earthquake; the Tinemaha short-period Benioff record is about twice the amplitude of the record for the Chilean earthquake and the Alaskan earthquake is the only event for which the

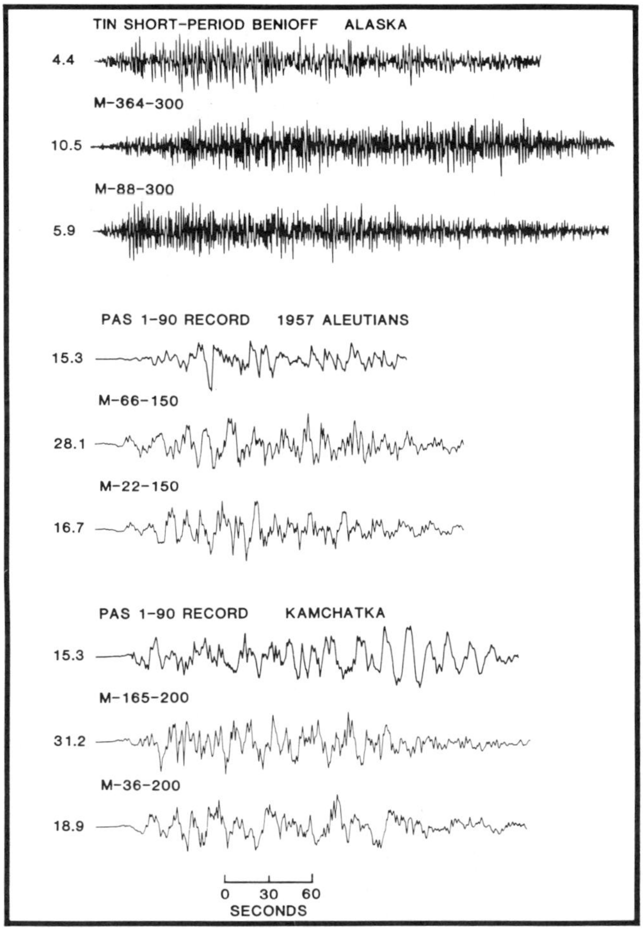

Figure 9

Comparison of observed teleseismic P-wave seismograms with records synthesized by summing Miyagi-Oki earthquakes records. In each case, the first model is the one for which enough Miyagi-Oki earthquakes were summed to match the long-period moment. In the second model, only enough records were summed to match the observed seismograms.

Pasadena long-period Benioff records were off scale. However, Alaska is relatively close to Pasadena, and when geometric spreading is considered, the Alaskan records are of comparable size to the Chilean records. The summation of 364 Miyagi-Oki records (the number indicated by the moment ratio) clearly overestimates the observed amplitude. Furthermore, the envelope of the waveform of model M–364–300 has a fairly uniform amplitude throughout the record, whereas the observed record seems to show a larger amplitude for the first 90 seconds than for the latter 90 seconds. This seems to provide corroborating evidence for the model of RUFF and KANAMORI (1983) in which they suggest that the earthquake initiated with the rupture of a large asperity with a diameter of about 200 km. In the model M–88–300, we have assumed that dislocations in the hypocentral area are about twice as large as those that occurred on the periphery of the rupture surface. In Figure 10, we show sketches of the spatial distributions of the dislocations that provided a reasonable match with the observed waveform envelopes for each of the earthquakes that we simulate. The model M–88–300, in which 88 Miyagi-Oki records are summed, gives a good overall fit to this data.

A comparison of synthetic long-period Benioff records for the 1957 Aleutian earthquake (M_w 9.1) with the observed is also shown in Figure 9. This earthquake has been assigned a very high energy magnitude because of its very long aftershock zone (SYKES, 1971; KANAMORI, 1977). However, direct measurements of the moment form very long-period surface waves are not currently available for this earthquake. The Pasadena long-period Benioff records are not particularly impressive and the summation of 66 Miyagi-Oki records (model M–66–150) results in significantly larger synthetics than the observed. The summation of only 22 Miyagi-Oki records (M–22–150) yields a more acceptable fit to the data. A fairly uniform dislocation distribution on a long, narrow fault (Figure 10) seems adequate to explain the Pasadena waveform.

Free oscillation recordings of the 1952 Kamchatka earthquake (M_w 9.0) provide fairly direct evidence that this event ranks among the giant historic earthquakes (KANAMORI, 1975). However, the summation of only 36 Miyagi-Oki records (M–36–200) provides an adequate fit to these records. The assumption of relatively uniform rupture along the fault plane seems adequate to explain the Pasadena records.

Comparisons of observed long- and short-period Benioff records from the 1965 Rat Island earthquake (M_w 8.7) are shown in Figure 11. The model M–18–150 has several asperities (Figure 10) that produce variations in the waveform envelope of the type that are seen in the observed long-period records.

In Figure 12, we show comparisons of the long- and short-period synthetics with the records observed for the 1968 Tokachi-Oki earthquake (M_w 8.2). A relatively detailed dislocation distribution has been deduced by KIKUCHI and FUKAO (1986) and this distribution (Figure 10) has been assumed in our model M–9–140. Summing the number of Miyagi-Oki records implied by the moment ratio of these two events results in a good match to the observed records.

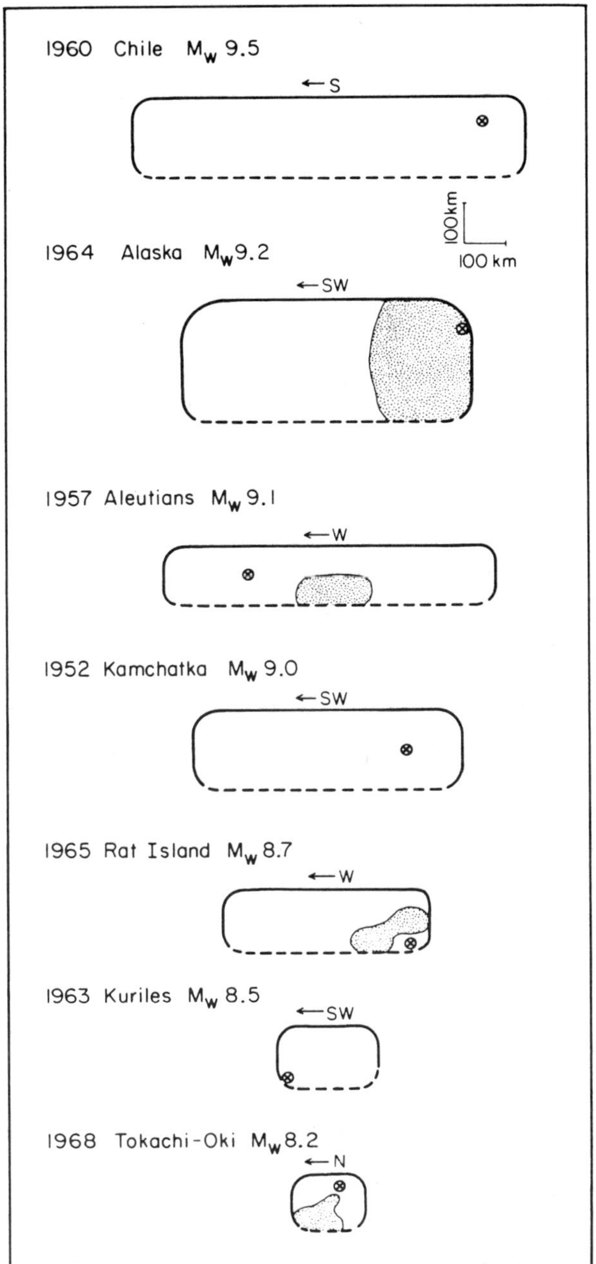

Figure 10

Schematic showing the assumed rupture characteristics of earthquakes modeled in this study. Although the bottom of the rupture surfaces are poorly determined, we considered the dashed line to define the bottom of the surface that radiated significant short-period energy. Hypocenters are shown by the ⊗ symbol and stippled regions designate areas that are thought to have relatively larger dislocations.

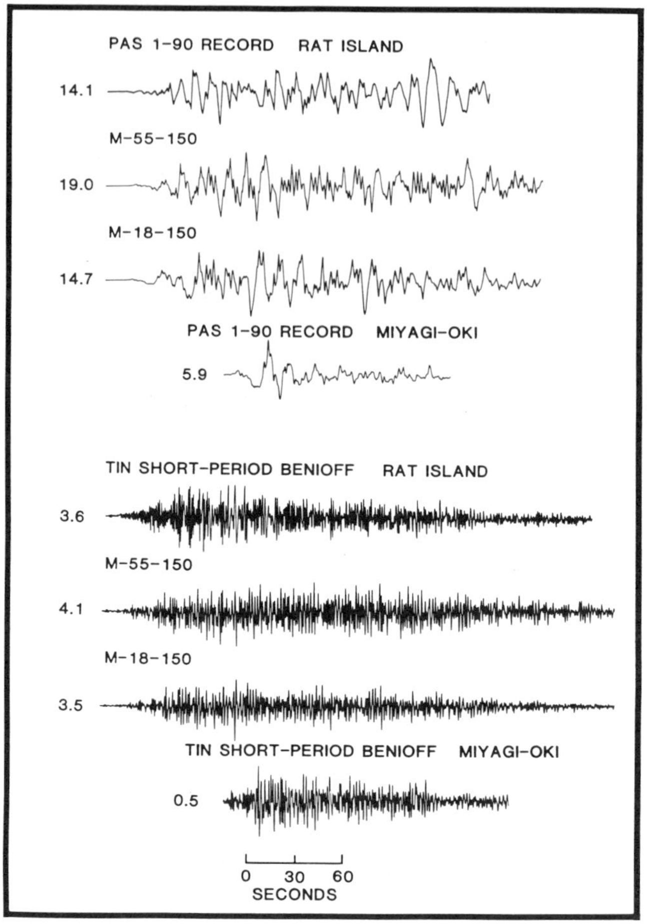

Figure 11
Comparison of observed teleseismic *P*-waveforms from the 1965 Rat Island earthquake with records synthesized by summing Miyagi-Oki earthquake records. Although the long-period moments of these earthquakes would indicate that 55 Miyagi-Oki records should be summed (M–55–150), a better match to the data is provided by the model in which only 18 records are summed (M–18–150).

The 1963 Kurile earthquake (M_w 8.5) is the last event that we consider. As can be seen in Figure 12, the observed records from this event are comparable in size to those observed from the Tokachi-Oki earthquake. The model M–12–150 assumes a relatively uniform dislocation distribution (Figure 10) and compares well with the observed records.

We have seen that the assumption that we can model giant earthquakes by randomly summing enough smaller ones to match the moments consistently overestimates the teleseismic *P*-waveforms. In several ways, this result is to be expected. As is pointed out by Joyner and Boore (1986) and also Appendix B, the random

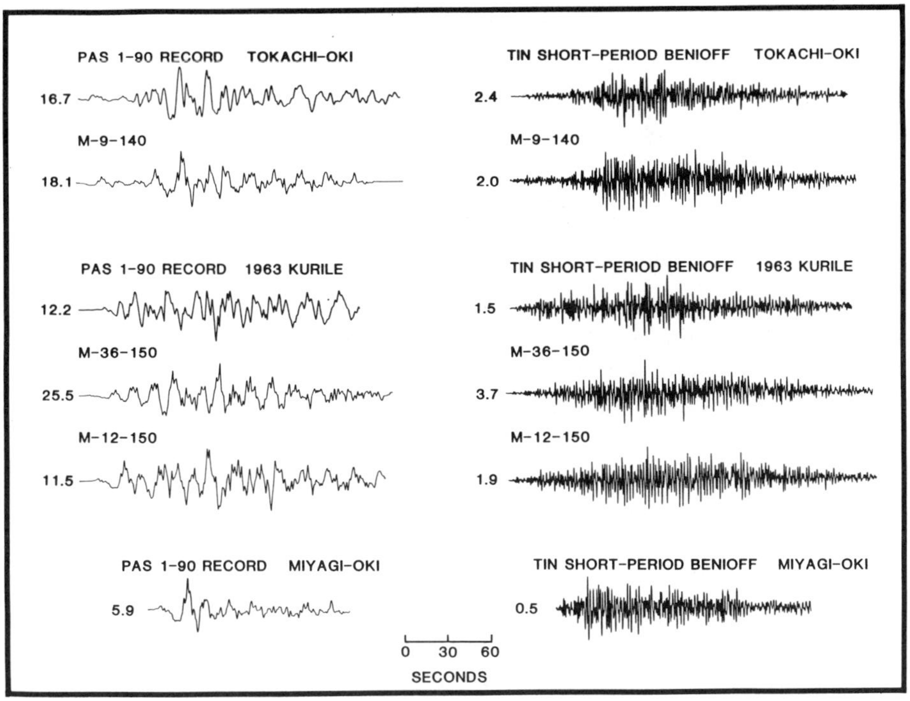

Figure 12

Comparison of observed teleseismic P-wave records from the 1968 Tokachi-Oki earthquake and the 1963 Kurile earthquake with records synthesized by summing Miyagi-Oki earthquake records. The summation of 9 Miyagi-Oki records (M–9–140), the number indicated from the moment ratio, provides a good match to the observed. However, for the 1963 Kurile earthquake, fewer records must be summed than is indicated by the moment ratios.

summation of N waveforms results in a $\sqrt{N}$ increase in high-frequency spectral levels; this is a larger increase than is produced by most spectral scaling laws that are based on the observed records of smaller earthquakes. Furthermore, HARTZELL and HEATON (1985) present clear evidence that teleseismic P-waves in the period range from 2 to 50 seconds saturate for earthquakes larger than about $8\frac{1}{4}$. This saturation effect can be seen both in the teleseismic time functions that they present and also in the Fourier amplitude spectra. In fact, the Fourier amplitude spectra presented by HARTZELL and HEATON (1985) seem to provide direct evidence that the rupture process of giant earthquakes is not self-similar to that of smaller ones. The increase in high-frequency spectral levels with moment appears to be less for earthquakes of $M_w > 8$ than for smaller earthquakes. This conclusion contrasts somewhat with the work of HOUSTON and KANAMORI (1986) who report that the amplitudes of high-frequency teleseismic P-waves grow at a constant rate for earthquakes varying from M_w 6 to M_w $9\frac{1}{2}$. The primary reason for this difference appears to be a difference in the giant earthquake data sample.

Results for Strong Ground Motions

Response spectra for the Chilean earthquake models described above and assuming the coastal ranges site located about 50 km from the coast are presented in Figure 13. The records that are used as Green's functions are found by cross-referencing Figures 4 and 5, Table 3, and Appendix A. Although the teleseismic motions for the models designated with an M were constructed by summing only Miyagi-Oki earthquake records, several of the Green's functions in the corresponding strong motion models were from the 1968 Hyuganada earthquake (M_w 7.4) and the 1979 St. Elias earthquake (M_w 7.8). These earthquakes have relatively simple far-field time functions and are similar in size to the Miyagi-Oki earthquake. We feel that including them in the Green's function set helps to provide synthesized ground motions that are less dependent on one particular set of records. The response spectra from the models designated T are obtained from summing records from the Tokachi-Oki earthquake only.

The response spectrum from each of these models is quite large. In particular, the spectra obtained by summing the number of records determined from the moment ratios (M–1140–200 and T–96–200) are alarmingly large. However, these models are inconsistent with the teleseismic data and thus we believe that they are not likely to be representative of the motions that may be encountered in giant earthquakes. It is not too surprising that our modeling procedure should produce large ground motions; the motions we use as Green's functions are already large. The sum of a number of these motions must result in a motion that is larger than the largest of the Green's functions.

In Figure 14, we show response spectra for the vertical ground motions produced by the models described above. These motions were synthesized using a procedure identical to the one used to sum the horizontal motions. In Figure 15, we show time histories for one horizontal component and the vertical component of ground motion for the model M–120–200 and for the coastal ranges site shown in Figure 5. We investigated many models in which we changed parameters (such as the hypocenter, the Green's function distribution, the number of Green's functions, the fault width) and the model M–120–200 represents a fair median to those models. Due to the assumption of random timing in our models, the hypocentral location

Figure 13
Synthesized pseudo-velocity response spectra (5% damping) for one component of horizontal ground motion from the models shown in Figures 5 and 6. The models in the top half are constructed by summing Miyagi-Oki-sized earthquake whereas the models in the bottom half are from summations of Tokachi-Oki earthquake records. Although the models M–1140–200 and T–96–200 are designed to match the long-period moment of the 1960 Chilean earthquake, they clearly overestimate the teleseismic P-waveforms and thus the spectra from these models are not considered to be plausible. The site is assumed to lie in the coastal ranges about 50 km from the coast.

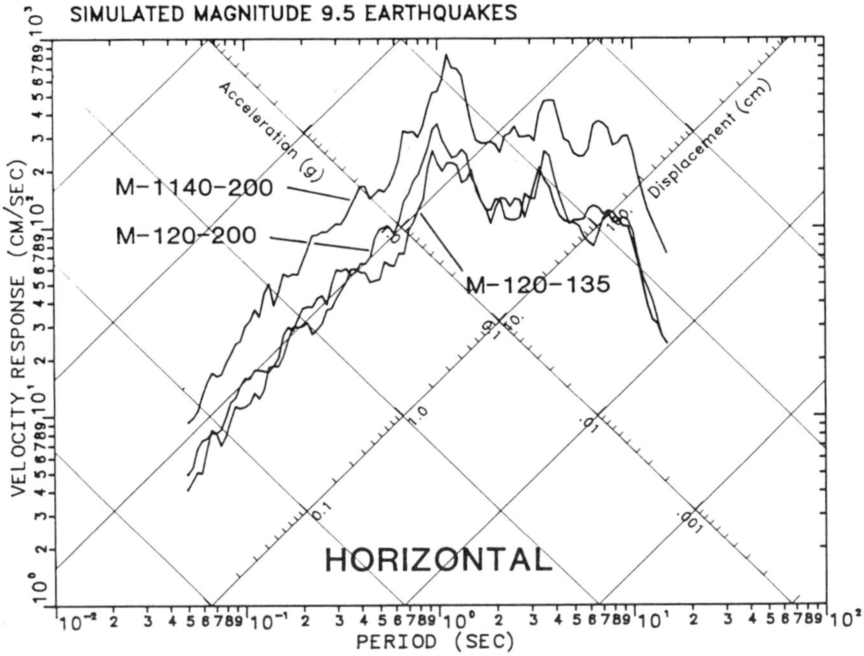
SIMULATED MAGNITUDE 9.5 EARTHQUAKES
M-1140-200
M-120-200
M-120-135
Acceleration (g)
Displacement (cm)
HORIZONTAL
VELOCITY RESPONSE (CM/SEC)
PERIOD (SEC)

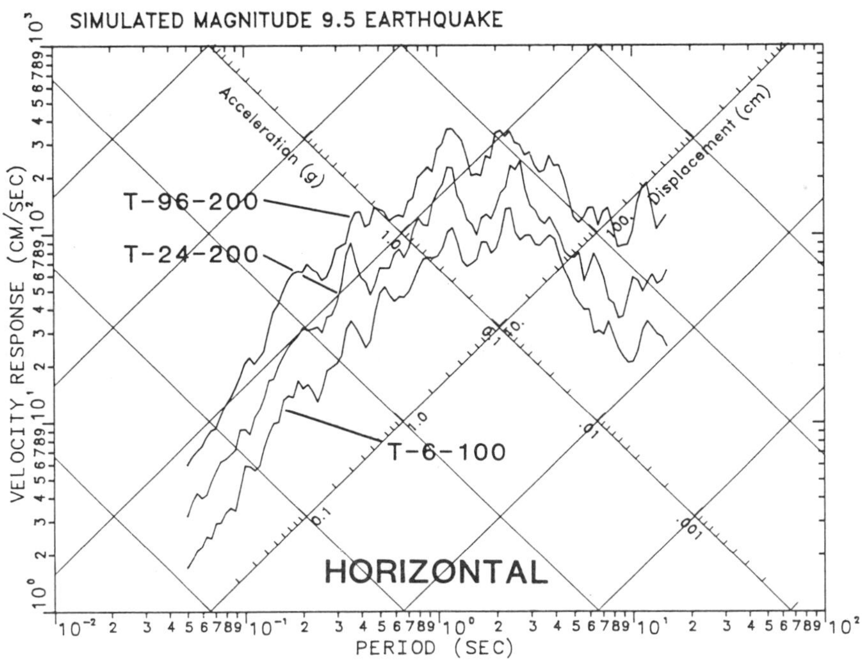
SIMULATED MAGNITUDE 9.5 EARTHQUAKE
T-96-200
T-24-200
T-6-100
Acceleration (g)
Displacement (cm)
HORIZONTAL
VELOCITY RESPONSE (CM/SEC)
PERIOD (SEC)

had little overall effect on either peak time domain or spectral amplitudes. Furthermore, the duration of strong shaking was fairly stable from model to model since it is mainly determined by the overall rupture length. The choice of individual records as Green's functions and the number of Green's functions to be summed appear to be the most important variables in our modeling procedure.

Since the individual records used as Green's functions have been filtered at periods longer than 10 seconds, their sum is also filtered in a similar manner. However in giant earthquakes, coastal regions may experience large static ground displacements; some coastal regions of Alaska were horizontally offset by 20 meters during the 1964 earthquake (PLAFKER, 1972). Obviously, the ground displacements given in Figure 15 do not account for static ground displacements. Recently, ANDERSON *et al.* (1986) presented stong motion recordings of the 19 September 1985 Mexico earthquake (M_s 8.1) in which there is convincing evidence for static offset of the recording sites. These offsets are on the order of 1 meter and occur in a linear ramp-like fashion over 10 to 20 seconds. In giant earthquakes, we anticipate that static offsets occur in a similar fashion, except that the displacements may exceed 10 meters and may occur over a duration of more than a minute. Ground motions of this nature can be modeled with acceptable accuracy using procedures similar to that described by HASKELL (1969). However, the accelerations associated with this very long-period motion are probably considerably smaller than the accelerations at higher frequencies that are presumably the result of irregularity of the rupture process. In some applications, such as the excitation of seiches, it may be important to consider the effects of the very long-period motions that give rise to static offsets.

In Figure 16, we show smoothed response spectra for the horizontal ground motions produced by the Chilean earthquake model M–120–200 as observed at sites located on the coast, in the coastal ranges 50 km inland, and in the Puget Sound region (see geometry in Figure 5). In this model, peak accelerations at the coast and in the Puget Sound are 0.89 g and 0.39 g, respectively. The time histories of ground motion are similar to those shown in Figure 15 for the coastal ranges site. The distance scaling relations of CROUSE *et al.* (1988) and KAWASHIMA *et al.* (1984) were used to scale the Green's functions in these calculations. The distance scaling relations of KAWASHIMA *et al.* (1984) indicate a somewhat stronger attenuation of motion with distance and their use results in motions that are about 10% higher and 10% lower at the coastal and Puget Sound sites, respectively. Although this model results in quite strong ground motions at Puget Sound, it should not be forgotten that Anchorage, Alaska lies at a similar point with respect to the Alaskan subduction zone. Although there are no strong motion recordings of the 1964 Alaskan earthquake, there was considerable damage in Anchorage and it seems clear that the ground motions were quite strong. If the zone of short-period seismic radiation is assumed to be narrower (model M–120–135), then the synthetic ground motions at the Puget Sound site drop by about 15%.

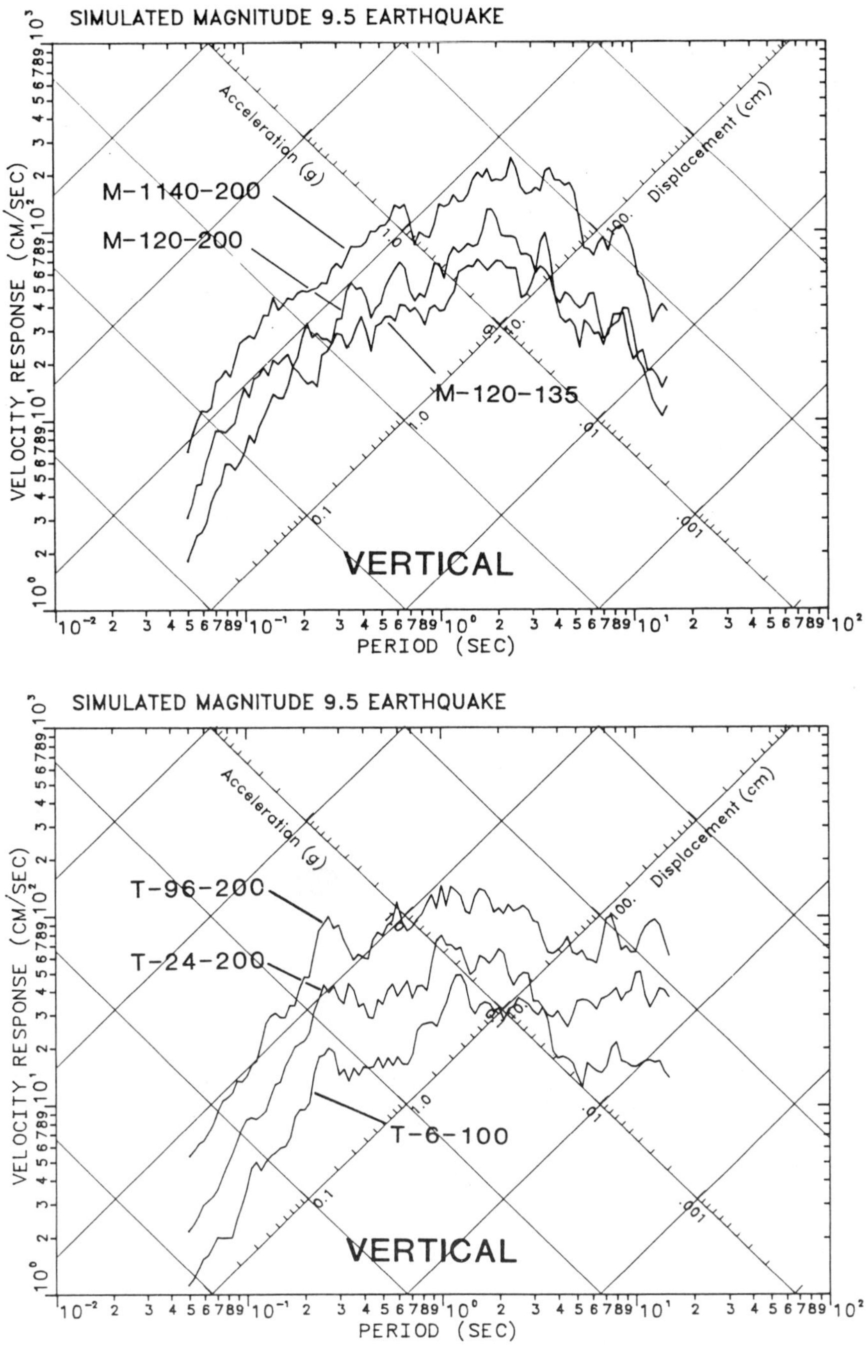

Figure 14

Same as Figure 13, except for the vertical component of synthesized ground motion.

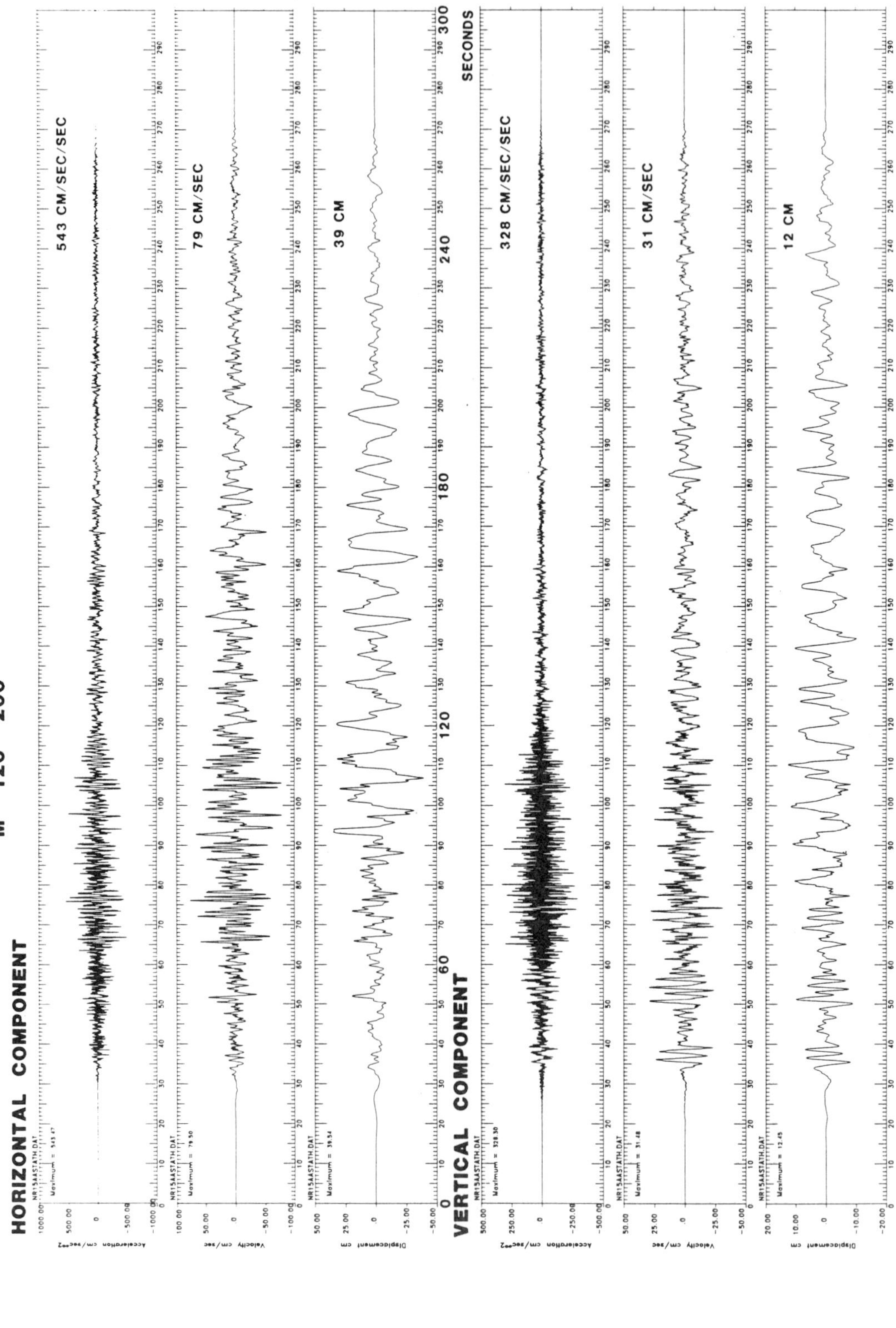
HORIZONTAL COMPONENT
M - 120 - 200
543 CM/SEC/SEC
79 CM/SEC
39 CM
SECONDS
VERTICAL COMPONENT
328 CM/SEC/SEC
31 CM/SEC
12 CM

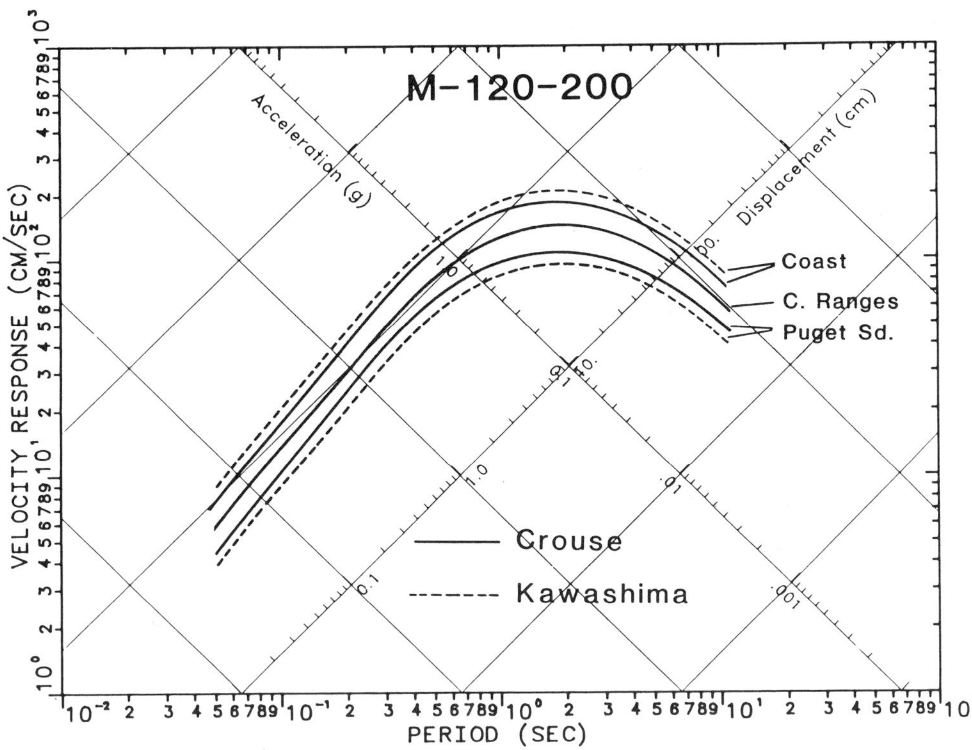

Figure 16

Comparison of synthetic response spectra as a function of site distance from the coast. Most of the models in this study are constructed using the CROUSE *et al.* (1988) distance attenuation law (solid curves). Use of the distance attenuation law of KAWASHIMA *et al.* (1984) results in motions that are nearly identical at the coastal ranges site and about 10% larger and 10% smaller at the coast and Puget Sound sites, respectively (dotted lines).

In addition to constructing models of the 1960 Chilean earthquake, we also constructed models of the 1964 Alaskan earthquake, the 1952 Kamchatka earthquake, and the 1957 Aleutian earthquake. Response spectra of horizontal and vertical components of synthetic ground motions are shown in Figure 17. The rupture parameters are those used to produce synthetic teleseismic *P*-waveforms that give a good overall match to the observed data (Figure 9). The synthetic

Figure 15

Synthetic ground motion for a giant earthquake in the Pacific Northwest that is similar to the 1960 Chilean earthquake. The site is assumed to be in the coastal ranges. This record was formed by summing records from Miyagi-Oki-sized earthquakes and is considered to be one of the larger motions that can be considered as reasonable for this site. Average ground motions for similar conditions may be somewhat smaller.

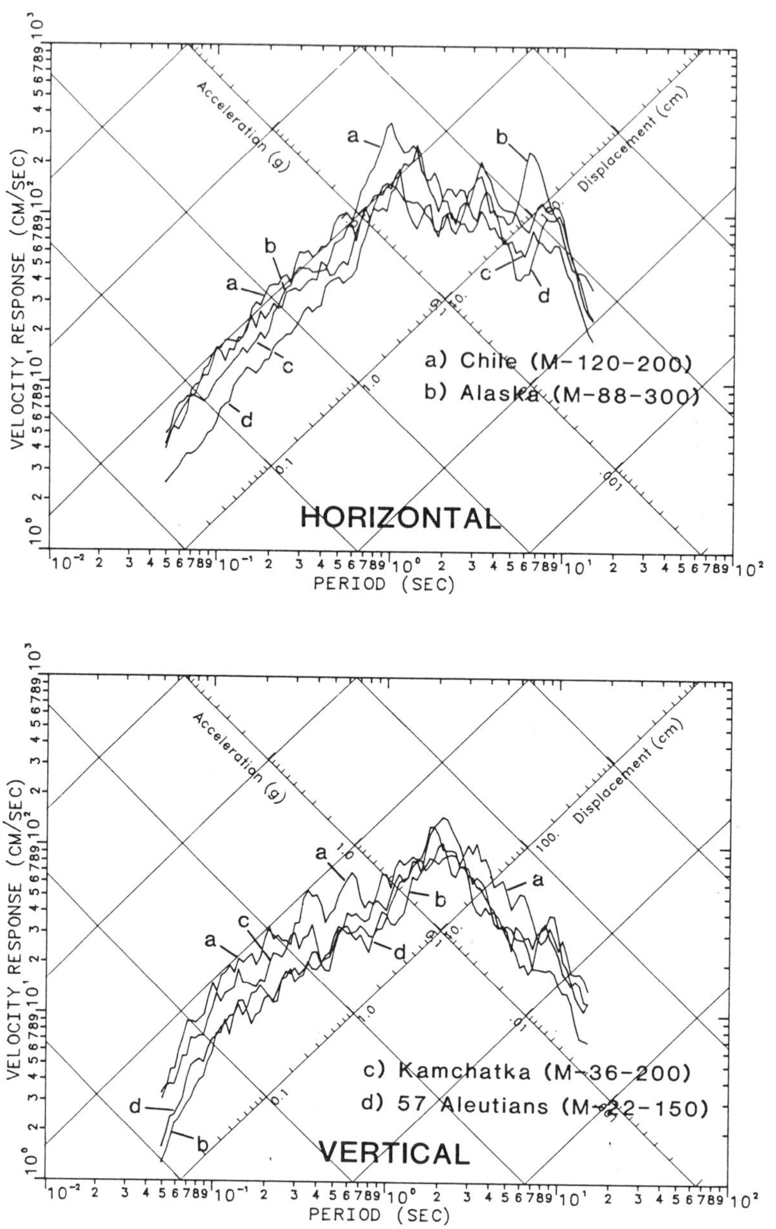

Figure 17

Comparison of synthetic response spectra (5% damping) for models of the four giant earthquakes ($M_w > = 9.0$) for which we have teleseismic P-wave data. Records from Miyagi-Oki-sized earthquakes are used as Green's functions and the motions are for a coastal ranges site. Schematics of the assumed fault geometries are shown in Figure 10. These motions are considered to be somewhat larger than the average motions that could be expected for sites at this distance.

motions are for the coastal ranges site and records from the Miyagi-Oki earthquake, the Hyuganada earthquake, and the St. Elias earthquake were used as Green's functions. All of the response spectra have similar shapes since the same set of records were used as Green's functions in each model. The number of records that are summed is the most important variable in determining the overall difference in the response spectral amplitudes for these different earthquakes.

Discussion and Conclusions

What is the overall significance of the synthetic ground motions that we have presented? Which, if any, of the motions represents the average motion and what is the scatter about the mean likely to be? In the case of actual data from smaller earthquakes, we noted the very large scatter for a given magnitude and distance range. There is little reason to expect that similar scatter does not exist for giant earthquakes as well. This scatter also helps to confuse the interpretation of our synthetic motions. Clearly the summation procedure assures that the final synthetics are at least as large as the largest of the motions that are summed. Thus Green's function records that fall within the large end of the scatter will dictate the size of the synthetic motion. To avoid this, it might be best to use only median records as Green's functions. However, the data set is very limited and too few (if any) records are available that could be confidently classified as median records. This situation is further complicated by some confusion about averaging procedures. That is, when we considered the spectra of earthquakes of $M_w < 8\frac{1}{4}$, we computed logarithmic averages (Figure 3). Logarithmic averages were used to de-emphasize the contribution of a few large records, if present. These logarithmic averages are about 80% as large as arithmetic averages for the same data. However, our synthetic ground motions result from the arithmetic sum of records within the data set. From these remarks, it does seem clear that most of our synthetic motions are likely to fall above the average motion for a giant earthquake.

In Figure 18, we present a schematic summary of the variation of idealized response spectra with energy magnitude for a coastal ranges site. The spectra for earthquakes of M_w $7\frac{1}{4}$ and 7.9 are smoothed interpretations of the averaged data presented in the first part of this paper. The spectrum shown for a M_w $9\frac{1}{2}$ earthquake is a compromise between the two models that produced the largest (M–120–200) and the smallest (T–6–100) ground motions and that were also compatible with the teleseismic waveforms of the 1960 Chilean earthquake. The response spectra for M_w 9 and $8\frac{1}{2}$ earthquakes were obtained by interpolation. The spectra presented in Figure 18 are only about two-thirds as large as the spectra that we obtained from most of our modeling using the records from Miyagi-Oki sized earthquakes as Green's functions. Because of the problems introduced by using a record set with very large scatter as Green's functions, we believe that our

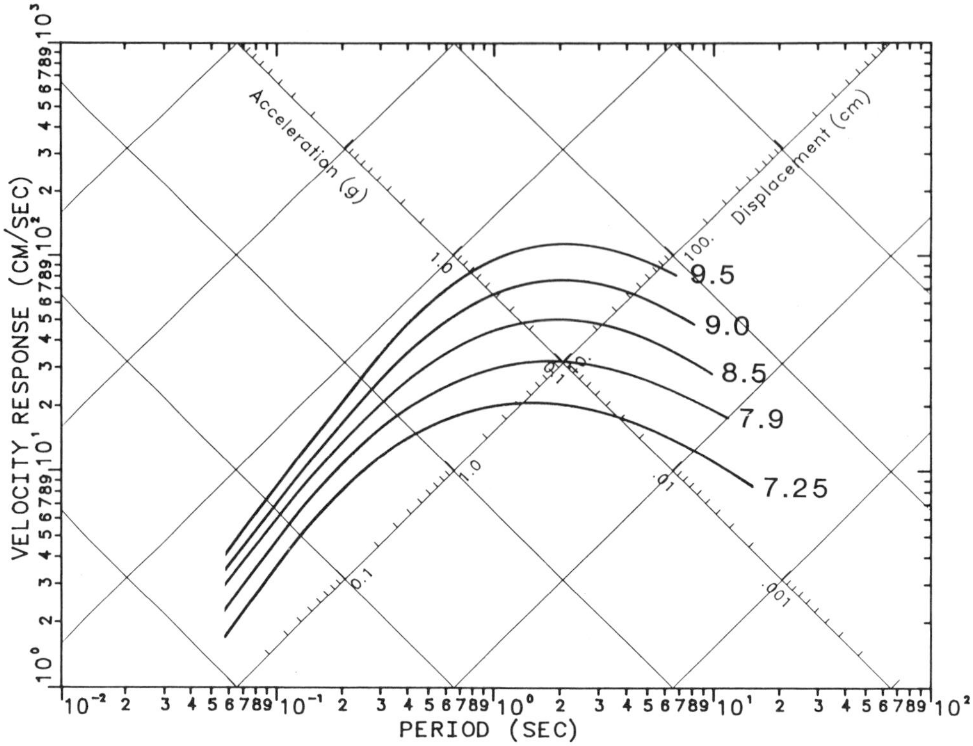

Figure 18
Estimates of the variation in average horizontal ground motion response spectra (5% damping) as a function of energy magnitude for a coastal range site. Scatter of actual data about mean values may be similar to that observed in the data displayed in Figure 1.

M-designated models are systematically larger than the mean. We believe that the spectra presented in Figure 18 are probably more representative of what average ground motions may be for giant earthquakes. However, it is important to recognize that we have not developed a rigorous methodology to produce average ground motions.

In Figure 19, we compare several of the response spectra that we produced for the coastal ranges site with the response spectrum of the S14°W component of motion at Pacoima Dam during the 9 February 1971 San Fernando, California earthquake (M_w 6.7), which is one of the largest motions recorded for any earthquake. The models M–120–200 and T–6–100 are shown since they produced the largest and smallest ground motions that were compatible with our teleseismic modeling. The model designated as "average" in Figure 19 is approximately equal to our "preferred" response spectrum for a M_w $9\frac{1}{2}$ that is shown in Figure 18. At high frequencies, all of our synthetic ground motions are clearly smaller than the Pacoima motion. However, at periods greater than 1 second, our synthetic motions are nearly comparable in strength to the Pacoima record.

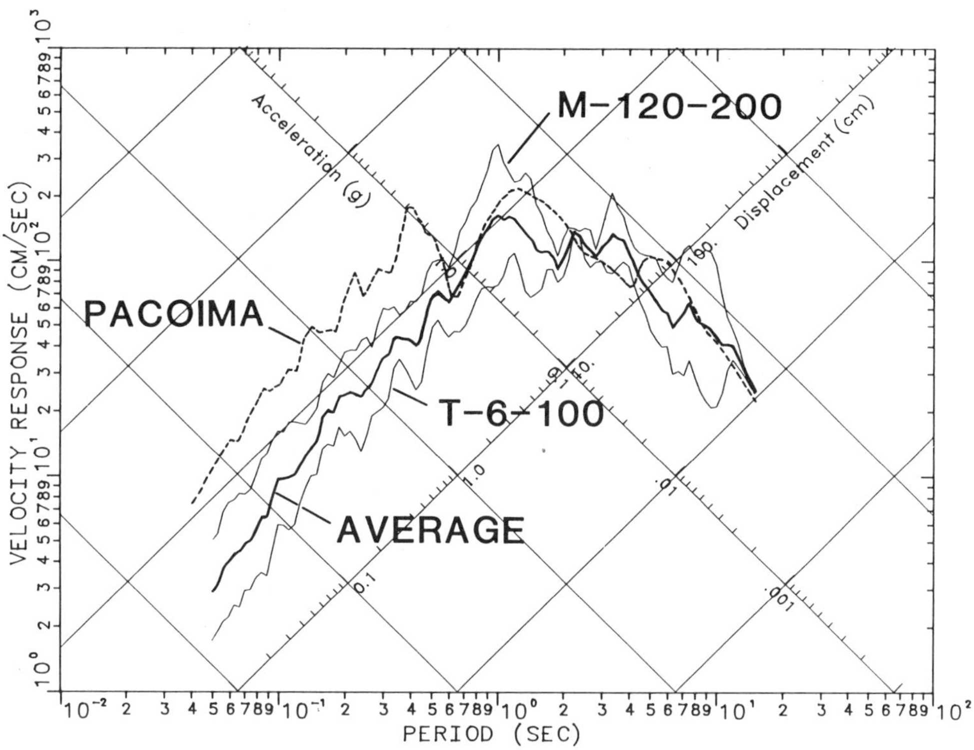

Figure 19
Comparison of the response spectrum of the S14W component of ground motion at Pacoima Dam from the 9 February 1971 San Fernando earthquake (M_w 6.7) with synthetic response spectra assuming a coastal ranges site and an earthquake similar to the 1960 Chilean earthquake. M–120–200 is one of the largest motions that is thought to be feasible and T–6–100 is one of the smallest motions produced in this study. The spectrum designated as average is similar to the one used in Figure 18.

Given that there are presently no strong motion recordings of giant earthquakes and that there are many uncertainties in our modeling procedure, there are still many uncertainties in estimating just how large and damaging the motions would be from a giant earthquake in the Pacific Northwest. We have seen that there is evidence that a large portion of the moment history of giant earthquakes is a very long period and cannot be seen in short- and intermediate-period teleseismic body-wave records. Nevertheless, this is small solace since the short-period records from such earthquakes are still among the largest ever recorded. Furthermore, data we do have suggests that relatively strong shaking occurs at surprisingly large distances from M_w 8 earthquakes. At this point, it is only natural to assume that M_w $9\frac{1}{2}$ earthquakes can produce even larger ground motions. In any event, it is clear that the 1960 Chilean and 1964 Alaskan earthquakes caused great damage over very large regions. The suggestion of a similar earthquake in an area as developed as the Pacific Northwest is a very disturbing notion.

Appendix A

Appendix A consists of figures that summarize the strong ground motion data set used in this study. The figures show the locations of strong motion recording sites, ISC mainshock and aftershock locations, the time histories of one horizontal component of ground acceleration and velocity for each site, and, in some cases, the teleseismic time function derived by HARTZELL and HEATON (1985). Common scales are used so that teleseismic time functions can be compared directly with strong ground motion time histories. The Japanese records were taken directly from the MORI and CROUSE (1981) catalogue (designated as Exxon) and also from a collection assembled by the Japanese Port and Harbor Research Institute (designated as PHRI). The Alaskan records are taken from BEAVAN and JACOB (1984), the Peruvian records are from BRADY and PEREZ (1977), and the Solomon Islands records are from DENHAM and SMALL (1971). All of these records, except for the ones from the Port and Harbor Research Institute, are available from the NOAA World Data Center in Boulder, Colorado. Additional information concerning the conditions under which these and other data were recorded have been tabulated by CROUSE *et al.* (1980).

Figures A1 to A20

Summary of strong ground motions from large ($M_w \geq = 7.0$) shallow subduction earthquakes considered in this study. Mainshock epicenters are designated with a star, aftershocks are shown as dots, and recording sites are shown as solid triangles. One horizontal component of ground acceleration and velocity are shown for each recording. Teleseismic far-field time functions reported by HARTZELL and HEATON (1985) are shown when available. All distances used in this study are epicentral distances, except in the case of the 16 May 1968 Tokachi-Oki earthquake (M_w 8.2) where distances are measured relative to the × symbol which is more central to the rupture surface.

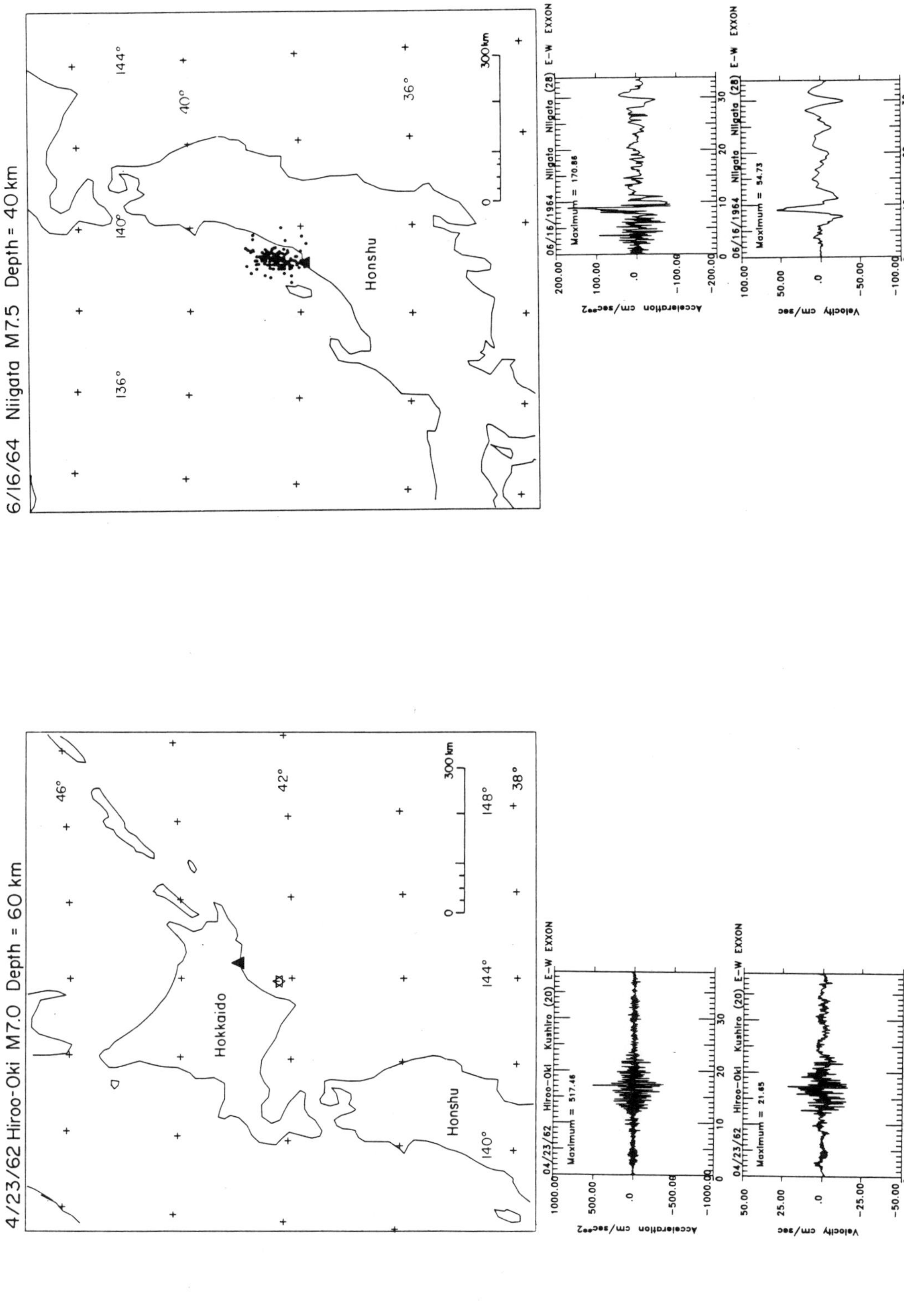
6/16/64 Niigata M7.5 Depth = 40 km
Honshu
06/16/1964 Niigata Niigata (28) E-W EXXON
Maximum = 170.86
Maximum = 54.73
Acceleration cm/sec**2
Velocity cm/sec
4/23/62 Hiroo-Oki M7.0 Depth = 60 km
Hokkaido
Honshu
04/23/62 Hiroo-Oki Kushiro (20) E-W EXXON
Maximum = 517.46
Maximum = 21.85

Fig. A1.

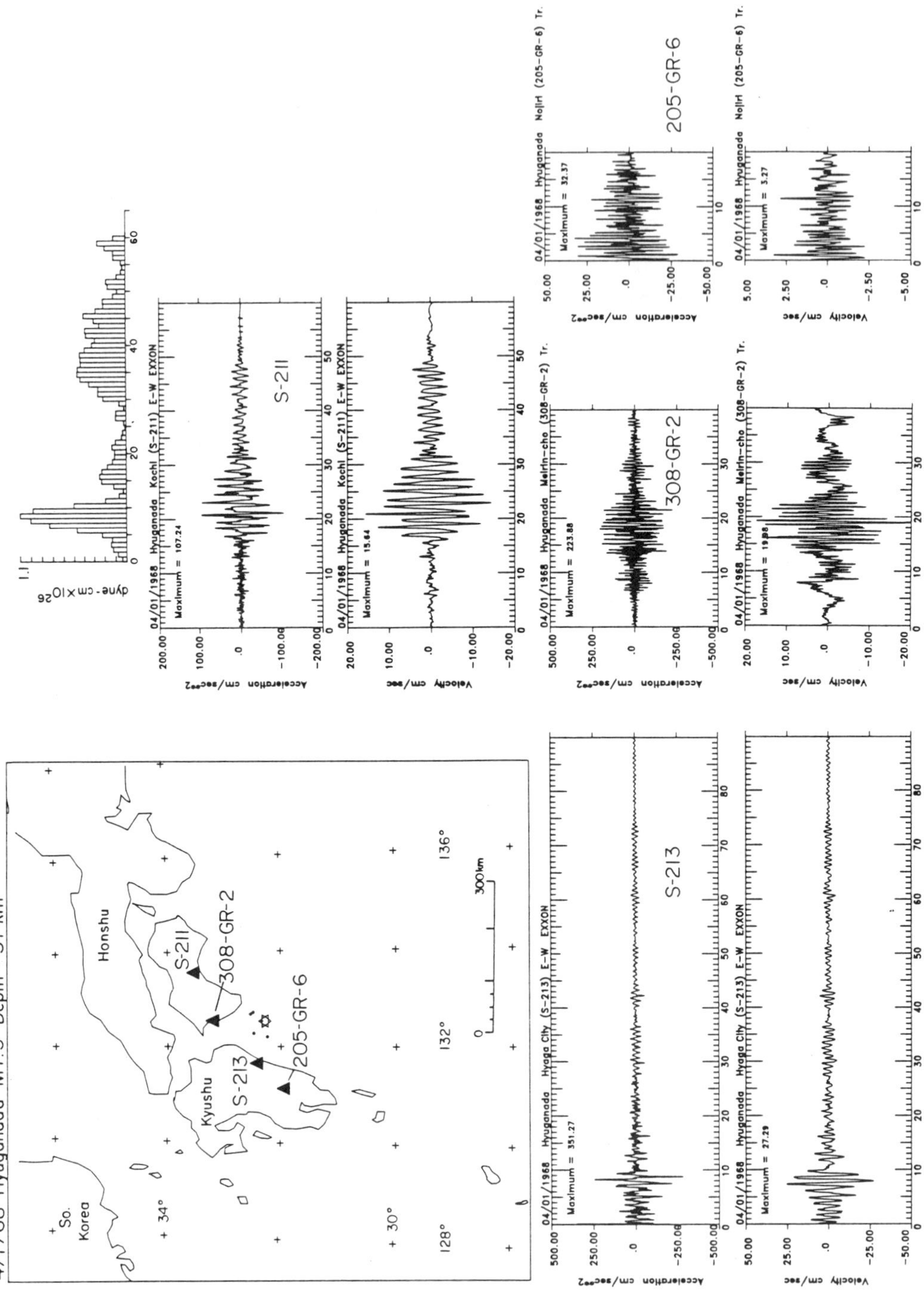
4/1/68 Hyuganada M7.5 Depth = 37 km
So. Korea
Honshu
Kyushu
S-211
308-GR-2
S-213
205-GR-6
300km
34°
30°
128°
132°
136°
dyne-cm x10^26
S-211
308-GR-2
205-GR-6
S-213

Fig. A2.

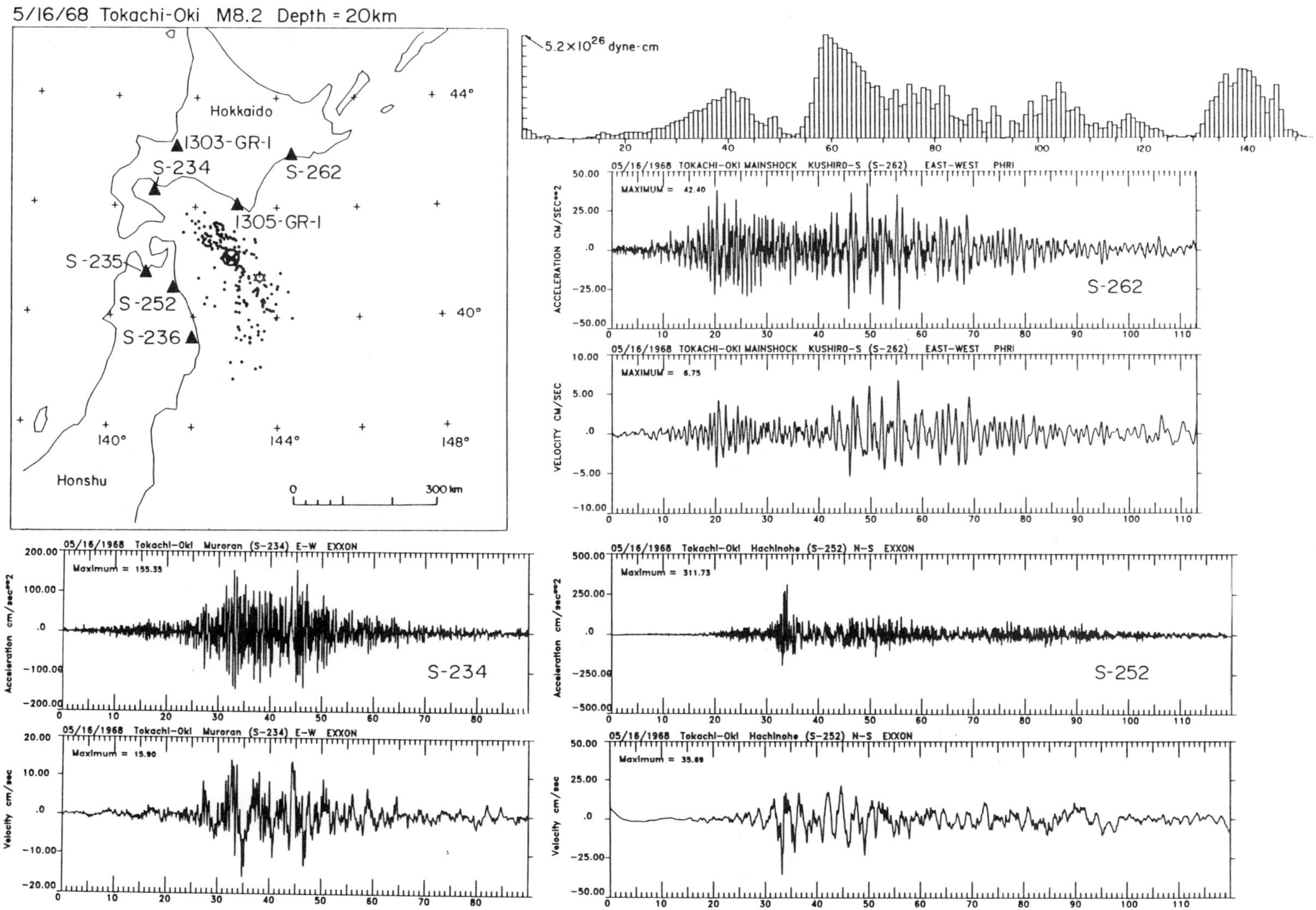

5/16/68 Tokachi-Oki M8.2 Depth = 20km
Hokkaido
1303-GR-I
S-234
S-262
1305-GR-I
S-235
S-252
S-236
Honshu
44°
40°
140°
144°
148°
0
300 km
5.2×10^26 dyne-cm
05/16/1968 TOKACHI-OKI MAINSHOCK KUSHIRO-S (S-262) EAST-WEST PHRI
MAXIMUM = 42.40
ACCELERATION CM/SEC**2
S-262
05/16/1968 TOKACHI-OKI MAINSHOCK KUSHIRO-S (S-262) EAST-WEST PHRI
MAXIMUM = 6.75
VELOCITY CM/SEC
05/16/1968 Tokachi-Oki Muroran (S-234) E-W EXXON
Maximum = 155.35
Acceleration cm/sec**2
S-234
05/16/1968 Tokachi-Oki Muroran (S-234) E-W EXXON
Maximum = 15.90
Velocity cm/sec
05/16/1968 Tokachi-Oki Hachinohe (S-252) N-S EXXON
Maximum = 311.73
Acceleration cm/sec**2
S-252
05/16/1968 Tokachi-Oki Hachinohe (S-252) N-S EXXON
Maximum = 35.89
Velocity cm/sec

Fig. A3.

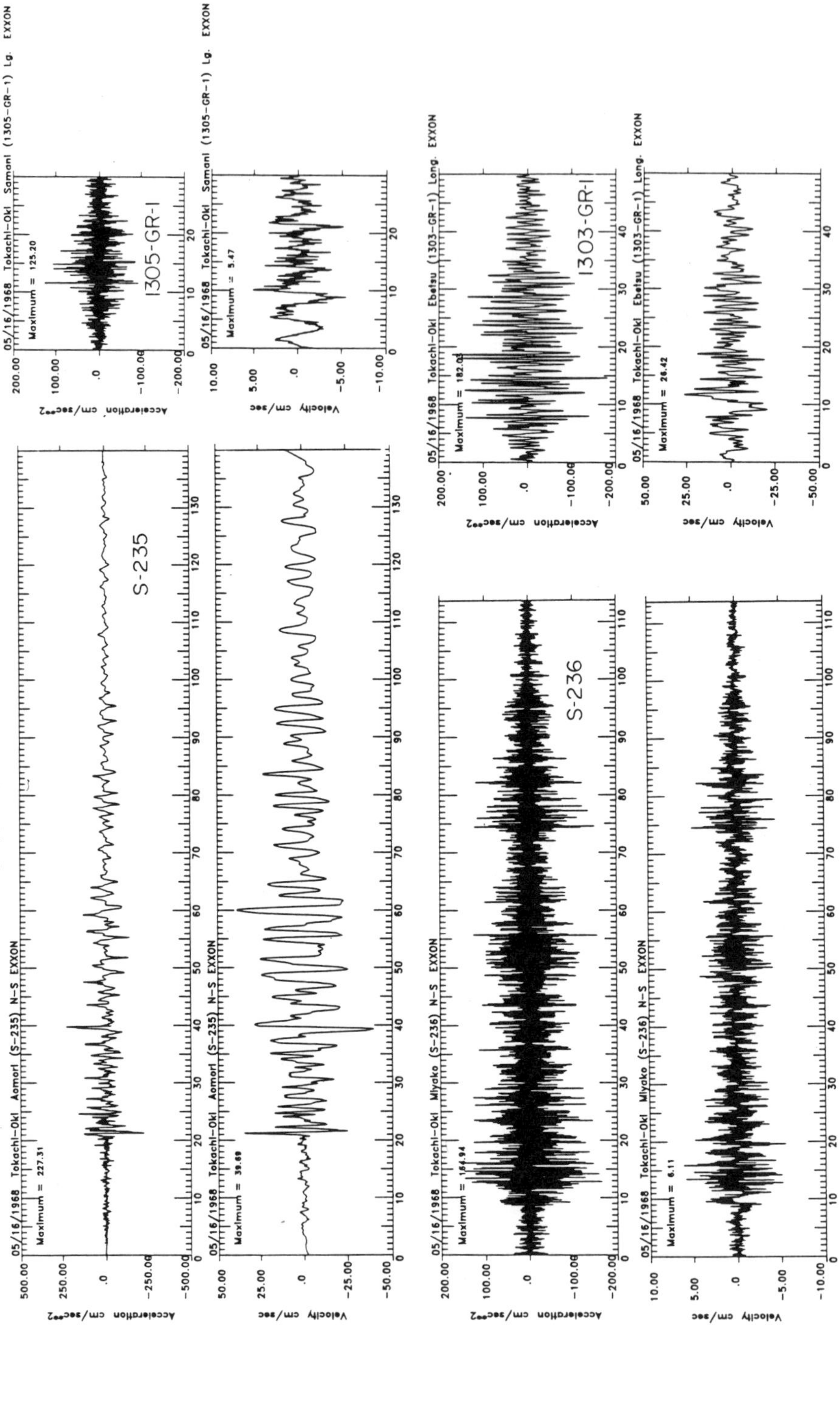

5/16/68 Tokachi-Oki M8.2
05/16/1968 Tokachi-Oki Aomori (S-235) N-S EXXON
Maximum = 227.31
05/16/1968 Tokachi-Oki Aomori (S-235) N-S EXXON
Maximum = 39.69
S-235
05/16/1968 Tokachi-Oki Samani (1305-GR-1) Lg. EXXON
Maximum = 125.20
05/16/1968 Tokachi-Oki Samani (1305-GR-1) Lg. EXXON
Maximum = 5.47
1305-GR-1
05/16/1968 Tokachi-Oki Miyako (S-236) N-S EXXON
Maximum = 184.94
05/16/1968 Tokachi-Oki Miyako (S-236) N-S EXXON
Maximum = 6.11
S-236
05/16/1968 Tokachi-Oki Ebetsu (1303-GR-1) Long. EXXON
Maximum = 182.03
05/16/1968 Tokachi-Oki Ebetsu (1303-GR-1) Long. EXXON
Maximum = 28.42
1303-GR-1
Acceleration cm/sec**2
Velocity cm/sec

Fig. A4.

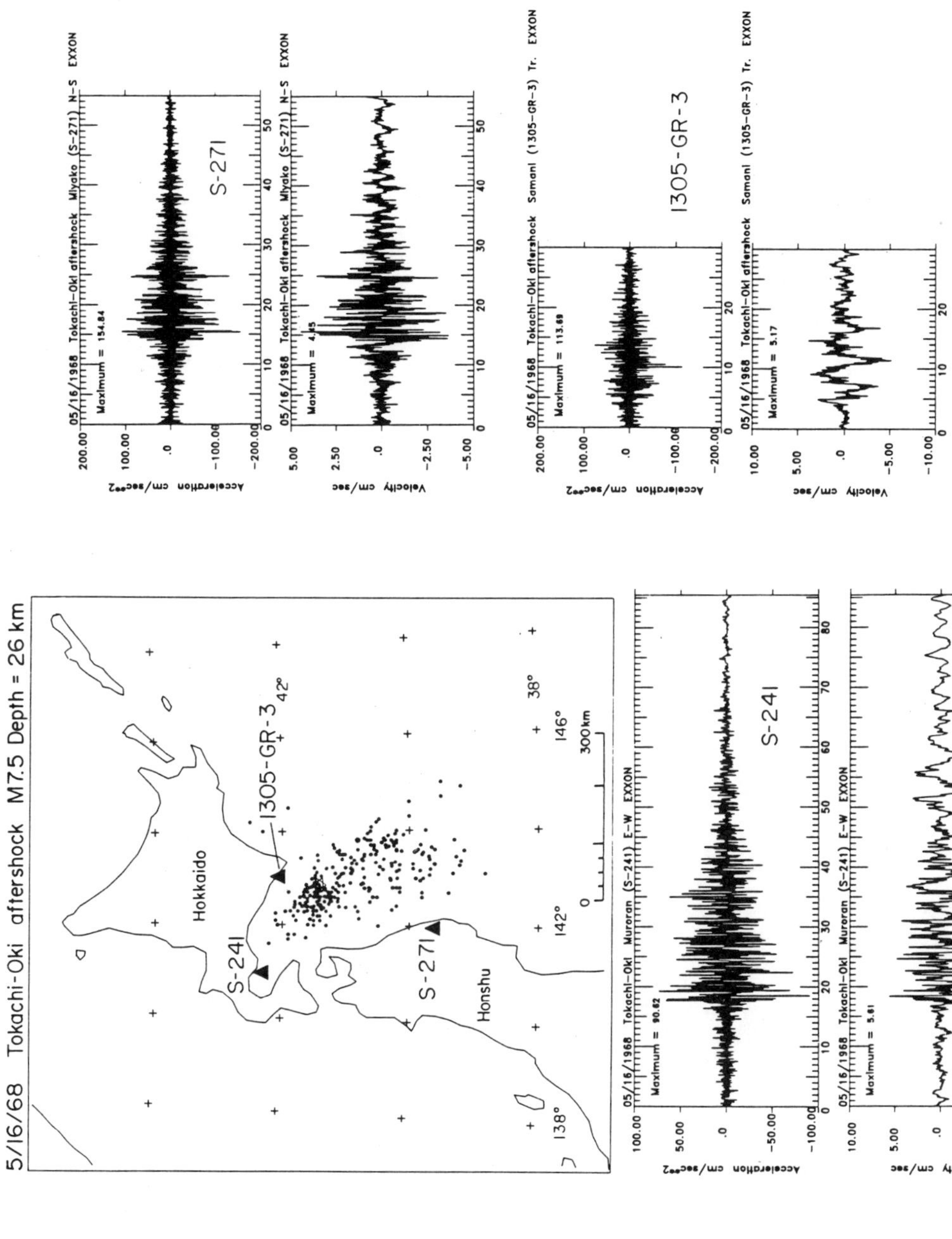
5/16/68 Tokachi-Oki aftershock M7.5 Depth = 26 km
Hokkaido
Honshu
S-241
S-271
1305-GR-3

Fig. A5.

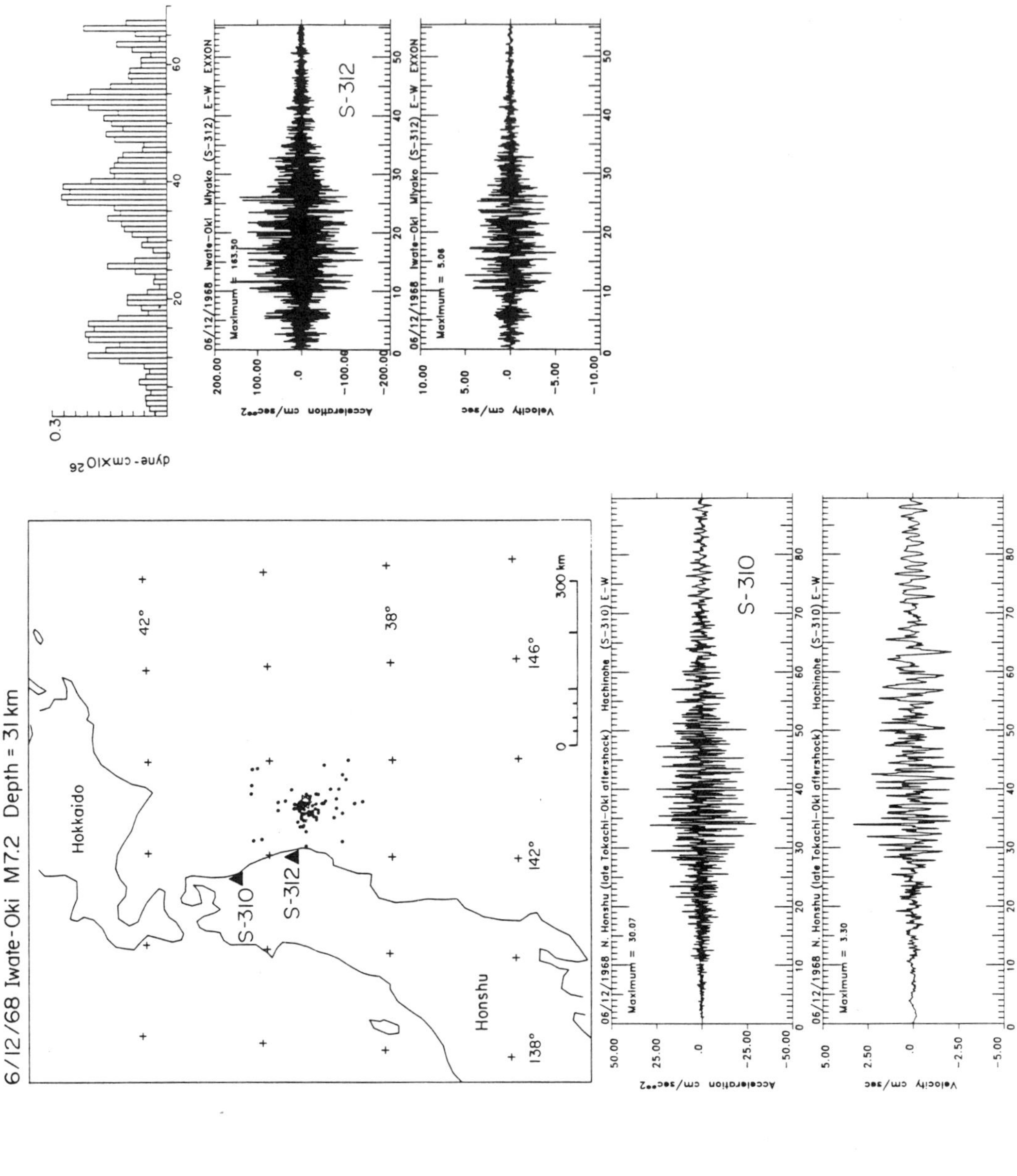
6/12/68 Iwate-Oki M7.2 Depth = 31 km
Hokkaido
Honshu
S-310
S-312
42°
38°
138°
142°
146°
0
300 km
0.3
dyne-cmX10 26
06/12/1968 Iwate-Oki Miyako (S-312) E-W EXXON
Maximum = 163.50
Maximum = 5.06
Acceleration cm/sec**2
Velocity cm/sec
06/12/1968 N. Honshu (late Tokachi-Oki aftershock) Hachinohe (S-310) E-W
Maximum = 30.07
Maximum = 3.30

Fig. A6.

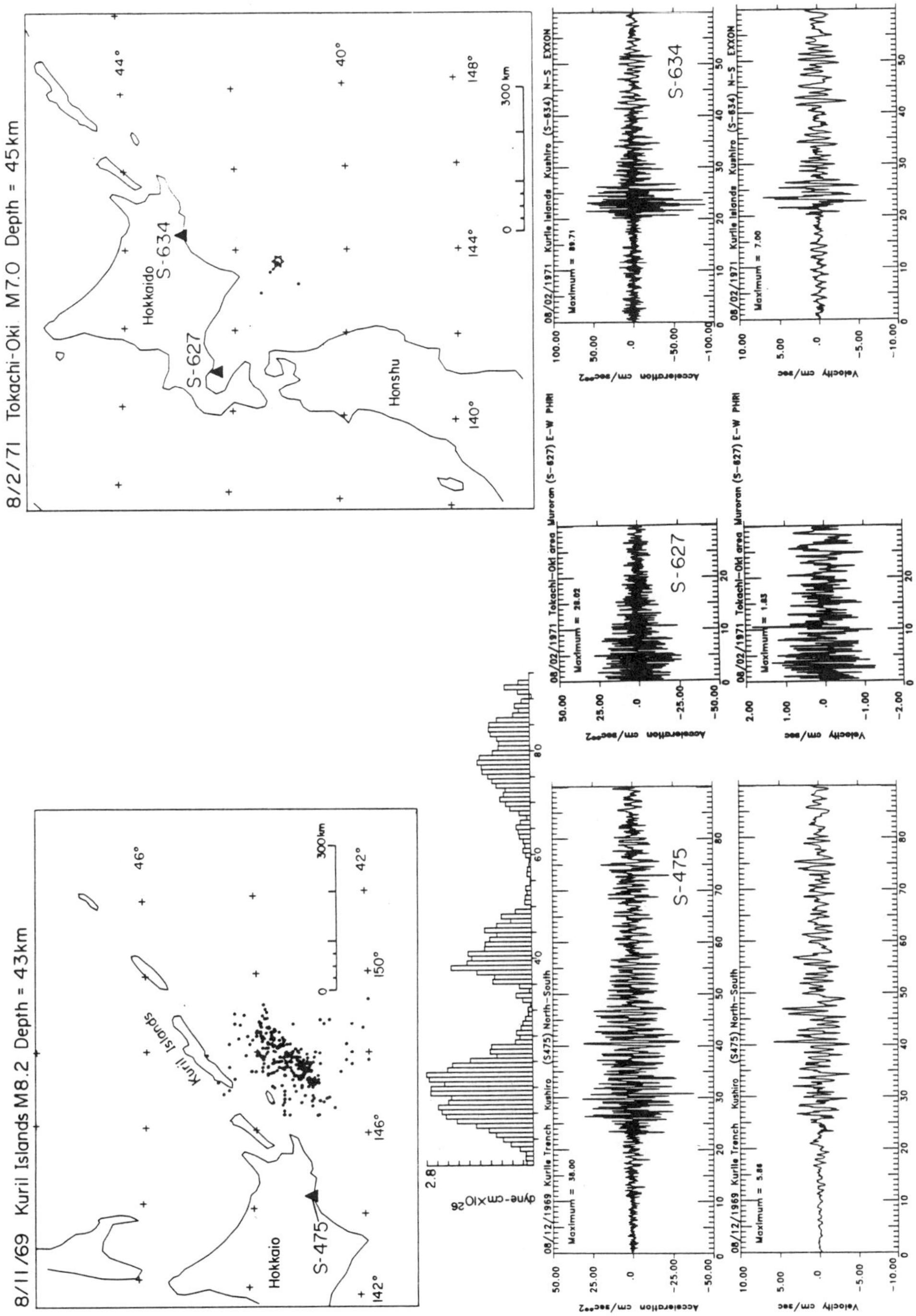
8/2/71 Tokachi-Oki M7.0 Depth = 45km
Hokkaido
Honshu
S-634
S-627
S-475
8/11/69 Kuril Islands M8.2 Depth = 43km
Kuril Islands
Hokkaio
300km
dyne-cm×10²⁶
2.8
Acceleration cm/sec**2
Velocity cm/sec

Fig. A7.

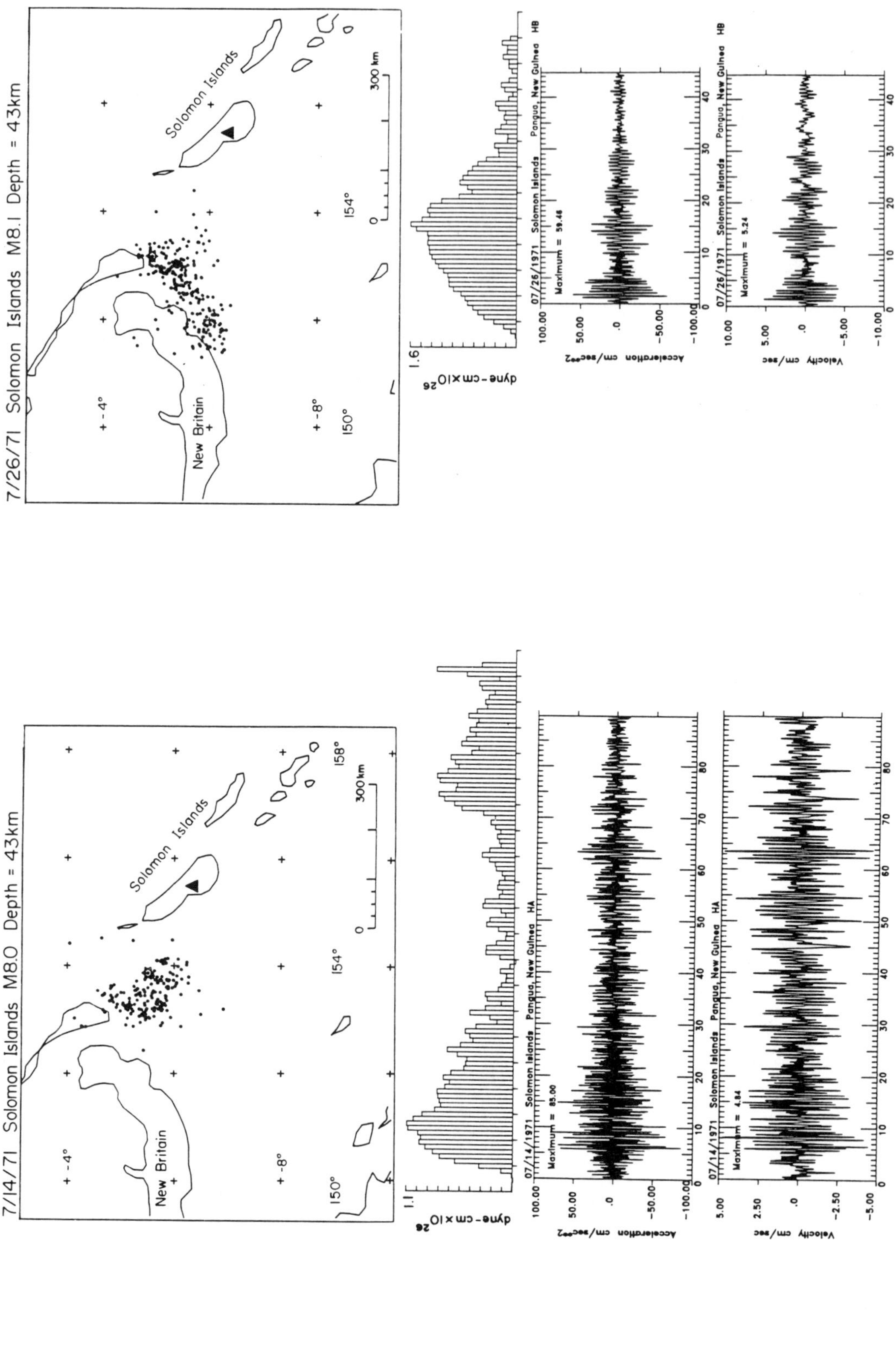
7/26/71 Solomon Islands M8.1 Depth = 43km
Solomon Islands
New Britain
300 km
07/26/1971 Solomon Islands Pangua, New Guinea HB
7/14/71 Solomon Islands M8.0 Depth = 43km
Solomon Islands
New Britain
300 km
07/14/1971 Solomon Islands Pangua, New Guinea HA
Acceleration cm/sec**2
Velocity cm/sec

Fig. A8.

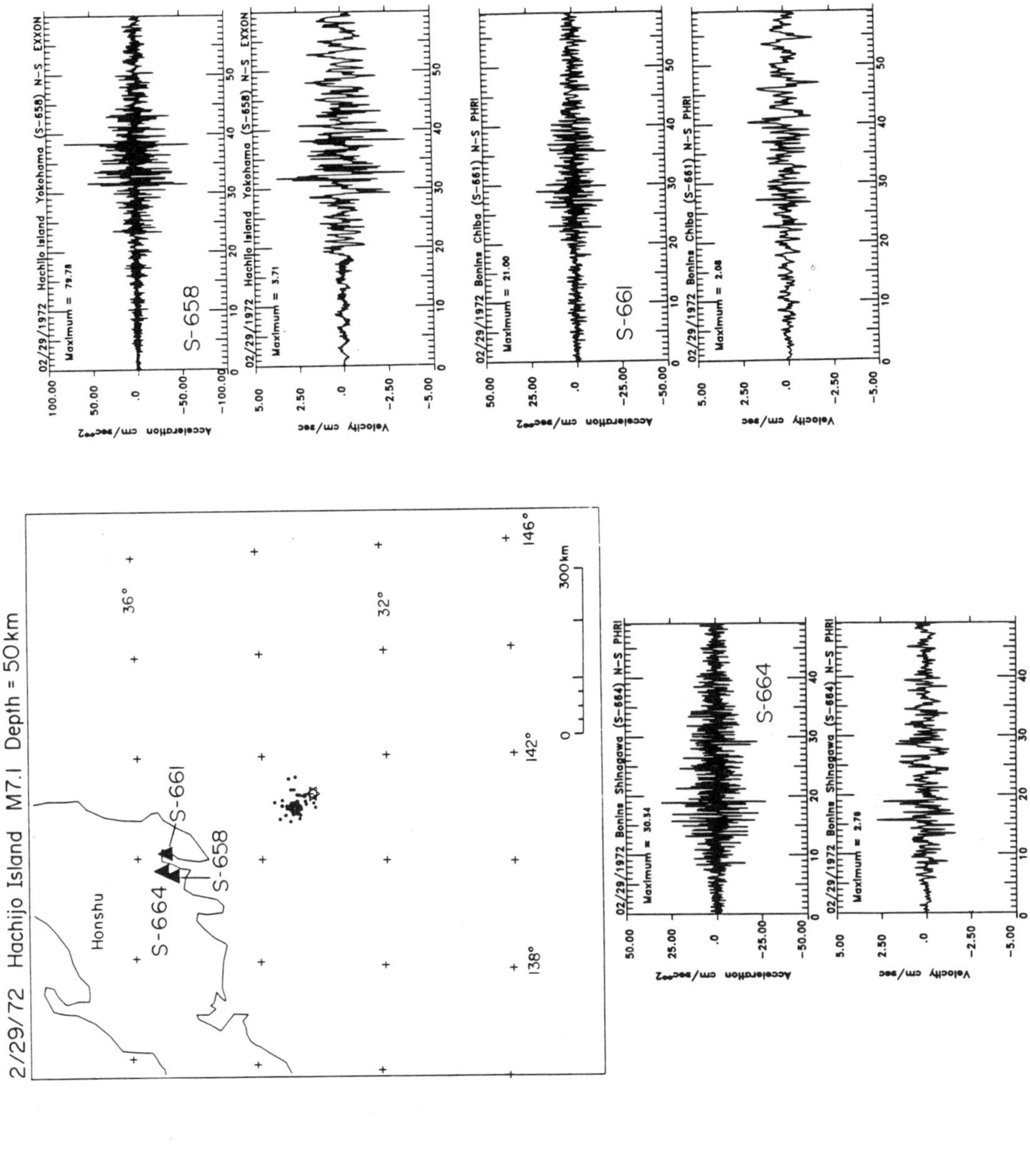
2/29/72 Hachijo Island M7.1 Depth = 50km
Honshu
S-664
S-658
S-661
36°
32°
138°
142°
146°
0
300km
02/29/1972 Hachijo Island Yokohama (S-658) N-S EXXON
Maximum = 79.78
Acceleration cm/sec**2
02/29/1972 Hachijo Island Yokohama (S-658) N-S EXXON
Maximum = 3.71
Velocity cm/sec
S-658
02/29/1972 Bonins Chiba (S-661) N-S PHRI
Maximum = 21.00
02/29/1972 Bonins Chiba (S-661) N-S PHRI
Maximum = 2.08
S-661
02/29/1972 Bonins Shinagawa (S-664) N-S PHRI
Maximum = 30.34
02/29/1972 Bonins Shinagawa (S-664) N-S PHRI
Maximum = 2.78
S-664

Fig. A9.

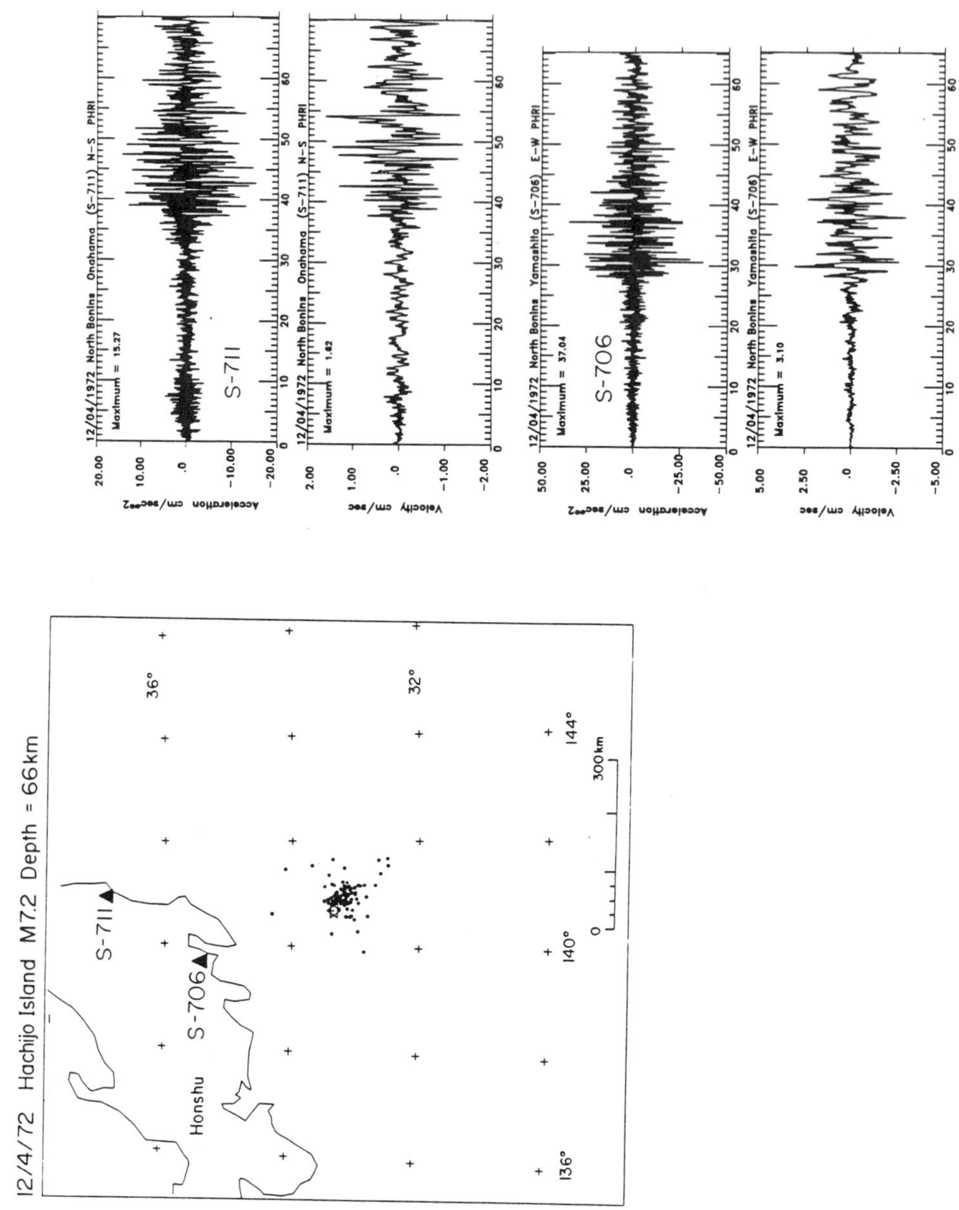

12/4/72 Hachijo Island M7.2 Depth = 66km
S-711
S-706
Honshu
36°
32°
136°
140°
144°
0
300km
12/04/1972 North Bonins Onahama (S-711) N-S PHRI
Maximum = 15.27
12/04/1972 North Bonins Onahama (S-711) N-S PHRI
Maximum = 1.82
12/04/1972 North Bonins Yamashita (S-706) E-W PHRI
Maximum = 37.04
12/04/1972 North Bonins Yamashita (S-706) E-W PHRI
Maximum = 3.10
Acceleration cm/sec**2
Velocity cm/sec

Fig. A10.

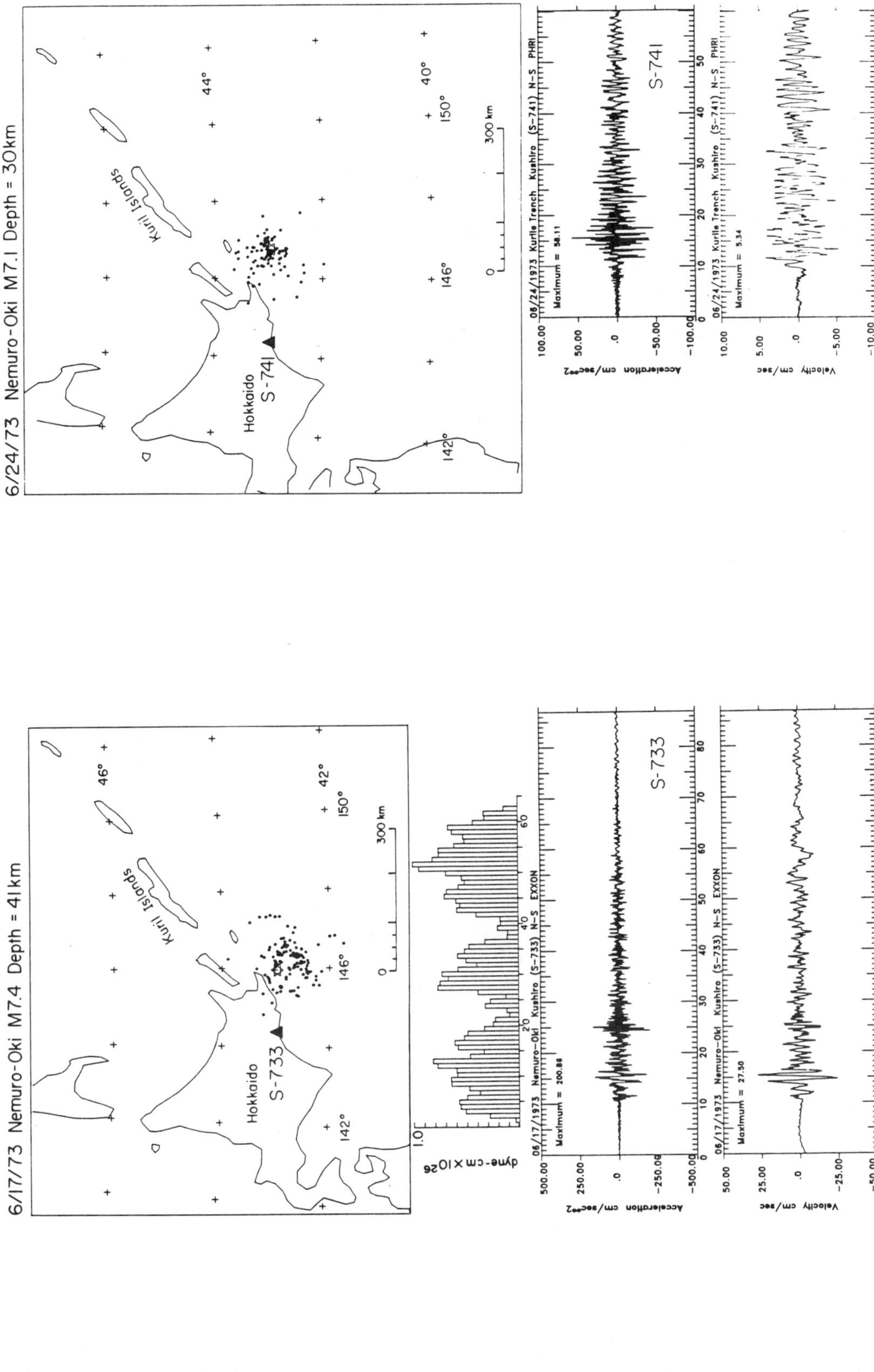
6/24/73 Nemuro-Oki M7.1 Depth = 30km
Kuril Islands
Hokkaido
S-741
44°
40°
150°
146°
142°
0
300 km
06/24/1973 Kurile Trench Kushiro (S-741) N-S PHRI
Maximum = 58.11
Acceleration cm/sec**2
S-741
06/24/1973 Kurile Trench Kushiro (S-741) N-S PHRI
Maximum = 5.34
Velocity cm/sec
6/17/73 Nemuro-Oki M7.4 Depth = 41km
Kuril Islands
Hokkaido
S-733
46°
42°
150°
146°
142°
0
300 km
dyne-cm x 10^26
1.0
06/17/1973 Nemuro-Oki Kushiro (S-733) N-S EXXON
Maximum = 200.86
Acceleration cm/sec**2
S-733
06/17/1973 Nemuro-Oki Kushiro (S-733) N-S EXXON
Maximum = 27.50
Velocity cm/sec

Fig. A11.

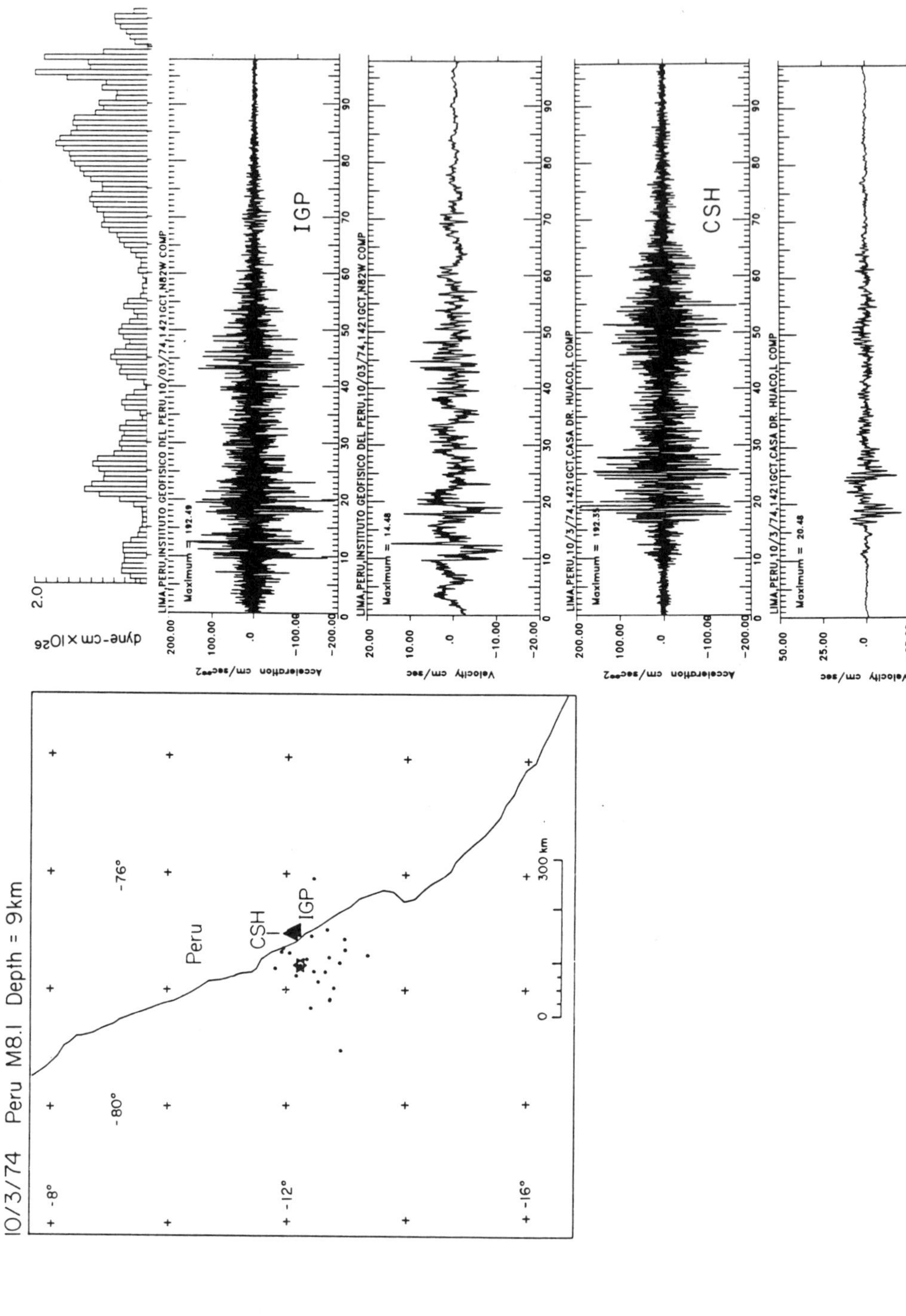
10/3/74 Peru M8.1 Depth = 9km
Peru
CSH
IGP
300 km
IGP
CSH
dyne-cm x 10^26
Acceleration cm/sec**2
Velocity cm/sec

Fig. A12.

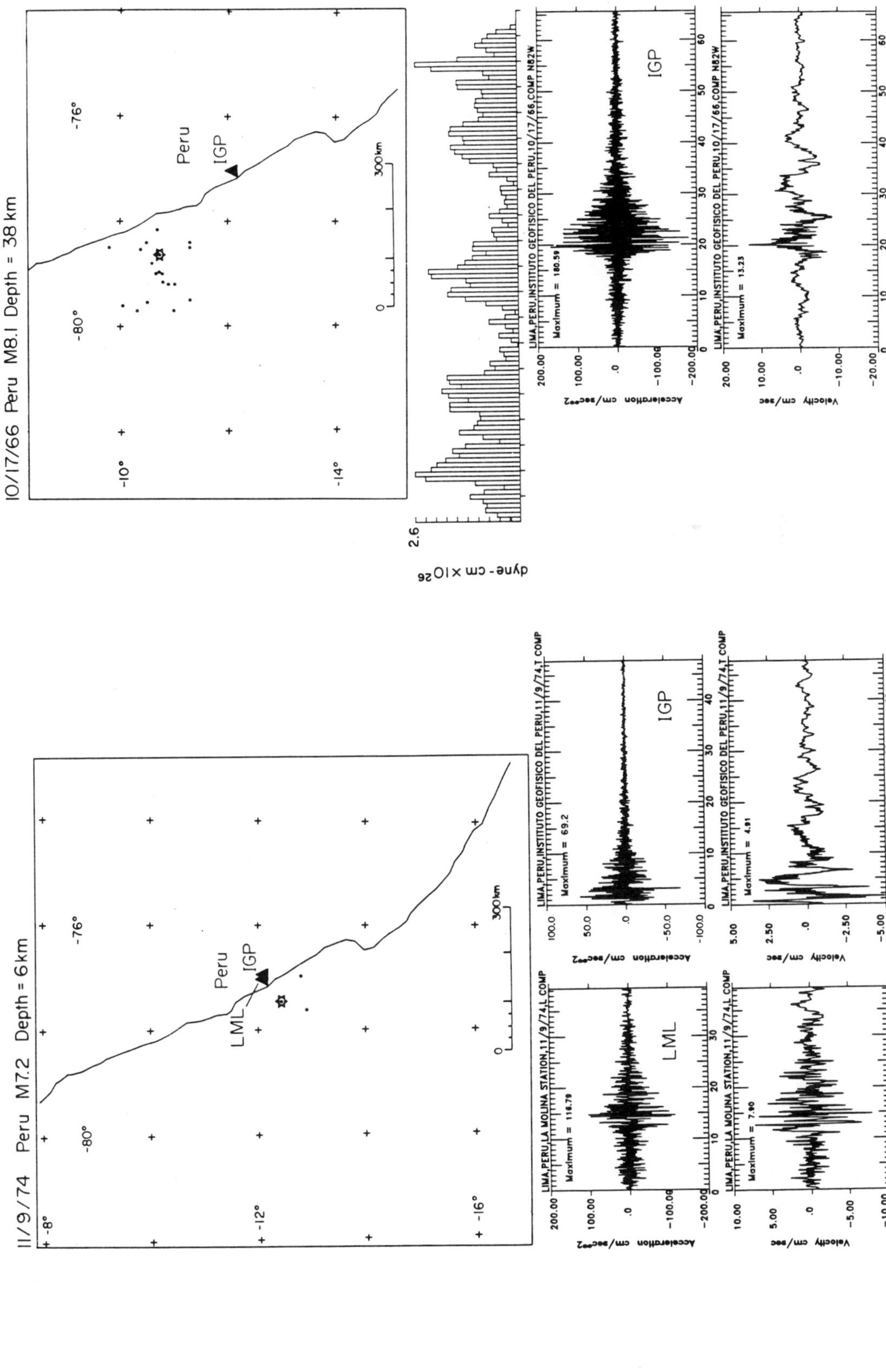
10/17/66 Peru M8.1 Depth = 38 km
Peru
IGP
300km
dyne-cm x10²⁶
LIMA,PERU,INSTITUTO GEOFISICO DEL PERU,10/17/66,COMP N82W
Maximum = 180.59
Maximum = 13.23
Acceleration cm/sec**2
Velocity cm/sec
11/9/74 Peru M7.2 Depth = 6 km
LML
LIMA,PERU,INSTITUTO GEOFISICO DEL PERU,11/9/74,T COMP
Maximum = 69.2
Maximum = 4.91
LIMA,PERU,LA MOLINA STATION,11/9/74,L COMP
Maximum = 118.79
Maximum = 7.90

Fig. A13.

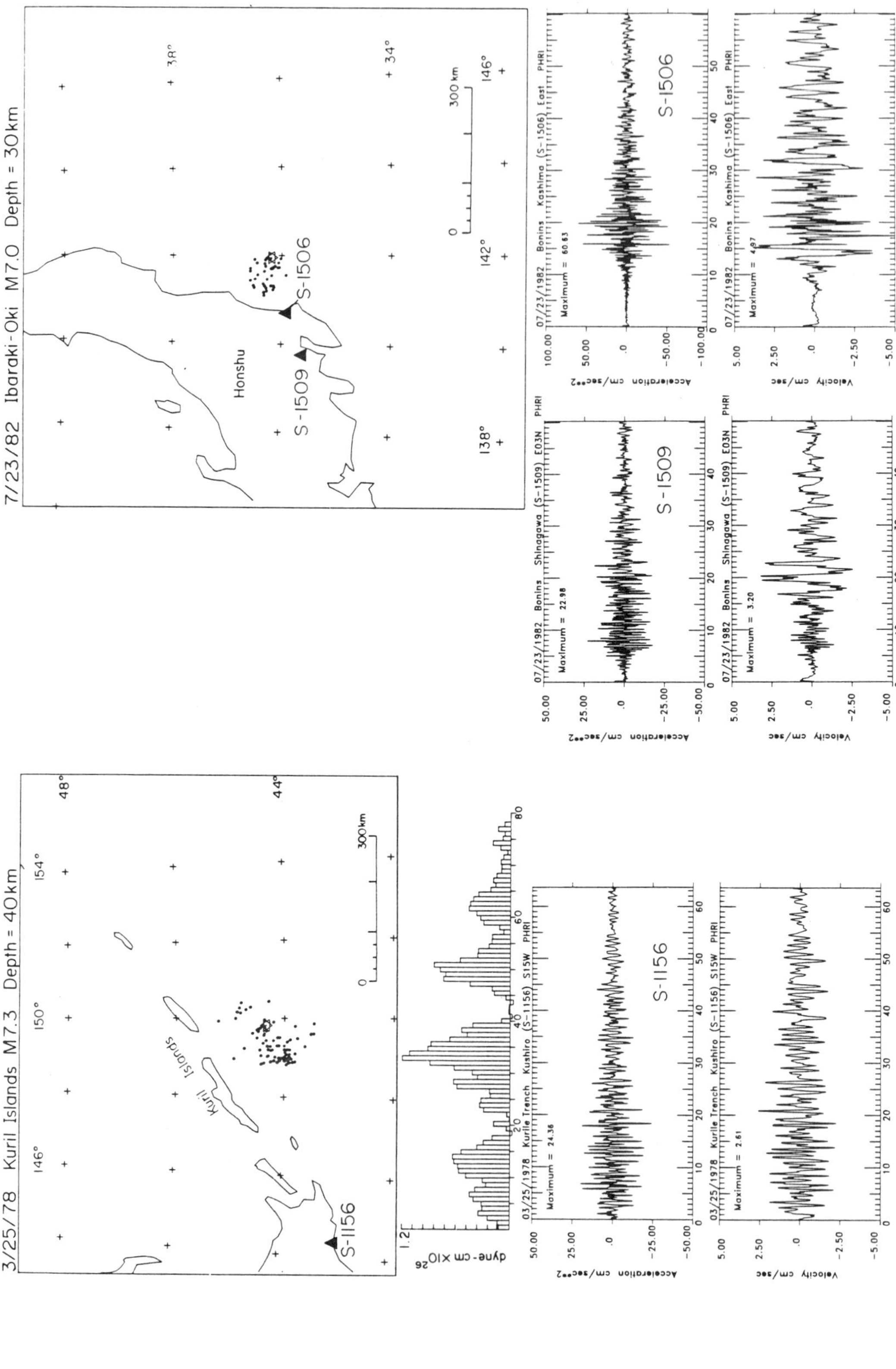
7/23/82 Ibaraki-Oki M7.0 Depth = 30km
Honshu
S-1509
S-1506
300 km
3/25/78 Kuril Islands M7.3 Depth = 40km
Kuril Islands
S-1156
300km
dyne-cm x10^26
07/23/1982 Bonins Shinagawa (S-1509) E03N PHRI
07/23/1982 Bonins Kashima (S-1506) East PHRI
03/25/1978 Kurile Trench Kushiro (S-1156) S15W PHRI
Acceleration cm/sec**2
Velocity cm/sec

Fig. A14.

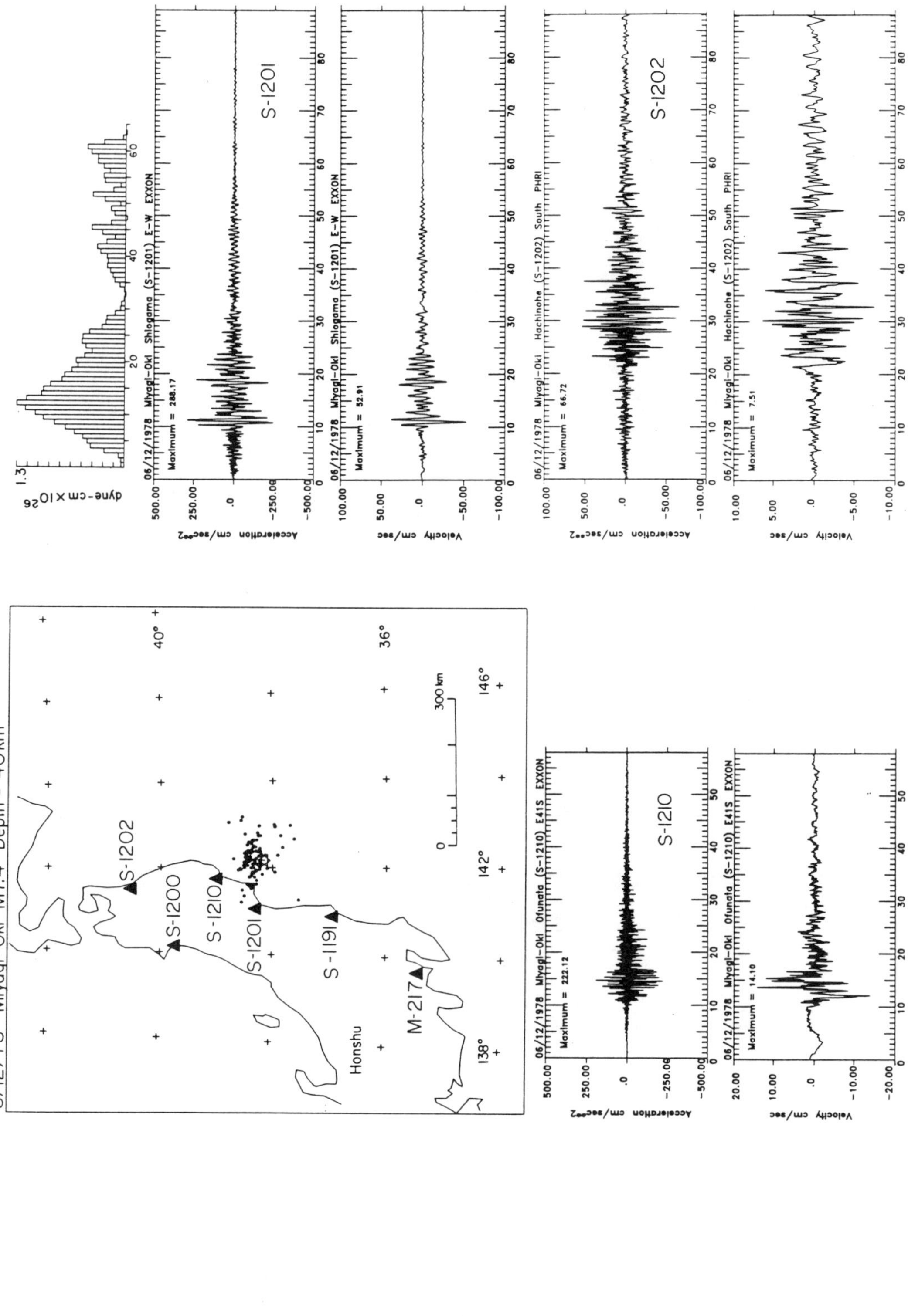

6/12/78 Miyagi-Oki M7.4 Depth = 40 km
S-1202
S-1200
S-1210
S-1201
S-1191
M-217
Honshu
40°
36°
138°
142°
146°
0
300 km
dyne-cm x 10^26
1.3
06/12/1978 Miyagi-Oki Shiogama (S-1201) E-W EXXON
Maximum = 288.17
Maximum = 52.91
S-1201
06/12/1978 Miyagi-Oki Hachinohe (S-1202) South PHRI
Maximum = 66.72
Maximum = 7.51
S-1202
06/12/1978 Miyagi-Oki Ofunato (S-1210) E41S EXXON
Maximum = 222.12
Maximum = 14.10
S-1210
Acceleration cm/sec**2
Velocity cm/sec

Fig. A15.

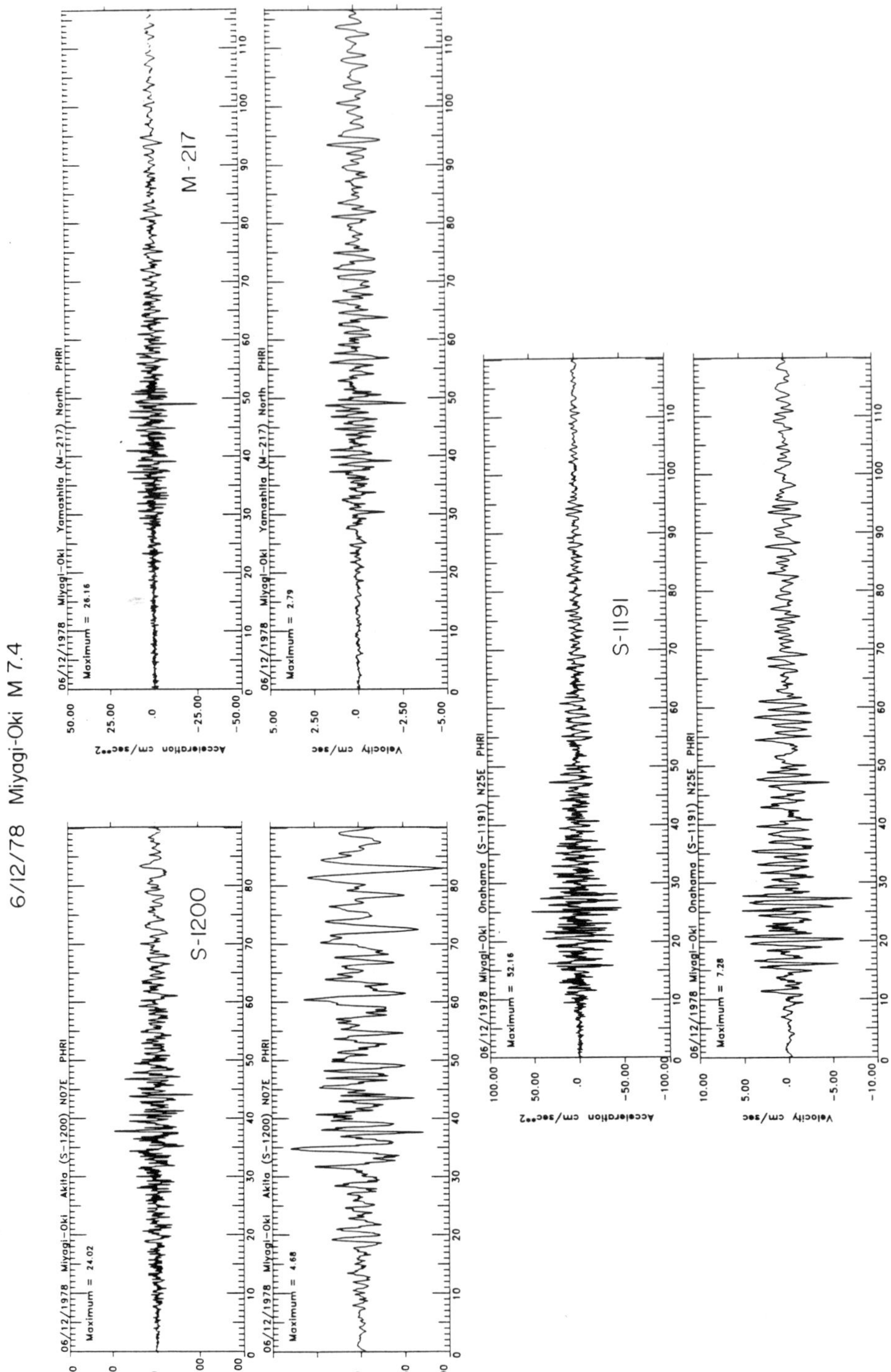
6/12/78 Miyagi-Oki M 7.4
06/12/1978 Miyagi-Oki Yamashita (M-217) North PHRI
Maximum = 26.16
M-217
06/12/1978 Miyagi-Oki Akita (S-1200) N07E PHRI
Maximum = 24.02
Maximum = 4.68
S-1200
06/12/1978 Miyagi-Oki Onahama (S-1191) N25E PHRI
Maximum = 52.16
Maximum = 7.28
S-1191
Acceleration cm/sec**2
Velocity cm/sec

Fig. A16.

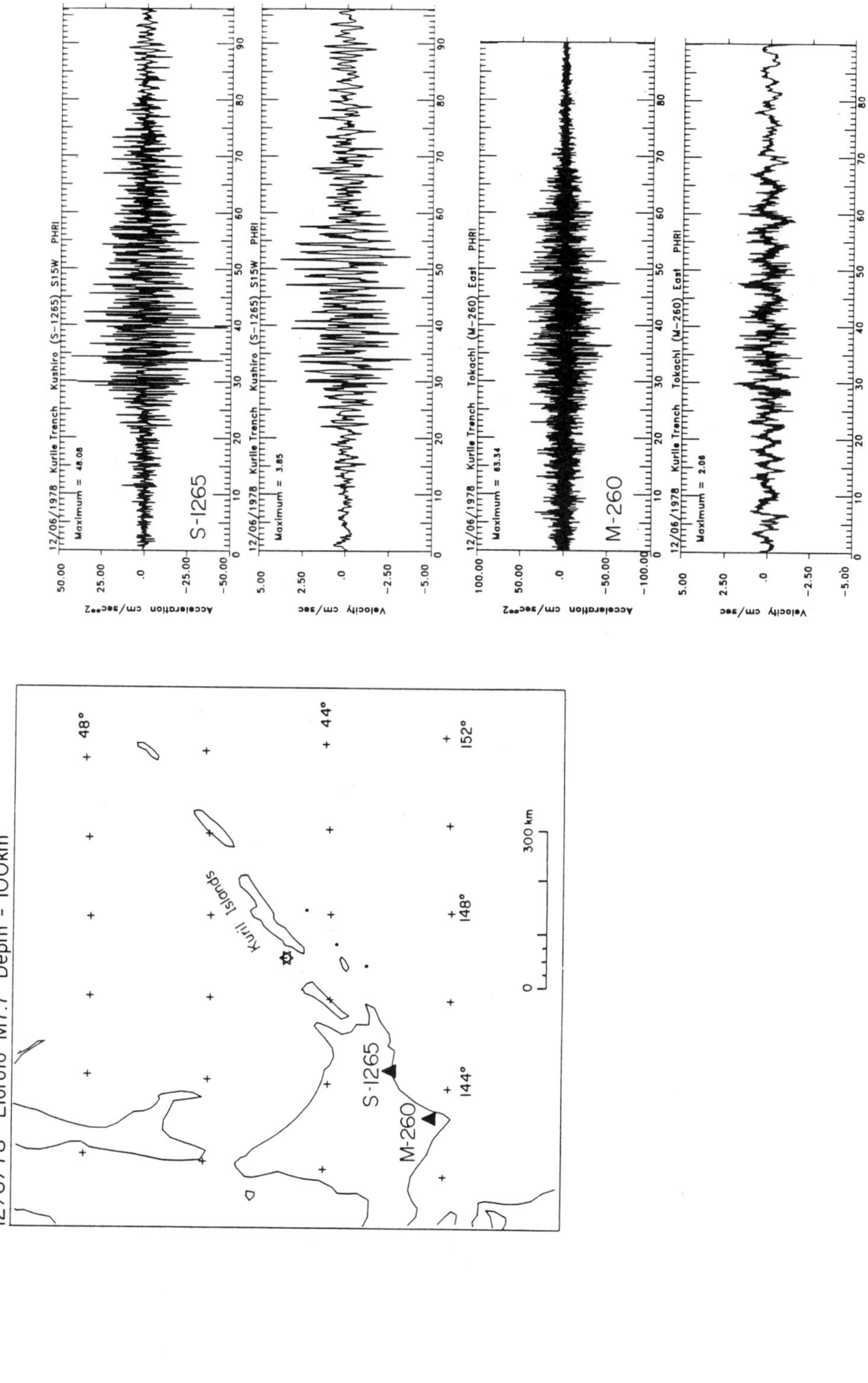
12/06/1978 Kurile Trench Kushiro (S-1265) S15W PHRI
Maximum = 48.08
S-1265
Acceleration cm/sec**2
12/06/1978 Kurile Trench Kushiro (S-1265) S15W PHRI
Maximum = 3.85
Velocity cm/sec
12/06/1978 Kurile Trench Tokachi (M-260) East PHRI
Maximum = 83.34
M-260
Acceleration cm/sec**2
12/06/1978 Kurile Trench Tokachi (M-260) East PHRI
Maximum = 2.06
Velocity cm/sec
12/6/78 Etorofo M7.7 Depth = 100km
Kuril Islands
S-1265
M-260
48°
44°
152°
148°
144°
0
300 km

Fig. A17.

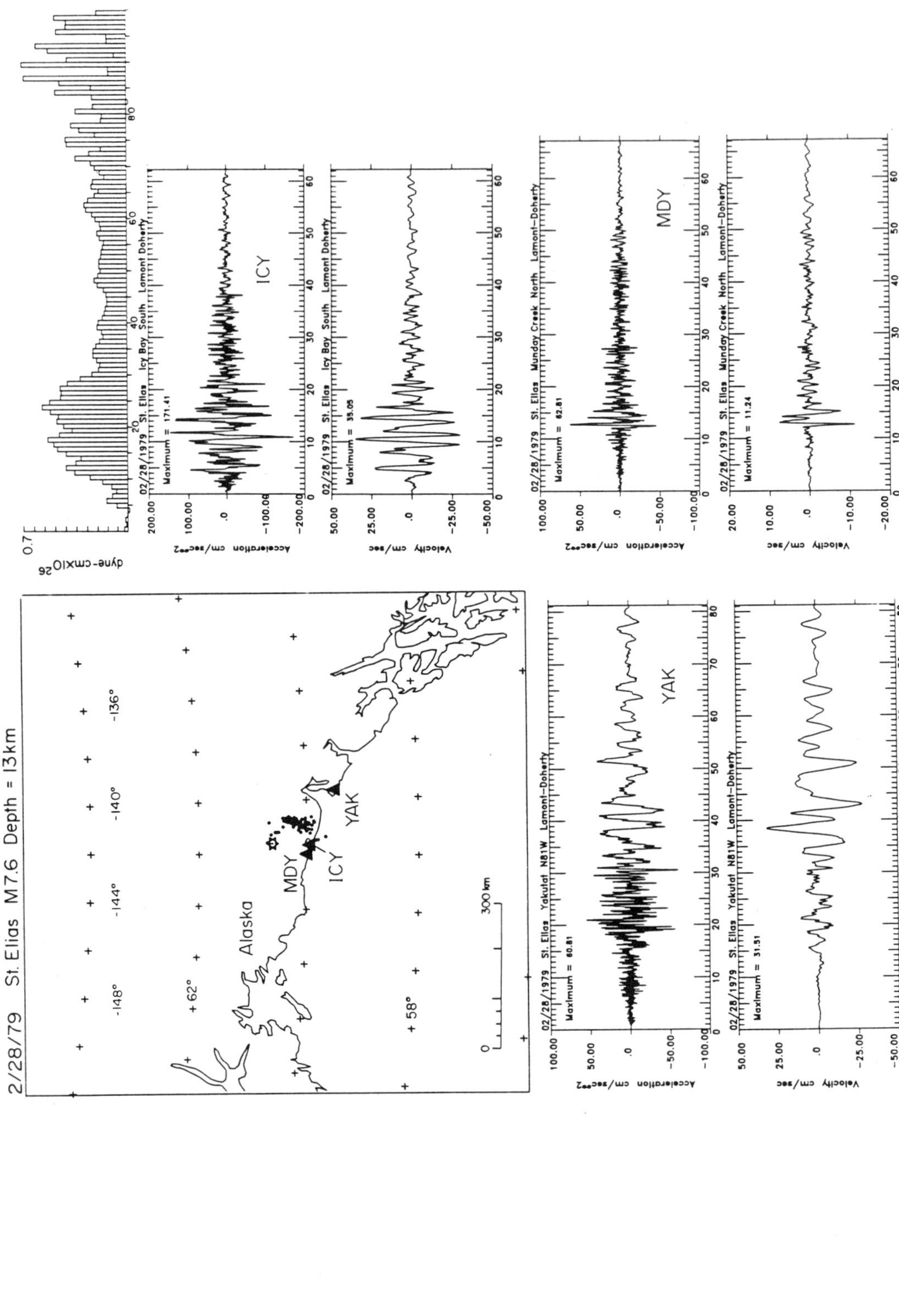

2/28/79 St. Elias M7.6 Depth = 13km
Alaska
MDY
ICY
YAK
-148°
-144°
-140°
-136°
+ 62°
+ 58°
0
300 km
dyne-cmX10^26
0.7
02/28/1979 St. Elias Icy Bay South Lamont Doherty
Maximum = 171.41
ICY
Acceleration cm/sec**2
02/28/1979 St. Elias Icy Bay South Lamont Doherty
Maximum = 35.05
Velocity cm/sec
02/28/1979 St. Elias Munday Creek North Lamont-Doherty
Maximum = 62.81
MDY
Acceleration cm/sec**2
02/28/1979 St. Elias Munday Creek North Lamont-Doherty
Maximum = 11.24
Velocity cm/sec
02/28/1979 St. Elias Yakutat N81W Lamont-Doherty
Maximum = 80.81
YAK
Acceleration cm/sec**2
02/28/1979 St. Elias Yakutat N81W Lamont-Doherty
Maximum = 31.51
Velocity cm/sec

Fig. A18.

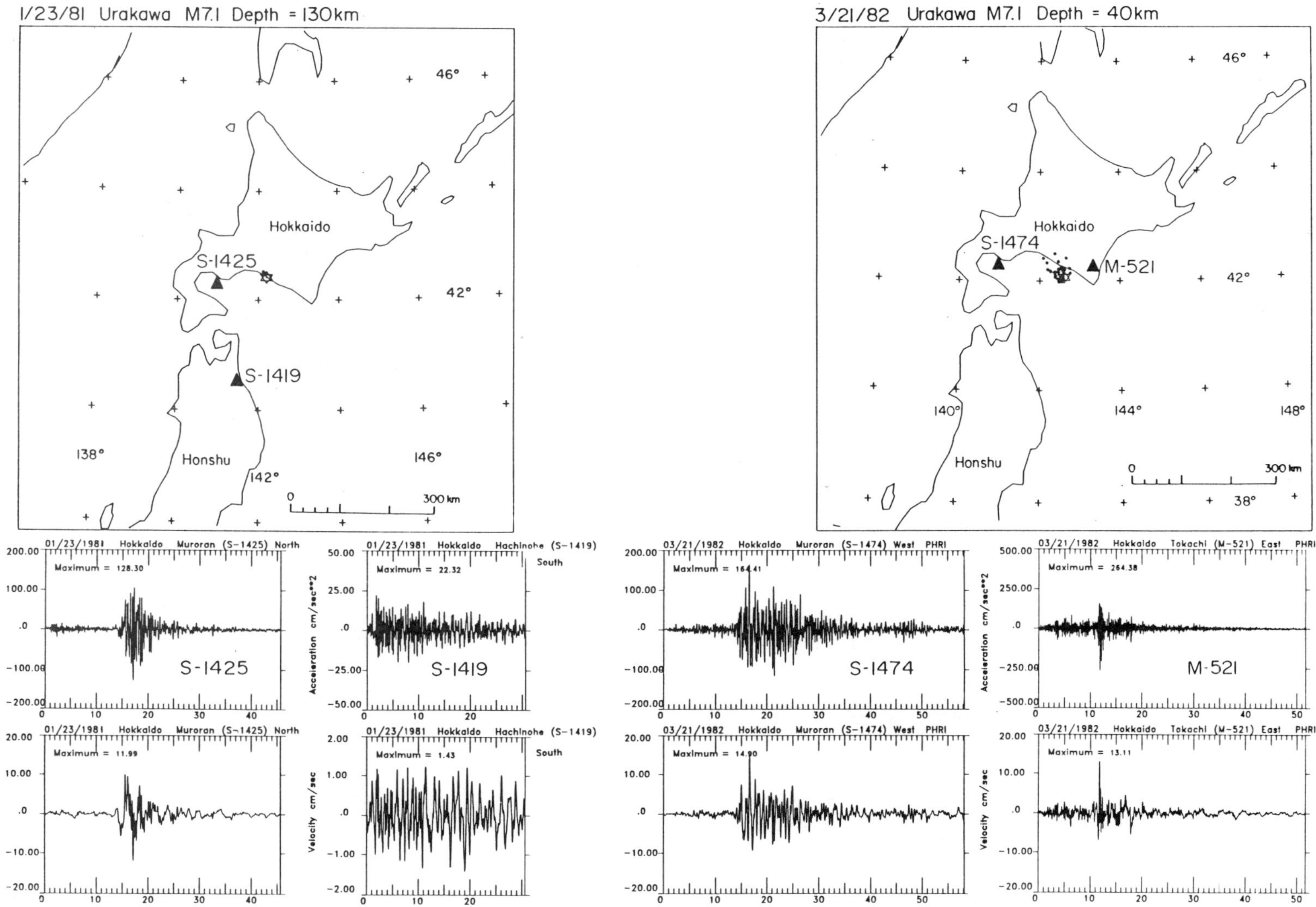
1/23/81 Urakawa M7.1 Depth = 130km
46°
Hokkaido
S-1425
S-1419
42°
138°
Honshu
142°
146°
0
300 km
3/21/82 Urakawa M7.1 Depth = 40km
46°
Hokkaido
S-1474
M-521
42°
140°
144°
148°
Honshu
0
300 km
38°
01/23/1981 Hokkaido Muroran (S-1425) North
Maximum = 128.30
S-1425
01/23/1981 Hokkaido Hachinohe (S-1419) South
Maximum = 22.32
S-1419
Acceleration cm/sec**2
03/21/1982 Hokkaido Muroran (S-1474) West PHRI
Maximum = 164.41
S-1474
03/21/1982 Hokkaido Tokachi (M-521) East PHRI
Maximum = 264.38
M-521
01/23/1981 Hokkaido Muroran (S-1425) North
Maximum = 11.99
01/23/1981 Hokkaido Hachinohe (S-1419) South
Maximum = 1.43
Velocity cm/sec
03/21/1982 Hokkaido Muroran (S-1474) West PHRI
Maximum = 14.90
03/21/1982 Hokkaido Tokachi (M-521) East PHRI
Maximum = 13.11

Fig. A19.

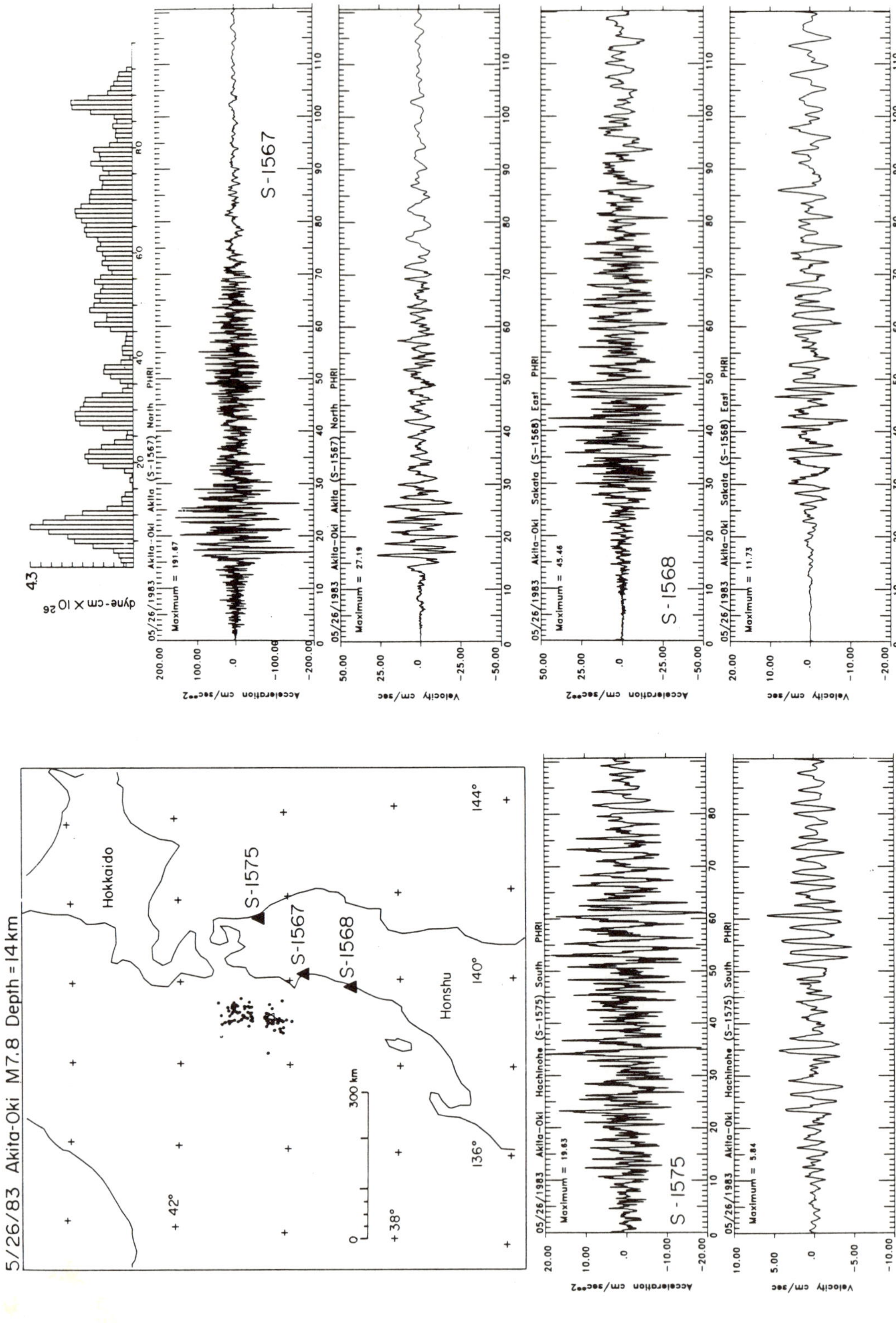
5/26/83 Akita-Oki M7.8 Depth = 14 km
Hokkaido
Honshu
S-1575
S-1567
S-1568
300 km
dyne-cm X 10^26
43

Fig. A20.

Appendix B

In this Appendix, we present a more quantitative discussion of the empirical Green's function technique. The ground motion $\mathbf{U}(t)$ that results from a complex distribution of dislocation time histories, $D(x, y, t)$, that are distributed on a planar fault of length L and width W can be written as

$$\mathbf{U}(t) = \int_0^L \int_0^W \dot{D}(x, y, t) * \mathbf{G}(x, y, t)\, dy\, dx, \tag{B1}$$

where x and y are cartesian coordinates along the fault strike and plunge, respectively, $\mathbf{G}(x, y, t)$ is the double-couple impulse response of the medium, and $\cdot$ and $*$ denote differentiation and convolution operators with respect to time. This expression is quite general and is valid in both the near- and far-fields provided that the earth is a linear system. Because the system is linear, we can always divide the solution into a sum over subfaults, or

$$\mathbf{U}(t) = \sum_{i=1}^{\ell} \sum_{j=1}^{m} \mathbf{U}_{ij}(t), \tag{B2}$$

where

$$\mathbf{U}_{ij}(t) = \int_{(i-1)\Delta L}^{i\Delta L} \int_{(j-1)\Delta W}^{j\Delta W} \dot{D}(x, y, t) * \mathbf{G}(x, y, t)\, dy\, dx, \tag{B3}$$

and where

$$\begin{aligned} \Delta L &= \frac{L}{\ell} \\ \Delta W &= \frac{W}{m}. \end{aligned} \tag{B4}$$

Now suppose that we already possess ground motions $\mathbf{u}_{ij}(t)$ that resulted from a smaller earthquake that ruptured the i, j-th subfault. Its motion would be given by

$$\mathbf{u}_{ij}(t) = \int_{(i-1)\Delta L}^{i\Delta L} \int_{(j-1)\Delta W}^{j\Delta W} \dot{d}_{ij}(x, y, t) * \mathbf{G}(x, y, t)\, dy\, dx, \tag{B5}$$

where $d_{ij}(x, y, t)$ is the dislocation time history distribution for the smaller earthquake. If a function $F_{ij}(t)$ exists such that

$$D(x, y, t) = F_{ij}(t) * d_{ij}(x, y, t) \tag{B6}$$

for $(i-1)\Delta L \leq x < i\Delta L$ and $(j-1)\Delta W \leq y < j\Delta W$, then

$$\mathbf{U}_{ij}(t) = F_{ij}(t) * \mathbf{u}_{ij}(t), \tag{B7}$$

and then

$$\mathbf{U}(t) = \sum_{i=1}^{\ell} \sum_{j=1}^{m} F_{ij}(t) * \mathbf{u}_{ij}(t). \tag{B8}$$

So in principle, if we have good data from smaller events and if we know what the functions $F_{ij}(t)$ are, then we can obtain the ground motions of a large earthquake from smaller ones. Of course, this assumes that the slip history of the larger event can be obtained from a linear combination of slip histories of the smaller events.

Unfortunately, determination of $F_{ij}(t)$ is somewhat problematic. It is customary to assume that $F_{ij}(t)$ exists and that it consists of a sequence of n Dirac-delta functions distributed over the dislocation rise time of the larger event, where n is the integral ratio of the dislocations for the larger and smaller events. This is actually an important assumption and we will discuss alternative assumptions later. For now, however, assume that $F_{ij}(t)$ has the form

$$F_{ij}(t) = \sum_{k=1}^{n} \delta(t - T_{ij} - \tau_k), \tag{B9}$$

where T_{ij} is the delay time for the rupture front to travel from the hypocenter to the ij-th subfault and the τ_k's are a yet undefined distribution of times between zero and the dislocation rise time. Combining (B8) and (B9), we obtain

$$\mathbf{U}(t) = \sum_{i=1}^{\ell} \sum_{j=1}^{m} \sum_{k=1}^{n} \mathbf{u}_{ij}(t - T_{ij} - \tau_k), \tag{B10}$$

where the ratio of the seismic moments of the larger event M_0 to the smaller event M_0' is

$$\frac{M_0}{M_0'} = \frac{LWD}{\Delta L \Delta W D'} = \ell mn = N. \tag{B11}$$

One simple distribution of the τ_k's is to assume that

$$\tau_k = (k-1)\frac{t_d}{n}, \quad k = 1, 2, 3, \ldots n \tag{B12}$$

where t_d is the duration of the dislocation of the large event. Unfortunately, this has the effect of convolving the empirical Green's functions with a picket fence function of periodicity t_d/n and the resulting synthetics ring at this period. In order to avoid this, a random time shift can be added to τ_k, or

$$\tau_k = (k - \chi_k)\frac{t_d}{n}, \tag{B13}$$

where χ_k's are uniform random numbers between 0 and 1. Combining (B13) and (B10), we obtain

$$\mathbf{U}(t) = \sum_{i=1}^{\ell} \sum_{j=1}^{m} \sum_{k=1}^{n} \mathbf{u}_{ij}\left[t - T_{ij} - (k - \chi_k)\frac{t_d}{n}\right]. \tag{B14}$$

Although the empirical Green's functions $\mathbf{u}_{ij}(t)$ must, in general, change from one subfault to another, in many respects they may be statistically quite similar to each other. This is particularly true in teleseismic cases since the observer-subfault geometry changes little from subfault to subfault. In order to study the general properties of the empirical Green's function technique, we approximate $\mathbf{u}_{ij}(t)$ by

$$\mathbf{u}_{ij}(t) \approx A_{ij}\mathbf{u}(t - \xi_{ij}), \tag{B15}$$

where $\mathbf{u}(t)$ is an observed record having general characteristics similar to other records, A_{ij} is a distance attenuation scaling factor, and ξ_{ij} is the source to receiver travel-time correction. Combining (B15) and (B14), we obtain

$$\mathbf{U}(t) \approx \mathbf{u}(t) * P(t), \tag{B16}$$

where

$$P(t) = \sum_{i=1}^{\ell} \sum_{j=1}^{m} \sum_{k=1}^{n} A_{ij}\delta\left[t - T_{ij} - \xi_{ij} - (k - \chi_k)\frac{t_d}{n}\right]. \tag{B17}$$

We can refer to $P(t)$ as a transfer function since it specifies how the record from a small earthquake can be transformed into the record of a larger earthquake. In Figure B1, we show $P(t)$ for the models M–1140–200 and M–120–135 (see Figure 5) and for both a coastal ranges strong motion site and also a teleseismic site. In the teleseismic case, the envelope of $P(t)$ is approximately constant in time since A_{ij} is

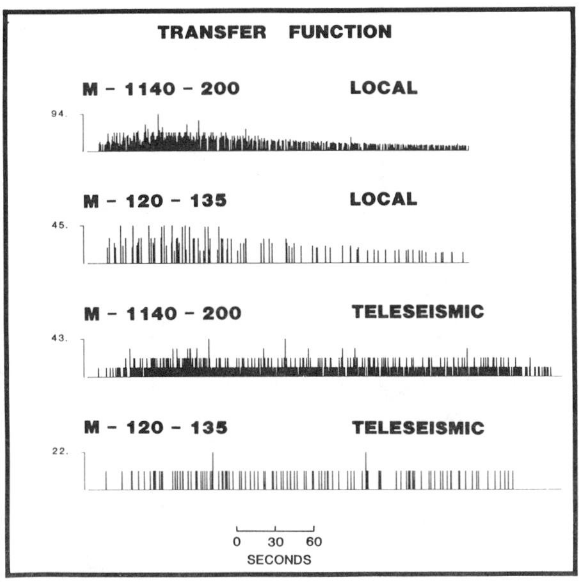

Figure B1

Examples of the form of the transfer function $P(t)$ defined by equation (B17) and assuming a coastal ranges site and two of the models of the 1960 Chilean earthquake shown in Figure 5. In the case of the teleseismic records, synthetic motions are determined by convolving the empirical Green's function with these transfer functions. Each spike represents an impulse function.

assumed to be 1.0 everywhere. In the strong motion case, the envelope has more character since the A_{ij} are determined from the CROUSE *et al.* (1988) distance attenuation relationship and also since the effects of directivity are more complex in the local field.

We can discover the spectral scaling implications of the empirical Green's function technique by studying the spectral characteristics of $P(t)$. In Figure B2, we show the Fourier amplitude spectra $\hat{P}(\omega)$ of the transfer functions shown in Figure B1. We studied many transfer functions and although the detailed shapes of their Fourier amplitude spectra are generally quite complex, we found that we could approximate their envelopes by

$$\hat{P}(\omega) \approx \begin{cases} N; & \omega < 2/T_f \\ \dfrac{2N}{T_f \omega^{\alpha}}; & \dfrac{2}{T_f} < \omega < \left(\dfrac{2}{T_f}\sqrt{N}\right)^{1/\alpha} \\ \sqrt{N}; & \left(\dfrac{2}{T_f}\sqrt{N}\right)^{1/\alpha} < \omega, \end{cases} \tag{B18}$$

where T_f is the total duration of $P(t)$, N is the total number of Green's functions that are summed, and α is a spectral fall-off parameter having a value between 1.0 and 3.0, depending on the nature of the fault geometry and rupture characteristics.

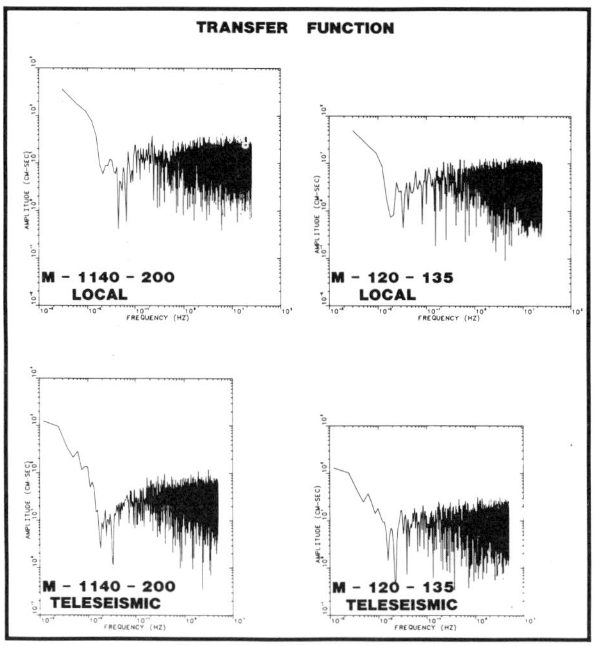

Figure B2
Fourier amplitude spectra of the transfer functions shown in Figure B1.

The fact that the high-frequency level of $\hat{P}(\omega)$ is $\sqrt{N}$ arises naturally (and inevitably) from the assumption that the individual Dirac-delta functions comprising $P(t)$ are shifted by random time delays.

If we assume that earthquake spectra obey certain simple similarity laws (and there is no good reason that they must), we can gain insight to the physical meaning of $\hat{P}(\omega)$. Suppose that we assume that $\hat{\mathbf{U}}(\omega)$ and $\hat{\mathbf{u}}(\omega)$ are self-similar and have a simple shape described by

$$\hat{\mathbf{U}}(\omega) = \begin{cases} CM_0; & \omega < \omega_c \\ CM_0\left(\dfrac{\omega_c}{\omega}\right)^{\nu}; & \omega > \omega_c, \end{cases} \tag{B19}$$

where C is a scaling constant, ν is a spectral fall off parameter, and ω_c is a corner frequency given by

$$\omega_c = DM_0^{\beta}, \tag{B20}$$

where D and β are also scaling constants. The spectral shape of $\hat{\mathbf{u}}(\omega)$ is described in a similar fashion as

$$\hat{\mathbf{u}}(\omega) = \begin{cases} CM_0'; & \omega < \omega_c' \\ CM_0'\left(\dfrac{\omega_c'}{\omega}\right)^{\nu}; & \omega > \omega_c', \end{cases} \tag{B21}$$

where

$$\omega_c' = DM_0'^{\beta}. \tag{B22}$$

We can now find $\hat{P}(\omega)$ such that it is compatible with our general similarity laws. From (B16) and (B19) through (B22) we have

$$\hat{P}(\omega) = \frac{\hat{U}}{\hat{\mathbf{u}}} = \begin{cases} N; & \omega < \omega_c \\ N\left(\dfrac{\omega_c}{\omega}\right)^{\nu}; & \omega_c < \omega < \omega_c', \\ N^{1+\beta\nu}; & \omega_c' < \omega \end{cases} \tag{B23}$$

where $N = M_0/M_0'$. Comparing (B23) with (B18), we conclude that the empirical Green's function technique that we have described will produce self-similar spectra if those spectra are characterized by

$$1 + \beta\nu = \tfrac{1}{2}. \tag{B24}$$

In many spectral scaling laws, it is assumed that the corner frequency scales with moment as

$$\omega_c = DM_0^{-1/3}, \quad \text{or} \quad \beta = -1/3. \tag{B25}$$

If (B25) is assumed, then from (B24),

$$\nu = 3/2. \tag{B26}$$

In other words, if we assume that the corner frequency scales inversely with the cube root of the seismic moment, then the empirical Green's function technique that we have described will produce self-similar spectra only if the records have spectral falloffs of $\omega^{-3/2}$. Unfortunately, such spectra are physically unreasonable since they result in infinite radiated energy at high frequencies.

A frequently used spectral scaling law is one that was introduced by BRUNE (1970), in which it is assumed that $\beta = -1/3$ and $\nu = 2$. If this is the case, then $\hat{P}(\omega)$ would have to have the following form

$$\hat{P}(\omega) = \begin{cases} N; & \omega < \omega_c \\ N\left(\dfrac{\omega_c}{\omega}\right)^2; & \omega_c < \omega < \omega_c'. \\ N^{1/3}; & \omega_c' < \omega \end{cases} \tag{B27}$$

Unfortunately, the summation of N records with a random phase lag inevitably leads to a synthetic having a high-frequency level that is larger than the original by a factor of $\sqrt{N}$. In order to obtain an empirical Green's function technique that is consistent with BRUNE'S (1970) spectral scaling law, JOYNER and BOORE (1986) propose that $N^{4/3}$ records should be randomly summed and that the final motion should be multiplied by $N^{-1/3}$. If $\alpha = 2$, $\beta = -1/3$, and if we replace N by $N^{4/3}$ in (B18) and then multiply the result by $N^{-1/3}$, then it is easy to see that we obtain a $\hat{P}(\omega)$ that is identical to the one that is required by the Brune spectral scaling law given by (B27).

The apparent inconsistencies that we have noted can be traced directly to the assumption given by (B9). It is this assumption that forces the high-frequency levels to scale as $\sqrt{N}$. If we assume that the stress drops and rupture aspect ratios of small and large earthquakes are identical (i.e., $\ell = m = n$), then we can reproduce the JOYNER and BOORE (1986) model by assuming that

$$F_{ij}(t) = \frac{1}{n} \sum_{k=1}^{n^2} \delta(t - T_{ij} - \tau_k). \tag{B28}$$

This is equivalent to saying that the dislocation of a large earthquake looks like the filtered sum of dislocations of smaller ones. In both assumptions (B9) and (B28), it is assumed that high-frequency energy is radiated from a patch throughout the duration of the dislocation on that patch. However, one could easily postulate that high-frequencies are only radiated during the initial phases of the rupture and that the dislocation is very smooth once the rupture front has broken through all adjacent areas. In this case, we might assume that $F_{ij}(t)$ may be approximated by

$$F_{ij}(t) = \delta(t - T_{ij}) + h(t), \tag{B29}$$

where $h(t)$ is a relatively smooth, positive function having a total area of $(n-1)$ and a duration of t_d. If we assume that $\ell = m = n$ and (B29), then it is not difficult to see that this assumption will lead to a $\hat{P}(\omega)$ in which the long-period levels grow as N, but the high-frequency levels grow as $N^{1/3}$ (provided that $h(t)$ is sufficiently smooth). Both assumptions (B28) and (B29) lead to the conclusion that the high-frequency energy radiated from a subfault is the same for both small and large earthquakes. This is undoubtedly an oversimplification of the real case, but aspects of this generality may well be true.

Self-similarity is a very useful concept since it provides rules for the way in which large earthquakes can be produced from smaller ones. However, it is difficult to argue that giant subduction earthquakes such as the 1960 Chilean earthquake must have rupture processes that are self-similar to those of smaller earthquakes. The aspect ratio of these giant earthquakes may be different from that of smaller ones. Furthermore, the rheology of the fault and surrounding rock is undoubtedly a function of depth and there is little reason to expect that the nature of the rupture process is the same for earthquakes that are confined to the shallow part of the interface and giant earthquakes which may rupture into the uppermost mantle. In their study of teleseismic P-waveforms from 60 of the largest subduction earthquakes, HARTZELL and HEATON (1985) find that the average spectra of giant earthquakes do not appear to be self-similar to the average of smaller earthquakes (see their Figure 11a).

Given these complications, we have chosen to apply the empirical Green's function technique in a simple way such that it matches the teleseismic P-waveforms in the frequency range that is relevant to strong ground motions. We simply sum enough records to match the teleseismic short-period amplitudes. As might be expected, this number is less than that given by the ratio of the moments of the larger and smaller earthquakes. Clearly, this procedure would seriously underestimate long-period (greater than 20 sec.) motions. However, since our strong motion empirical Green's functions have already been filtered so that they have no information at these periods, we believe that little is lost in our application of the procedure.

Acknowledgments

We thank Hiroo Kanamori and C. B. Crouse for helpful discussions and we thank C. B. Crouse for his help in collection of the digitized strong motion data used in this report. We thank Bob Page, Bill Joyner, and Heidi Houston for their comments on this manuscript. This work was partially supported by the United States Nuclear Regulatory Commission.

References

ANDERSON, J. G., P. BODIN, J. N. BRUNE, J. PRINCE, S. K. SINGH, R. QUAAS, and M. ONATE (1986), *Strong ground motion from the Michoacan, Mexico, earthquake*, Science *233*, 1043–1049.

BEAVAN, J. and K. H. JACOB (1984), *Processed strong motion data from subduction zones: Alaska*, in *Lamont-Processed Strong Motion Data, Report 1* (Lamont-Doherty Geological Observatory, Palisades, NY) 249 p.

BRADY, A. G. and V. PEREZ (1977), *Strong-motion earthquake accelerograms, digitization and analysis, records from Lima, Peru: 1951 to 1974*, U.S. Geol. Surv. Open-File Report, 77–587.

BRUNE, J. N. (1970), *Tectonic stress and the spectra of seismic shear waves from earthquakes*, J. Geophys. Res. *75*, 4997–5009.

CROUSE, C. B., J. A. HILEMAN, B. E. TURNER, and G. R. MARTIN (1980), *Compilation assessment and expansion of the strong earthquake ground motion data base*, U.S. Nuclear Regulatory Commission, NUREG/CR–1660.

CROUSE, C. B., Y. K. VYAS, and B. A. SCHELL (1988), *Ground motions from subduction zone earthquakes*, Bull. Seism. Soc. Am. *78*, 1–25.

DENHAM, D. and R. G. SMALL (1971), *Strong motion data centre: Bureau of Mineral Resources, Canberra*, Bull. New Zealand Soc. for Earthq. Eng. *4*, March.

GUZMAN, R. A. and P. C. JENNINGS (1976), *Design spectra for nuclear power plants*, Journal of the Power Division, ASCE *102*, Po2, Proc. paper 12521, 165–178.

HARTZELL, S. H. (1978), *Earthquake aftershocks as Green's functions*, Geophys. Res. Letters *5*, 1–4.

HARTZELL, S. H. (1985), *The use of small earthquakes as Green's functions*, in Proceedings: *Strong-Ground-Motion Simulation and Earthquake Engineering Applications* (eds. R. E. Scholl and J. L. King) Electric Power Research Institute, NP–4299, Project 2556–1, pp. 22–1 to 22–8.

HARTZELL, S. H. and T. H. HEATON (1985), *Teleseismic time functions for large shallow subduction zone earthquakes*, Bull. Seism. Soc. Am. *75*, 965–1004.

HASKELL, N. A. (1969), *Elastic displacements in the near-field of a propagating fault*, Bull. Seism. Soc. Am. *59*, 865–908.

HEATON, T. H. and S. H. HARTZELL (1986), *Source characteristics of hypothetical subduction earthquakes in the northwestern United States*, Bull. Seism. Soc. Am. *76*, 675–708.

HEATON, T. H. and H. KANAMORI (1984), *Seismic potential associated with subduction in the northwestern United States*, Bull. Seism. Soc. Am. *74*, 933–941.

HEATON, T. H., F. TAJIMA, and A. W. MORI (1986), *Estimating ground motions using recorded accelerograms*, Surveys in Geophysics *8*, 25–83.

HOUSTON, H. and H. KANAMORI (1986), *Source spectra of great earthquakes: Teleseismic constraints on rupture process and strong motion*, Bull. Seism. Soc. Am. *76*, 19–42.

IRIKURA, K. (1983), *Semi-empirical estimation of strong ground motions during large earthquakes*, Bull. Disas. Prev. Res. Inst., Kyoto Univ. *33*, 63–104.

JOYNER, W. B. and D. M. BOORE (1981), *Peak horizontal acceleration and velocity from strong-motion records including records from the 1979 Imperial Valley, California, earthquake*, Bull. Seism. Soc. Am. *71*, 2011–2038.

JOYNER, W. B. and D. M. BOORE (1982), *Prediction of earthquake response spectra*, U.S. Geol. Survey, Open-File Report, 82–977.

JOYNER, W. B. and D. M. BOORE (1986), *On simulating large earthquakes by Green's function addition of smaller earthquakes*, Proceedings of the 5th Maurice Ewing Symposium on Earthquake Source Mechanics (eds. S. Das, J. Boatwright and C. Scholz), AGU, 269–274.

KANAMORI, H. (1970), *Synthesis of long-period surface waves and its application to earthquake source studies—Kurile Islands earthquake of October 13, 1963*, J. Geophys. Res. *75*, 5011–5027.

KANAMORI, H. (1975), *Re-examination of the Earth's free oscillations excited by the Kamchatka earthquake of November 4, 1952*, Phys. Earth Plane. Int. *11*, 216–216.

KANAMORI, H. (1977), *The energy release in great earthquakes*, J. Geophys. Res. *82*, 2981–2987.

KANAMORI, H. (1979), *A semi-empirical approach to prediction of long-period ground motions from great earthquakes*, Bull. Seism. Soc. Am. *69*, 1654–1670.

KAWASHIMA, K., K. AIZAWA, and K. TAKAHASHI (1984), *Attenuation of peak ground motion and absolute acceleration response spectra*, in Proceedings of the Eighth World Conference on Earthquake Engineering Vol. II, San Francisco, 257–264.

KELLEHER, J., J. SAVINO, H. ROWLETT, and W. MCCANN (1974), *Why and where great thrust earthquakes occur along island arcs*, J. Geophys. Res. *79*, 4889–4899.

KIKUCHI, M. and Y. FUKAO (1987), *Inversion of long-period P-waves from great earthquakes along subduction zones*, Tectonophysics *144*, 231–247.

LIU, H. L. and T. H. HEATON (1984), *Array analysis of the ground velocities and accelerations from the 1971 San Fernando, California, earthquake*, Bull. Seism. Soc. Am. *74*, 1951–1968.

MORI, A. W. and C. B. CROUSE (1981), *Strong motion data from Japanese earthquakes*, World Data Center A, Report SE-29, National Oceanic and Atmospheric Administration, Boulder, CO.

OHTA, Y., H. KAGAMI, N. GOTO, and K. KUDO (1978), *Observation of 1- to 5- second micrometers and their application to earthquake engineering. Part I: Comparison with long-period accelerations at the Tokachi-Oki earthquake of 1968*, Bull, Seism. Soc. Am. *68*, 767–779.

PLAFKER, G. (1972), *Tectonics*, in *The Great Alaskan Earthquake of 1965, Seismology and Geodesy* (National Academy of Sciences, Washington, D.C.).

RUFF, L. and H. KANAMORI (1983), *The rupture process and asperity distribution of three great earthquakes from long-period diffracted P-waves*, Phys. Earth Planet. Inter. *31*, 202–230.

SYKES, L. (1971), *Aftershock zones of great earthquakes, seismicity gaps, and earthquake prediction for Alaska and the Aleutians*, J. Geophys. Res. *76*, 8021–8041.

TRIFUNAC, M. D. (1976), *Preliminary analysis of the peaks of strong ground motion—dependence of peaks on earthquake magnitude, epicentral distance, and recording site conditions*, Bull. Seism. Soc. Am. *66*, 189–219.

WU, F. T. and H. KANAMORI (1973), *Source mechanism of February 4, 1965, Rat Island earthquake*, J. Geophys. Res. *78*, 6082–6092.

(Received August 5, 1986, accepted July 14, 1987)

PAGEOPH, Vol. 129, Nos. 1/2 (1989) 0033–4553/89/020203–29$1.50 + 0.20/0

Subduction and Back-Arc Activity at the Hikurangi Convergent Margin, New Zealand

EUAN G. C. SMITH,[1] TIM STERN[1] and MARTIN REYNERS[1]

Abstract—The Hikurangi Margin is a region of oblique subduction with northwest-dipping intermediate depth seismicity extending southwest from the Kermadec system to about 42°S. The current episode of subduction is at least 16–20 Ma old. The plate convergence rate varies along the margin from about 60 mm/a at the south end of the Kermadec Trench to about 45 mm/a at 42°S. The age of the Pacific lithosphere adjacent to the Hikurangi Trench is not known.

The margin divides at about latitude 39°S into two quite dissimilar parts. The northern part has experienced andesitic volcanism for about 18 Ma, and back-arc extension in the last 4 Ma that has produced a back-arc basin onshore with high heatflow, thin crust and low upper-mantle seismic velocities. The extension appears to have arisen from a seawards migration of the Hikurangi Trench north of 39°S. Here the plate interface is thought to be currently uncoupled, as geodetic data indicate extension of the fore-arc basin, and historic earthquakes have not exceeded $M_s = 7$.

South of 39°S there is no volcanism and a back-arc basin has been produced by downward flexure of the lithosphere due to strong coupling with the subducting plate. Heatflow in the basin is normal. Evidence for strong coupling comes from historic earthquakes of up to about $M_s = 8$ and high rates of uplift on the southeast coast of the North Island.

The reason for this division of the margin is not known but may be related to an inferred increase, from northeast to southwest, in the buoyancy of the Pacific lithosphere.

Key words: Subduction, back-arc basin, lithospheric coupling, historical seismicity, Hikurangi Margin, New Zealand.

Introduction

New Zealand largely owes its existence to being at the boundary of the Pacific and Australian plates. The nature of this boundary has changed rapidly over the last 25 Ma from transcurrent to obliquely convergent, shown by the southern migration of Australian/Pacific rotation pole positions (e.g., STOCK and MOLNAR, 1982; SMITH, 1981). The current boundary consists of three segments (Figure 1): the Hikurangi Margin—a subduction zone extending from the southern end of the Kermadec Trench to about latitude 42°S, the largely transcurrent Alpine Fault of the central South Island and the Fiordland subduction zone extending from about

[1] Geophysics Division, Department of Scientific and Industrial Research, Wellington, New Zealand.

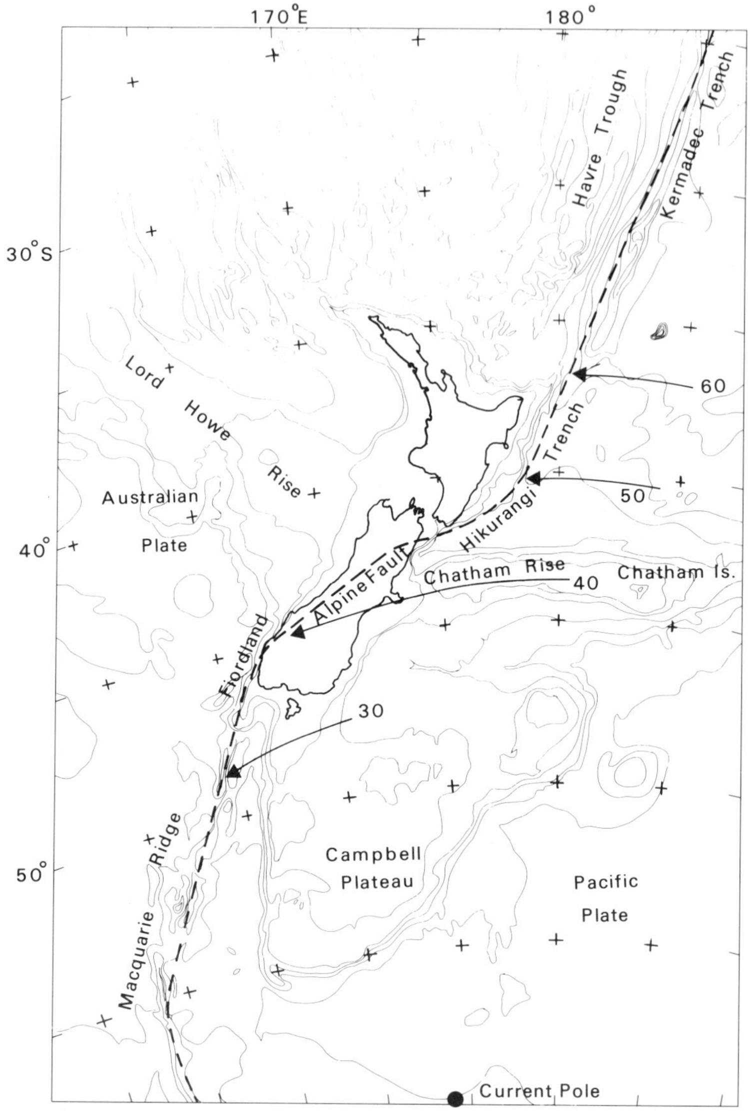

Figure 1
The location of the Hikurangi Margin section of the Australian/Pacific plate boundary. Pole and rotation rates (mm/a) after MINSTER and JORDAN (1978). Bathymetry after CARTER (1980).

latitude 45°S to the Macquarie Ridge (DAVEY and SMITH, 1983a; SMITH and DAVEY, 1984).

This paper is concerned with the Hikurangi Margin and those features that make it both similar to and different from other subduction zones. The data that have been gathered will be briefly described, and then an attempt will be made to relate the subduction and back-arc processes of the Hikurangi Margin to those at other subduction zones. RUFF and KANAMORI (1980) demonstrated a convincing relationship between age, convergence rate and magnitude of the largest earthquakes at various subduction zones. Measurements of the age of sea-floor that is currently subducting at the Hikurangi Margin have not been made, so the Hikurangi Margin's place in this relationship is not immediately evident. Moreover, it will be argued that other features of the Hikurangi Margin make direct comparison with other systems difficult. The various data will be used to infer that, for the purpose of making this comparison, the Hikurangi Margin should be divided into two segments. In addition, the consequent evolution of the Hikurangi Margin, particularly the continuation of back-arc processes, will be considered.

At the Hikurangi Margin, standard subduction zone features include a well-defined zone of intermediate depth seismicity, arcs of andesite volcanoes, a fore-arc basin exhibiting negative gravity anomalies and low heatflow, and a growing accretionary prism. Behind the arc lies an area of high heatflow, active volcanism and extensional deformation. Unusual features include a trench (Hikurangi Trench) that is so shallow as to have been often termed a trough (e.g., VAN DER LINGEN, 1982), a transcurrent component of plate motion equal to the compressive component, and the existence in the north of an active back-arc basin onshore. Data that serve to characterise the Hikurangi Margin are: a comprehensive set of gravity observations, a useful quantity of heatflow data especially from the back-arc region, a catalogue of historical earthquakes that is largely complete for onshore events of $M \geq 7$ since about 1840 and $M \geq 5$ since 1940, geodetic and palaeomagnetic observations for comparison with strain rates predicted by plate tectonics, some geologically derived uplift data, comparatively well observed segmentation of the subducted plate, offshore and onshore seismic reflection data that reveal the structure of the accretionary prism and the subducted plate dipping under it, and an incomplete but useful collection of seismic velocity determinations, principally obtained from arrival times of regional earthquakes or similar data from a number of microseismic surveys, but including some refraction profiles.

However, a number of important measurements have not yet been made. There is very little data on return times or characteristic size of earthquakes; few seismic moments or mechanisms of historical events have been determined; and not only the age, but also the structure and likely buoyancy of the lithosphere east of the Hikurangi Trench, and hence that of the lithosphere recently subducted, arc poorly known.

Morphology of the Subducted Pacific Plate

The nationwide network of seismographs in New Zealand has enabled a detailed study of intermediate depth earthquakes associated with subduction of the Pacific plate at the Hikurangi Margin (e.g., ADAMS and WARE, 1977). However, the large station spacing of the network (typically 140 km) provides poor depth control for shallower earthquakes, and only in the last decade, with the advent of microearthquake surveying, has it been possible to study the shallow part of the dipping seismic zone in detail.

Microseismicity studies have been made of most of the onshore part of the Hikurangi Margin between latitudes 38.5°S and 42.5°S. From northeast to southwest: Hawke's Bay (latitudes 38.5°S to 39.5°S), BANNISTER (1986); southern Hawke's Bay (latitudes 39°S to 40°S), CHONG (1982); southern Hawke's Bay, including a 200 km SE-NW traverse, REYNERS (1980); Wairarapa (latitudes 40°S to 41°S), ARABASZ and LOWRY (1980) and KAYAL (1984); Wellington (latitudes 40.5°S to 42°S), ROBINSON (1986); Marlborough—northern South Island (latitudes 41°S to 42.5°S), ARABASZ and ROBINSON (1976). For locations, see Figure 2.

The subducted plate consists of three largely planar zones striking N45°E with different dips. Between the trench and the coast, the plate dips at 3–5° (DAVEY *et al.*, 1986; BANNISTER, 1986). From just onshore, the dip increases to 20–30°. The dip then changes to about 50° beneath the current volcanic front at a depth of about 70 km (REYNERS, 1980; ADAMS and WARE, 1977). Maximum depths of intermediate events shallow from more than 300 km at latitude 37°S to about 200 km at latitude 41°S (ANSELL and ADAMS, 1986). Directly beneath the deepest of the intermediate depth events at latitude 39°S there have occurred a few events

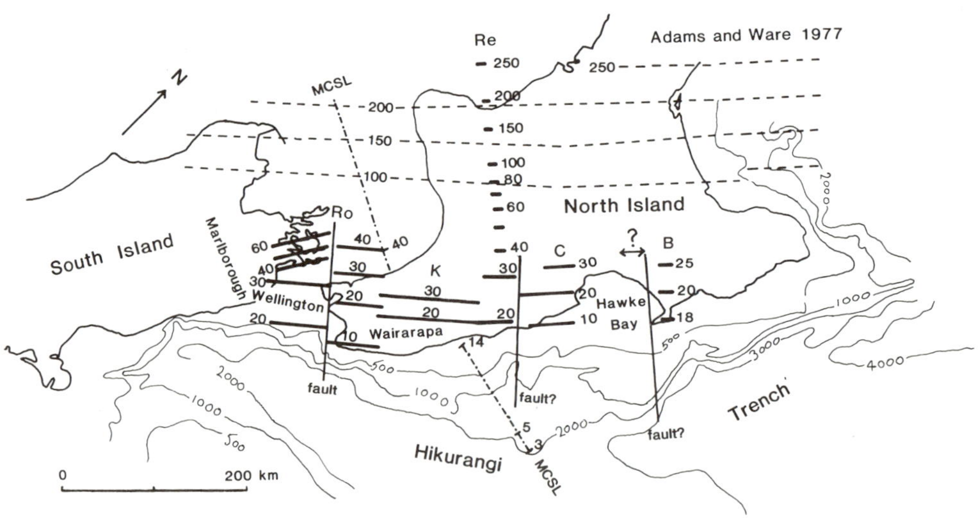

Figure 2
Morphology of the subducted Pacific plate at the Hikurangi Margin. Dashed lines show the median depths of intermediate depth seismicity (ADAMS and WARE, 1977). Bold lines are the plate interface as interpreted by: BANNISTER (1986)—B; CHONG (1982)—C; REYNERS (1980)—Re; KAYAL (1984)—K; ROBINSON (1986)—Ro. Dot-dash lines are multi-channel seismic lines (MCSL, DAVEY *et al.*, 1986 and DAVEY, 1987). Bathymetry in meters. For latitude and longitude grid see Figure 4.

at 600 km (e.g., ADAMS and FERRIS, 1976). These are possibly the remains of a previous subduction episode.

The subducted plate is also divided laterally by tears or faults that accommodate the varying deformation along the margin. The best defined tear in the plate occurs at the south end of the North Island (ROBINSON, 1986). Its strike is NW-SE (the dip direction of the subducted plate) with the southwest side downthrown an average of 7 km. Also, the strike of the deepest part of the southwestern portion of the plate is more northerly (see Figure 2). These distortions in the subducted plate show clear correlations with seismicity in the overlying Australian plate, with topography, and with gravity anomalies. Similar down-dip tears in the subducted plate have been suggested 200 km to the northeast of Wellington (REYNERS, 1983), and in northern Hawke Bay (BANNISTER, 1986).

Stress Regime

Seismic activity beneath Wellington defines two parallel zones separated by about 20 km, the lower zone being much less active and perhaps merging with the upper zone at a depth of 80 to 90 km (ROBINSON, 1986). The upper envelope of the upper zone is interpreted as the plate interface, because its position is similar to that of a strong reflector identified under Wellington (DAVEY and SMITH, 1983b), and focal mechanisms of earthquakes in the upper zone indicate intraplate stresses, rather than shallow-angle interplate thrusting. A similar double-zoned structure is seen farther north (latitude 40°S; REYNERS, 1980) and in the following discussion of the stress regime of the Pacific plate, events have been ascribed to the upper or lower zone, depending on whether they are less or more than 15 km below the plate interface.

The dominant feature of the fault-plane solutions of earthquakes in the upper zone is that the T-axis is oriented down the dip of the subducted plate. The P-axis direction is more variable, though there is a general tendency for it to move from being perpendicular to the plate interface (normal faulting) to being parallel to it (strike-slip faulting in the plane of the subducted plate) as the plate deepens. Earthquakes in the lower zone also consistently show the T-axis oriented approximately down the dip of the subducted plate, while the P-axis again shows greater variability.

Stresses deeper in the subducted plate have been investigated by HARRIS (1982a,b), who studied the mechanisms of intermediate depth events between latitudes 38°S and 40°S. Most of these events fit a mechanism which has the P- and the T-axis in the plane of the dipping seismic zone, the T-axis being rotated clockwise 30° from directly down-dip. Twenty percent of events have a down-dip T-axis and a P-axis perpendicular to the earthquake zone. The similarity of these mechanisms to those found in the lower and upper zones under Wellington has led

ROBINSON (1986) to suggest that the double seismic zone may extend to depths of 200 km or so, although ANSELL and SMITH (1975) have shown that the whole zone is only about 9 km thick at depths below 100 km.

In the overlying Australian plate, the stress regime is dominated by strike-slip faulting with subhorizontal compression in an E-W to SE-NW direction except that in the central North Island the stress regime is modified by active back-arc opening in the Central Volcanic Region. This compression tends to confirm the suggestion (WALCOTT, 1978) that the plate interface is currently locked, at least in its shallowest portion, in the southwest. The pervasive down-dip tension in the subducted plate is assumed to be due to slab pull. In contrast to other subduction zones, very few interplate thrusting events have been identified. In the northeast, however, a possible interplate thrust event ($M_L = 6$) has been observed (WEBB *et al.*, 1985).

Velocity Structure from Inversions

Arrival-time data from microearthquake surveys have been used by BANNISTER (1986), CHONG (1982) and ROBINSON (1986) to perform velocity inversions in northern and southern Hawke's Bay and Wellington, respectively (see Figure 2 for their survey areas). Chong and Bannister used *Sp* converted phases to constrain the positions of some of the interfaces between regions of different velocity, while Robinson inferred the positions of these interfaces from the seismicity distribution. The results of Bannister and Robinson are summarised in Figure 3.

At Wellington, after further subdividing the two uppermost dipping layers (Figure 3a), Robinson estimates the crust of the subducted plate to be about 15 km thick. Bannister's results are more complicated. In the south of his survey region, the model is similar to that of Chong and the central part of Robinson's model. The subducted plate crustal thickness is about 11–12 km (Figure 3b). North of about 39°S, however, the data require a velocity reversal at the plate interface: two possible interpretations are shown (Figures 3c, 3d). Bannister discusses the possibility that the 3 km thick, low-velocity layer in Figure 3d represents sediment subducted with the Pacific plate.

Historical Seismicity and Large Shallow Earthquakes

The historical period of New Zealand's seismic history is normally taken to begin about 1840 from which time European colonisation proceeded apace. The most recent catalogue of large historical events was published by SMITH and BERRYMAN (1983). Details of how magnitudes were assigned to the pre-1900 (pre-instrumental) shocks are given in SMITH and BERRYMAN (1986). ABE (1981),

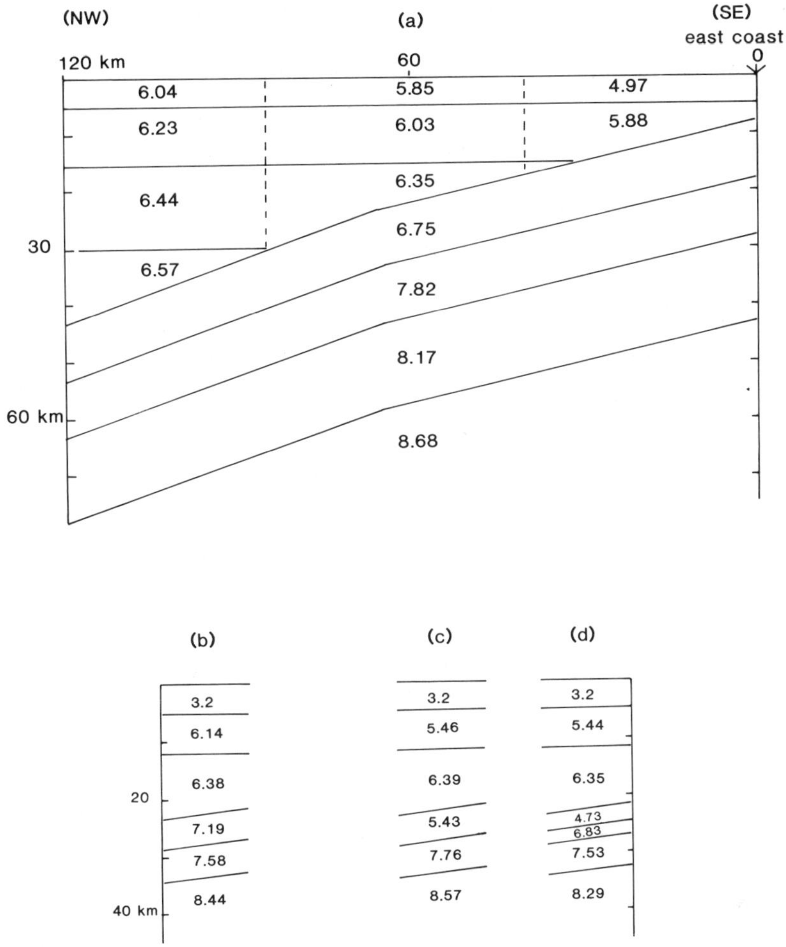

Figure 3
Arrival-time inversion models. (a) ROBINSON (1986)—Wellington; (b) BANNISTER (1986)—central Hawkes Bay (south of 39°S); (c) and (d) BANNISTER (1986)—alternative models for northern Hawkes Bay. Natural scale. P-wave velocities in km/s.

ABE and NOGUCHI (1983a,b) and ABE (1984) have published self-consistent magnitudes of large shallow earthquakes and thereby provide a catalogue of post-1900 shocks of $M_s \geq 7$ which appears to be complete for New Zealand.

Figure 4 shows the distribution of earthquakes of $M_s \geq 7$ since 1840 at the Hikurangi Margin. An attempt will be made to estimate the total coseismic slip per unit length of the plate boundary produced by these events. To do this, a number of assumptions must be made. There is little information available on the style and extent of faulting of these events. Few, if any, appear to be simple interplate thrusting events. Some primary faulting on-shore has been identified for about half

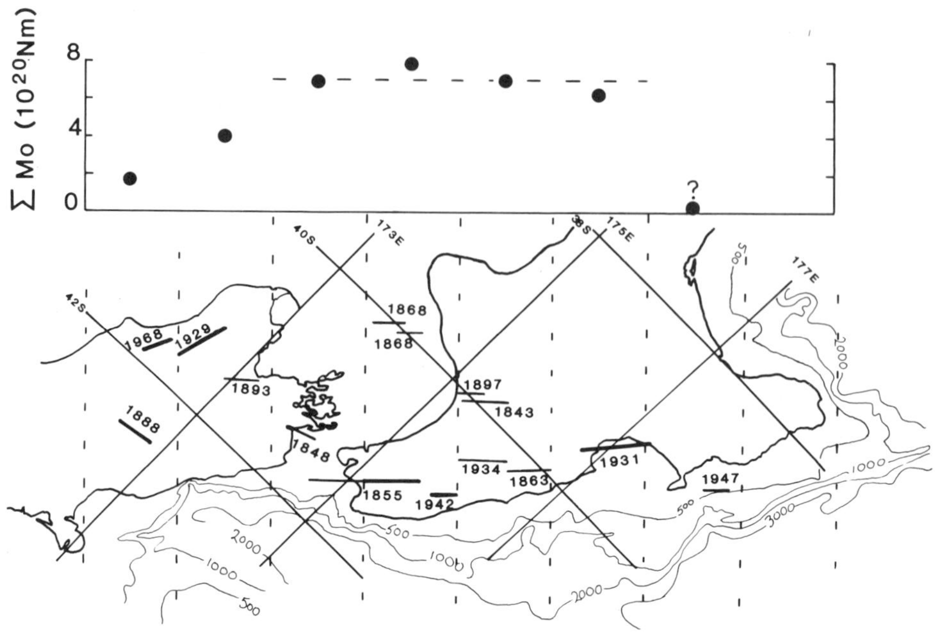

Figure 4
Historical seismicity $M_s \geq 7$. Bold lines indicate where faulting was observed, and thin lines indicate inferred faulting. Also shown is accumulated moment in 100 km wide strips perpendicular to the plate boundary. For data sources and treatment, see text.

the events (Figure 4), and this typically shows both thrust and strike-slip components. For example, the largest event (1855) apparently produced 12 m of dextral and 2.7 m of vertical displacement on a NE striking fault (STEVENS, 1974, p. 62). An area of at least 3000 km^2 and perhaps more than 6000 km^2 was uplifted, suggesting a significant component of low-angle thrusting.

Unpublished modeling of level changes produced by the 1929 and 1931 events (pers. comm. A. J. Haines, 1987) shows that these were dominantly thrusts with significant strike-slip components, on faults dipping at about 60°. The slip vector for the 1931 event is approximately parallel to the plate convergence direction, as is that for the 1888 event which produced a maximum of 2.4 m of dextral displacement and negligible uplift (FREUND, 1971). In contrast, the 1968 M_s 7.1 event was almost a pure thrust.

Mechanisms of other events are completely unknown. It is strongly suspected from its position and macro-seismic effects (BULLEN, 1938) that at least one, 1934, may have been a large normal faulting event within the subducted plate, but there is no direct evidence for this. Similarly, the events in the west (1843, 1868 twice, and 1897) may have been normal faulting events associated with the postulated down-buckling of the overlying plate (see below: "Crustal structure of the back-arc region").

It follows that the subsequent analysis is highly uncertain, but the principal conclusion that will be drawn, namely that there was markedly less coseismic strain release in the NE part of the margin, could be obtained qualitatively by inspection of Figure 4. To quantify this difference, magnitudes have been converted into seismic moments and the moments summed. In the absence of direct evidence that any particular event should be excluded, all historical shallow events of $M_s \geq 7$ will be included in the summation. Published M_s values have been used where available, otherwise SMITH and BERRYMAN'S (1983) values have been used. M_s has been converted into moment using

$$\log M_0 = 1.5M_s + 9.1, \quad M_0 \text{ in Newton-metre (Nm)} \tag{1}$$

(after HANKS and KANAMORI, 1979). We also calculate pseudo-fault lengths from

$$\log M_0 = 1.5 \log S + \log(16\,\Delta\sigma/7\pi^{3/2}), \tag{2}$$

(KANAMORI and ANDERSON, 1975) where S is the fault area (m^2) and $\Delta\sigma$ is the static stress drop (N/m^2). This is valid for circular faults and a good approximation for elliptical faults with moderate length to width ratios (ESHELBY, 1957). The circular fault approximation is assumed for events of $M_s \leq 7.8$. For the one event, 1855, that was larger, the whole brittle-elastic crust must have been broken and a constant width model is assumed. We also take $\Delta\sigma = 6 \times 10^6$ N/m^2 (60 Bar), appropriate for an "average" event (KANAMORI and ANDERSON, 1975). Equations (1) and (2) can be combined to give fault area, and hence a representative fault length L, in terms of M_s:

$$L = 2[10^{M_s - 4.2}/\pi]^{1/2} \text{ km}, \quad \text{for } M_s \leq 7.8,$$

and

$$L = 2[10^{M_s - 4.2}/35\pi] \text{ km}, \quad M_s > 7.8. \tag{3}$$

In Figure 4, bold lines indicate where primary faulting was observed. Where none was observed, the pseudo-faults have been arbitrarily oriented NE-SW. The upper part of Figure 4 shows the summed scalar moment in 100 km wide strips. Note that some events have been deliberately and arbitrarily placed to make the resulting graph of accumulated moment appear uniform. Events repositioned to achieve this are 1855 (for the unobserved portion of faulting), 1843, 1863, 1897 and 1934. In no case is this repositioning inconsistent with previously published positions for the events (e.g., SMITH and BERRYMAN, 1983).

The graph of accumulated moments shows that the central 400 km of the margin could have experienced a very uniform rate of coseismic strain dissipation in the last 150 years. Taking the crustal thickness to be 35 km and $\mu = 4 \times 10^{10}$ N/m^2 for the whole crust, the average of 7×10^{20} Nm per 100 km in this region is equivalent to a slip of 5 m. This figure is exaggerated because no allowance has been made for vertical deformation which, as mentioned above, was significant or dominant in the largest three events (1855, 1929, 1931). On the other hand, on the assumption of a b-value of 1.0, events of $M_s < 7$ would have contributed about 20% to the total slip, and this will tend to offset the former effect.

From geodetic data spanning almost 100 years, BIBBY (1981) calculated a slip rate of 53 ± 9 mm/a (azimuth $83° \pm 10°$) for the northeastern end of the South Island, i.e., the southwestern end of the region that has experienced 7×10^{20} Nm accumulated moment per 100 km. Also, the Australian/Pacific pole of MINSTER and JORDAN (1978) implies a convergence rate of 47–57 mm/a (azimuth 90°) at the Hikurangi Margin (Figure 1). The seismicity thus accounts for about 100 years worth of accumulated strain.

At the ends of the central region, strain dissipation was less. At the southwestern end there appears to be a steady reduction in seismic strain dissipation, marking the transition from the region where there is subduction to the central South Island region where continental rocks collide at the plate boundary (Alpine Fault). On this portion of the boundary, no historical earthquake has reached $M_s = 7$, but there is abundant geological evidence of movements of several metres on the Alpine Fault in the last few thousand years (e.g., BERRYMAN, 1979; HULL and BERRYMAN, 1986).

At the northeastern end there is an abrupt diminution of seismic strain release. Only one event of $M_s \geq 7$ has been observed in historical times, although SMITH and BERRYMAN (1983) judge an event in this region, on 1914 October 6, to have been larger than the $M_s = 6\frac{1}{2}$ calculated by RICHTER (1958). It is possible that significant pre-instrumental shocks could have occurred off-shore in this region and produced sufficiently low felt intensities to be considered only moderately sized events. Alternatively, like the central South Island, the region may have suffered no major shocks at all since 1840.

There are two other pieces of evidence that support the idea that the region northeast of the 1931 event might be different in its seismic behaviour from the adjacent region to the southwest. First, while the rest of the Hikurangi Margin shows a large number of recently active faults, there are almost none in the northeastern portion (Officers of the N.Z. Geological Survey, 1983). This supports the idea that here the seismic deformation occurs principally offshore. Second, the relative seismic quiescence of the northeastern region and back-arc processes to the west may be causally related. Both terminate at about latitude 39°S. This relationship will be further pursued later.

Earth Deformation

Repeated triangulations cover a substantial part of the Hikurangi Margin and a number of workers have inferred horizontal strain rates from these data. In a recent study, WALCOTT (1984) discusses the kinematics of the plate boundary zone and includes a comprehensive bibliography. His interpretation of the data is shown in Figure 5. WALCOTT (1978) has shown that the pattern of horizontal strain within the Hikurangi Margin has shown changes following large earthquakes. Since about 1920, the southwestern portion of the region, including the source region of the 1855 earthquake (Figure 4) has been subjected to compression in a WNW-ESE direction with a maximum shear strain rate of about 0.45×10^{-6} radian/a. In the northeast, however, the principal axis of compressive horizontal strain is oriented NE-SW, with a maximum shear strain rate of about 0.34×10^{-6} radian/a. Walcott has further shown that in the region northeast of the 1931 shock, the principal axis of horizontal compression was oriented roughly NW-SE before 1931 and adopted its current orientation following the 1931 event. He has concluded that following the large earthquakes of the 1930s, the northeastern portion of the plate interface has been essentially unlocked and therefore has dissipated the relative plate motion both seismically (through small-to-moderate earthquakes), and aseismically, while the southwestern portion has been locked and accumulating strain.

Recent vertical deformation determined from geological observations is also shown in Figure 5. At the southwestern end of the Hikurangi Margin, uplift rates of 5 to 7 mm/a at the axes of the Kaikoura Ranges are exceeded in New Zealand only by the 10 mm/a rate in the Southern Alps (WELLMAN, 1979). The high uplift rate of the Kaikoura Ranges supports the idea that there is considerable resistance to subduction at the southwestern end of the margin.

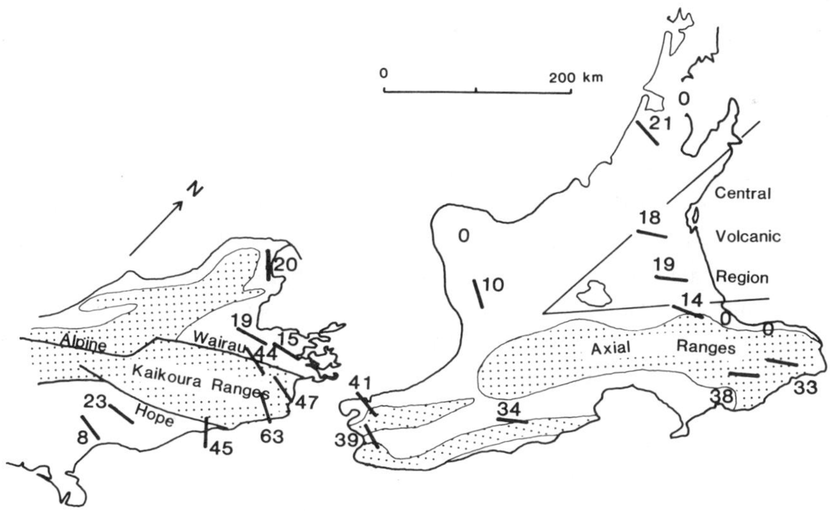

Figure 5

Deformation. Bars indicate direction of principal axis of horizontal compression, accompanying numbers are maximum shear strain rates (10^{-8} radian/a) (from WALCOTT, 1984). Shading indicates regions with uplift of more than 1 mm/a (WALCOTT, 1987).

In the North Island, a maximum uplift rate of 4 mm/a has been found at the southern end (WELLMAN, 1972; GHANI, 1978), and PILLANS (1986) found rates of more than 3 mm/a in the axial ranges in the northeast, where the geodetically observed relative extension is perpendicular to the axis of the ranges. WALCOTT (1987) shows that it is feasible that the Australian plate is underplated with a proportion of the large volume of sediments in the Hikurangi Trench (next section) and that the uplift could consequently result from isostatic adjustment. The subduction of large quantities of sediment in the northeast is supported by BANNISTER'S (1986) velocity model (Figure 3d). An alternative explanation for the uplift invoking plutonism would appear to be ruled out by the low average heatflow of about 40 mW/m^2 in the forearc basin and the axial ranges (PANDEY, 1981).

The Hikurangi Trench and Accretionary Prism

Marine seismic and other data have provided information about the structure and development of the off-shore part of the Hikurangi Margin. In particular, the southwestern portion (south of latitude 39°S) has been well surveyed and a recent multi-channel reflection profile at about latitude 41°S has shown sediment-covered Pacific crust dipping northwest under the accretionary prism between the trench and the coast, where the Pacific crust is about 14 km deep (DAVEY *et al.*, 1986). This profile and other data (e.g., COLE and LEWIS, 1981; LEWIS and BENNETT, 1985) reveal that the prism contains numerous southwest striking thrusts separating partially sedimented basins, and antithetic normal faults. The sedimentation rate has probably increased throughout the Cenozoic (WALCOTT, 1987) and the thickness of sediments lying in the Hikurangi Trench is more than 3.5 km at latitude 40°S (LEWIS and BENNETT, 1985, their Figure 13). The sediment thickness further north is not known but is probably not greater than 1.5 km.

The width of the prism varies from about 70 km at the southwestern end to a maximum width of 160 km at latitude 41°S (LEWIS and BENNETT, 1985). Thereafter, the prism narrows to about 70 km north of 39°S. An increase in maximum water depth parallels the decrease in sediment thickness: from 2800 m at 42°S (CARTER, 1980) to about 3500 m north of 39°S. From isostasy, each kilometre of extra sediment (density 2.2 Mgm/m^3) can be expected to decrease the water depth by about half a kilometre, so the increase in water depth is roughly in accord with the decrease in sediment thickness. The changing width of the accretionary prism is equivalent to the strike of the trench slowly changing from an average of about N60°E south of 41°S to N20°E north of 39°S. Note that a much smaller change in the strike of the Benioff zone below 100 km is observed between latitudes 37°S and 42°S (ADAMS and WARE, 1977; isobaths Figure 2). The variation in the strike of the trench has been attributed to variations in deformation styles and rates in different segments of the subducted plate (KATZ, 1982; REYNERS, 1983).

Little is known of the lithospheric structure east of the Hikurangi Trench; in particular, crustal thickness and age are not known. The seismic *Pn* and *Sn* velocities have been estimated using earthquake arrival times by HAINES (1979) for the region between the east coast of the North Island north of 39°S and the Chatham Islands (Figure 1), i.e., over a range of more than 700 km. The results, 8.0 ± 0.1 km/s (*Pn*) and 4.55 ± 0.1 km/s (*Sn*) are similar to the values Haines found for the central South Island and the *Pn* velocity is similar to that observed (unpublished data) for the region south of the Chatham Rise (Figure 1). Haines and others (e.g., KAYAL and SMITH, 1984; ROBINSON, 1986) have found upper mantle *Pn* velocities of 8.5–8.7 km/s for ray paths along the east coast of the North Island, which have usually been interpreted as evidence for the presence of "standard" oceanic upper mantle under the east coast (KOSMINSKAYA and KAPUSTIAN, 1976; HAINES, 1979; ROBINSON, 1986). There are three possible ways to reconcile the difference in seismic velocity between the east coast North Island and the area east of the Hikurangi Trench:

(i) the two regions have essentially different lithosphere, so that the lithosphere now arriving at the trench is different from that which immediately preceded it;
(ii) the difference is due to anisotropy;
(iii) phase changes at shallow (< 50 km) depth have occurred and produced the high-velocity mantle layer.

It is difficult to reconcile suggestions (i) and (ii) with plate reconstruction models. Prior to its subduction, the material currently exhibiting intermediate depth seismicity under the North Island was immediately west of the lithosphere currently at the Hikurangi Trench (ANSELL and ADAMS, 1986). Plate reconstructions for the Southwest Pacific show isochrons generally trending northeast (e.g., MOLNAR and ATWATER, 1978; SCHROEDER, 1984) so that although the age and evolution of the Pacific plate between the Louisville Ridge and Chatham Rise are poorly known, it is likely that the pre- and post-subduction lithosphere had the same genesis.

Similarly, if the region were to have spread from a northeast striking ridge, one might expect higher *P*-wave velocities in a NW-SE direction than NE-SW (e.g., CLOWES and AU, 1982), which is the opposite of what has been observed.

The other source of information about the structure of the subducting lithosphere comes from water depths. Between the Hikurangi Trench south of 40°S and the Chatham Rise, water depth is less than 3000 m, and for the adjacent region north of 40°S it is about 3500 m. By way of comparison, the water depth is 5500 m on the Pacific side of the Kermadec Trench. To account for the difference in water depth by sedimentation would require > 4000 m of sediments, implying a sediment volume eight times that of supermarine New Zealand (REILLY, 1965). It is likely, therefore, that some of the shallowness of the water is due to a crust that is thicker, even without sediment, than normal oceanic crust (6–7 km). A 1 km thickening of ocean crust can be expected to produce about $\frac{1}{4}$ km shallowing of the water depth. Thus the water depth east of the Hikurangi Trench implies a crustal thickness of, for example, 14–17 km with no sediment or, with 2 km of sediment, 10–13 km exclusive of the sediment.

These values are similar to ROBINSON's (1986) and BANNISTER's (1986) estimates of 15 km and 11–12 km for the thickness of the crust of the subducting plate at latitudes 41°S and 39°S respectively, and it must be concluded that not only is thicker than normal oceanic crust adjacent to the trench but from Robinson's results it follows that such crust has been subducting for at least the last 2–3 Ma.

Back-arc Structures Associated with the Hikurangi Margin

Behind the northeast section of the Hikurangi Margin (north of 39°S) there is a linear calc-alkaline volcanic arc, and behind the arc there is a wedge-shaped basin of predominantly Quaternary rhyolitic volcanism—the Central Volcanic Region (CVR—Figure 6). Its western boundary largely coincides with the apparent position of a 4 Ma old arc of andesitic volcanoes (CALHAEM, 1973) and its eastern boundary coincides with the line of active or recently active andesitic to dacitic volcanoes—the Active Volcanic Front of ADAMS and HATHERTON (1973). Sharing this boundary with the CVR is the Taupo Volcanic Zone which makes up the eastern half of the CVR and contains most of the rhyolitic volcanism active within the last 1 Ma, and nearly all the economically important geothermal fields. However, various geophysical studies of crust and upper mantle structure indicate that the CVR is the principal volcano-tectonic structure in the central North Island (STERN, 1985; STERN and DAVEY, 1987).

KARIG (1970) described the CVR as an apparent extension of the Havre Trough, the oceanic back-arc basin of the Kermadec system, into New Zealand. Geodetic data show that the CVR is currently widening at about 12 mm/a at the coast (WALCOTT, 1987). After a presentation of crustal structure and heatflow data, STERN (1985) argued that the CVR is an active back-arc basin that has formed within continental lithosphere. By virtue of the CVR being above sea level, and containing economically important geothermal resources that are being extensively exploited, it is one of the better studied back-arc basins of the world.

Although subducted plate is present as far south as 42°S, no volcanic arc or extensional volcanic graben exists behind the fore-arc south of 39°S. Instead, there is a broad basin contining 4–5 km of shallow marine Plio-Pleistocene sediments (Figure 6). This is the South Wanganui Basin described by HUNT (1980) and ANDERTON (1981). As will be discussed, the development of this basin in both space and time appears to be controlled by the underlying subducted Pacific plate. Thus the basin qualifies as a back-arc basin by the definition of TAYLOR and KARNER (1983)—"a marginal basin located behind an active or inactive trench system and whose origin is inferred to be subduction related."

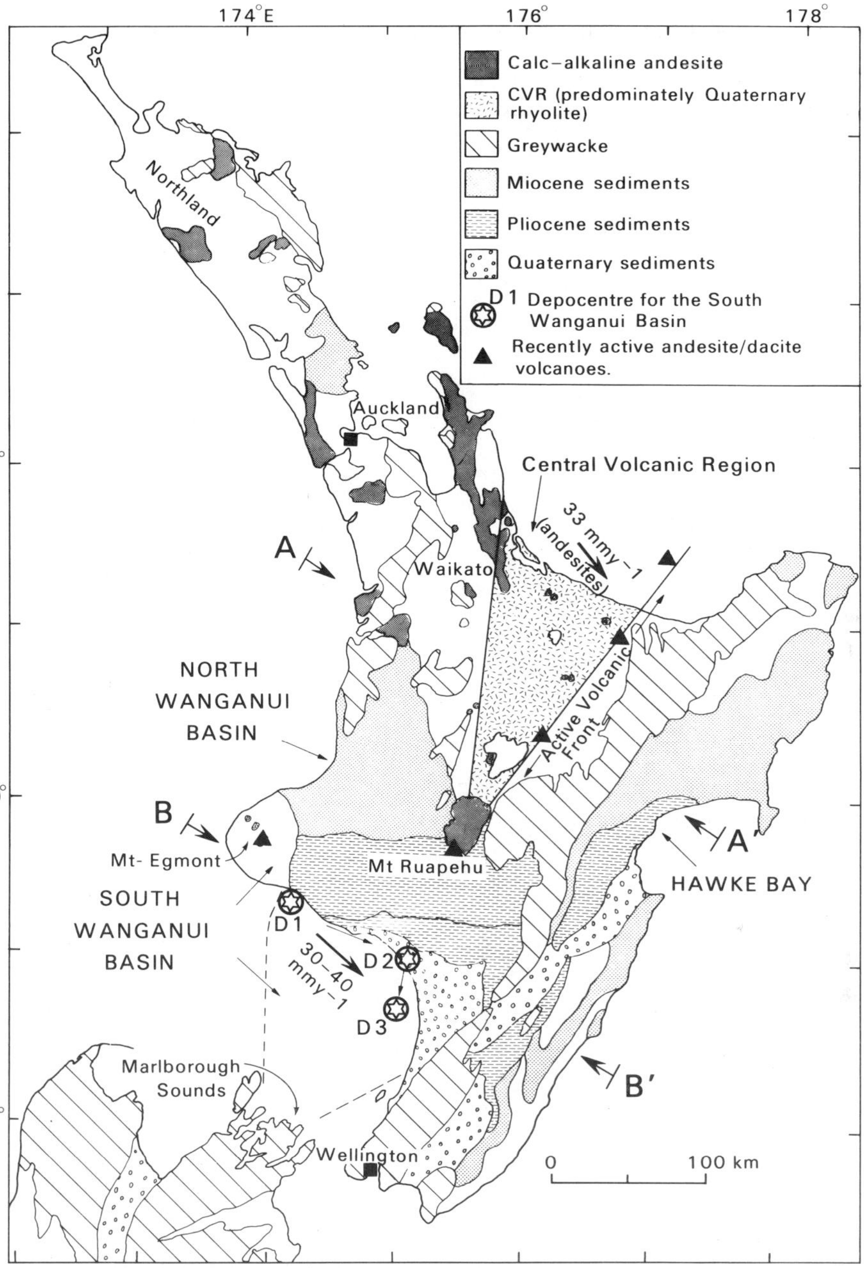

Figure 6

Map of the North Island showing the distribution of andesites, Quaternary rhyolite and principal sedimentary basins. Also shown are three depocentres for the South Wanganui Basin (after ANDERTON, 1981) with the time spans for each centre being: D1 = 5–3.5 Ma B.P. D2 = 3.5–1 Ma B.P. and D3 = 1 Ma–present. The resultant vector represents depocentre migration from 5 Ma B.P. to the present. The vector representing low-potash andesite migration is after CALHAEM (1973) and covers the period 4 Ma to the present.

Volcanic Basins

Figure 6 shows the distribution of Cenozoic volcanism and major Plio-Pleistocene sedimentary basins of the North Island. Within Northland outcrops of calc-alkaline andesite and basaltic volcanism are found interspersed among older volcanics of an obducted ophiolite sequence (BROTHERS and DELALOYE, 1982) and Permian-Triassic basement greywacke. A number of workers have noted an apparent migration of the low-potash andesites southeastwards through the Northland-Auckland peninsula with time (e.g., KEAR, 1959; HATHERTON, 1969). With the availability of K-Ar dates of these andesites, various plate reconstructions have been developed that either take the plate boundary during the Tertiary as being adjacent and parallel to Northland (e.g., CALHAEM, 1973; COLE and LEWIS, 1981; WALCOTT, 1987), or that a substantially shallower dipping Pacific plate extended beneath Northland from the current plate boundary (BROTHERS and DELALOYE, 1982).

Regardless of what the exact orientation for the plate boundary was, two points should be noted. First, the Northland-Waikato area has been subjected to up to 18 My of volcanism and, presumably, thermal weakening of the lithosphere, and second, there has been a rapid migration of the andesite arc across the CVR compared to that within Northland—CALHAEM (1973) estimates the present-day rate of movement at the Bay of Plenty coast to be about 33 mm/a.

Drilling for exploration and exploitation of geothermal power within the CVR over the past 20 years has provided a great deal of stratigraphic data. About 10 drill holes have gone deeper than 2 km, and these show predominantly rhyolitic material in the form of ignimbrite sheets within the upper 1500 m or so, and then an increasing amount of dense volcanic rocks, particularly andesite, below 1500 metres. A west-east cross-section of drill hole stratigraphy across the CVR is given by STERN (1986). Mesozoic greywacke has been encountered at the base of some holes that are within 6 km of the eastern boundary of the CVR but no sedimentary basement rocks have been encountered in drill holes elsewhere within the CVR. The deepest hole to date terminated in a quartz diorite at a depth of about 2.8 km (WOOD, 1986).

Sedimentary Basins

Southwest of the CVR there is the Miocene-Quaternary Wanganui sedimentary basin (Figure 6). The Wanganui basin is divided stratigraphically into northern and southern sections. Data from numerous multichannel seismic reflection surveys in the South Wanganui Basin have been collated, analysed and summarized by ANDERTON (1981). His analysis shows the basin has developed by progressive subsidence and on-lap to the south while emerging and off-lapping to the north. The coast to the south of Mt Ruapehu (Figure 6) is currently being uplifted (PILLANS,

1983). Further south, the Marlborough Sounds (Figure 6) have recently subsided. The maximum sediment thickness within the South Wanganui basin is 4–5 km (ANDERTON, 1981; HUNT, 1980).

Anderton also mapped three depocentres, on the basis of characteristic seismic horizons, showing a movement of the depocentres over the last 5 My that parallels the motion of the Active Volcanic Front (Figure 6). This association of depocentre migration with the low-potash andesite axis suggests that the development of the Wanganui Basin is intimately related to movement of the underlying subducted plate.

Heatflow

Figure 7 displays a comparison of heatflow (PANDEY, 1981) and *Pn* velocity (HAINES, 1979) provinces for the North Island. A heatflow profile across *A*-*A*′ (of Figure 6) for the North Island would show the classic subduction zone pattern of low heatflow between the trench and the volcanic arc, then high heatflow over and

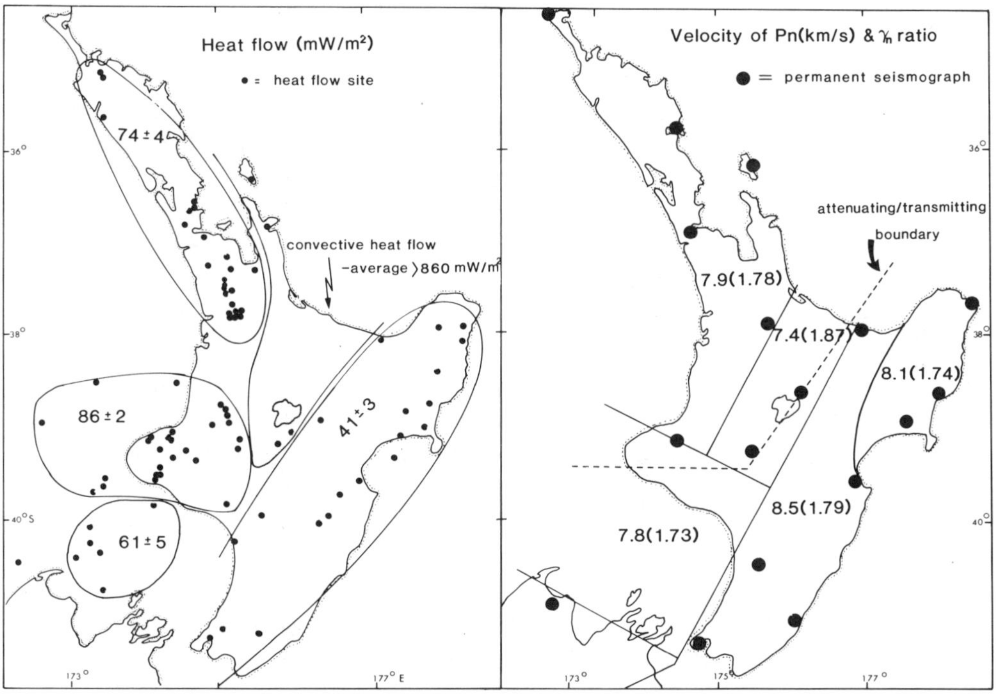

Figure 7
A comparison of heatflow (after PANDEY, 1981) and *Pn*-velocity (after HAINES, 1979) provinces of the North Island. γn refers to ratio of *Pn* to *Sn* seismic velocity. The attenuating/transmitting boundary is explained in the text (after MOONEY, 1970).

behind the arc (WATANABE *et al.*, 1977; ZIAGOS *et al.*, 1985). However, a distinguishing feature of the North Island profile is the remarkably high heat-flow in the CVR.

Heat output from the CVR is almost totally expressed in the discharge of hot water and steam from hydrothermal systems (STUDT and THOMPSON, 1969; DAWSON and DICKINSON, 1970). Adding 400 MW of natural heat output from the recently discovered Mokai field (BIBBY *et al.*, 1984) to the natural heat output of 3.9×10^9 watts for the other hydrothermal systems (CALHAEM, 1973), and taking an area of convection for the hydrothermal systems of 5000 km^2 (STERN, 1985), a value of 860 mW/m^2 for the average heatflow is obtained. This is necessarily a minimum value as it is possible that other "hidden" geothermal fields, similar to Mokai, exist.

An average heatflow value of 860 mW/m^2 is, as far as we are aware, one of the highest reported for a back-arc basin. This may be because most back-arc basins are oceanic and therefore covered by a shallow ocean that effectively masks the true heat output, thus resulting in an underestimate of the heatflow (SCLATER *et al.*, 1980).

Another feature of the CVR is the asymmetry in the distribution of both ages of low-potash andesites and centres of heat discharge; this asymmetry is demonstrated in Figure 8 which shows the apparent positions of the Active Volcanic Front at 4 Ma, 2 Ma and currently. Also shown is the distribution of geothermal fields on the eastern side of the CVR and the proposed downflow region as roughly outlined by heatflow sites that show a zero geothermal gradient (STUDT and THOMPSON, 1969; THOMPSON, 1977; STERN, 1985). Other geophysical properties of the CVR such as micro- and macroseismicity (EVISON *et al.*, 1976; SMITH and WEBB, 1986) and resistivity (HATHERTON *et al.*, 1966), also display an asymmetry. This inherent asymmetry suggests that the back-arc opening is not taking place symmetrically about a central spreading centre, which is the commonly adopted model for oceanic back-arc basins (WEISSEL, 1981). Moreover, the complexity is further demonstrated by the large strike-slip component in the focal mechanisms of small earthquakes in the crust of the CVR (SMITH and WEBB, 1986), and a delay-time and attenuation profile at the southern end of the CVR by ROBINSON *et al.* (1981) that found no extensive volume of partial melt in the mantle.

Pn Velocities and Attenuation

A comparison between the various heatflow and *Pn*-velocity provinces (Figure 7) demonstrates a broad similarity in province boundaries and an apparent inverse relationship between heatflow and *Pn* velocity. Within continental United States, BLACK and BRAILLE (1982) note such a relationship between heatflow and *Pn* velocity and therefore suggest some dependance of *Pn* velocity in temperature.

Also shown on Figure 7 is the attenuating/transmitting boundary found by

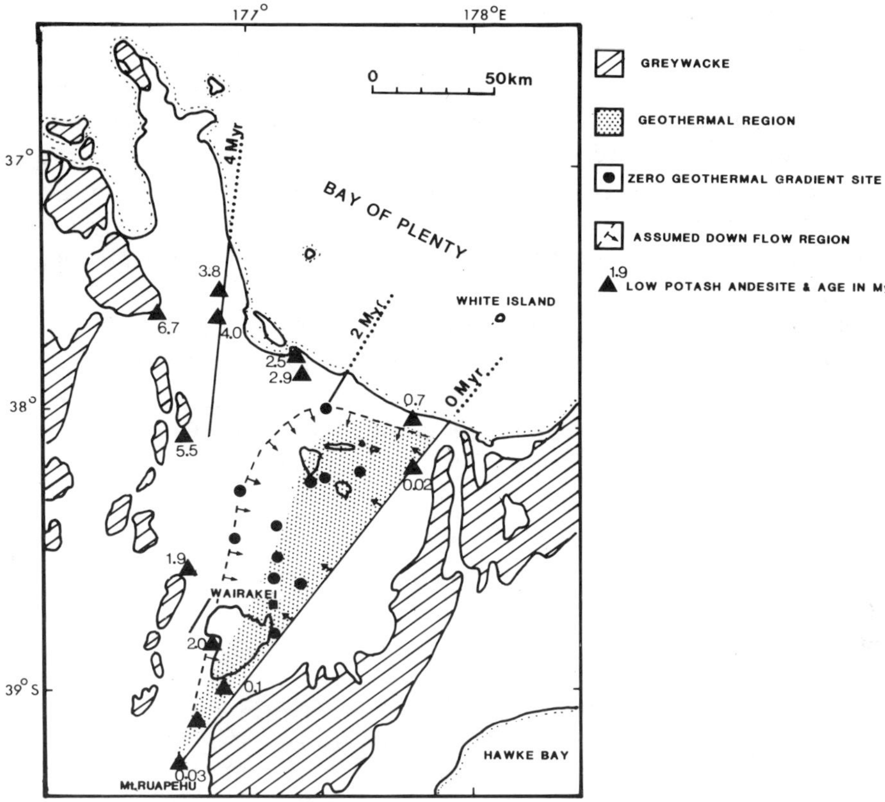

Figure 8
Map showing the overall asymmetry in the distribution of low-potash andesite ages and geothermal areas (modified from Stern, 1985). K–Ar ages of andesites adjacent to and within the Central Volcanic Region are from Calhaem (1973). Apparent positions of the Active Volcanic Front at 4, 2 and 0 Ma B.P. are shown.

Mooney (1970) and supported by Haines (1981). At this boundary there are major changes in the geophysical properties here considered—volcanism, heatflow, seismic velocity—and also in the gravity anomaly field (Hatherton, 1970). Mooney found that to the north and west of the boundary, earthquake waves that had traversed the upper mantle had their high frequency content attenuated, while south and east of the boundary attenuation was much less. Thus the upper mantle in the northwest North Island has a classic back-arc low-Q zone (Oliver and Isacks, 1967), while the southwestern part does not.

Crustal Structure of the Back-arc Region

The state of knowledge of crustal structure within the North Island is summarised in Stern *et al.* (1986). More recently the two multichannel seismic reflection profiles shown in Figure 2 have significantly extended our understanding of the crustal structure in these areas.

Figure 9 is a cartoon profile across *A-A′* of Figure 6 that summarizes this knowledge. In the CVR, the crust is interpreted to be only 15 km thick and the *Pn* velocity is correspondingly low (7.4–7.5 km/s). West of the CVR the crust is about 25 km thick, also thinner than the continental norm (35 km). Upper mantle *P*-wave velocities are 7.6 km/s at a depth of 25 km increasing steadily to 7.9 km/s at a depth of about 45 km (Stern *et al.*, 1987). Haines (1979) also found a *Pn* velocity of 7.9 km/s in this region.

Two gravity interpretation profiles along *A-A′* and *B-B′* of Figure 6 are shown in Figures 10 and 11. The areas of thin crust and low upper mantle seismic velocities (and hence densities) of the central and western North Island in profile *A-A′* are constrained by the seismic refraction data already discussed. The regions with thin

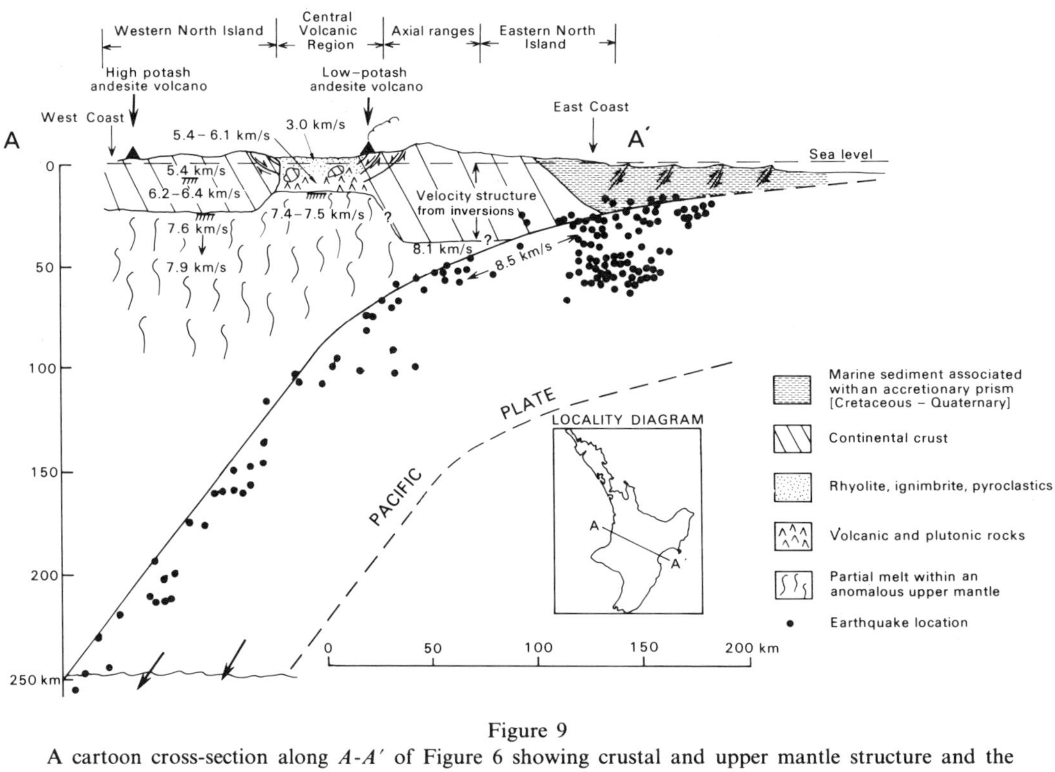

Figure 9

A cartoon cross-section along *A-A′* of Figure 6 showing crustal and upper mantle structure and the inferred position of the subducted Pacific plate. Crustal and upper mantle structure for the CVR and the western North Island are from Stern and Davey (1987) and Stern *et al.* (1987). Velocity structure beneath the eastern North Island is given by the velocity inversions shown in Figure 3 with the 8.1 km/s *Pn* velocity coming from Haines (1979). Earthquake hypocentres and position of the subducted plate are from Reyners (1980). Note that Reyners profile is 50 km SW of *A-A′*.

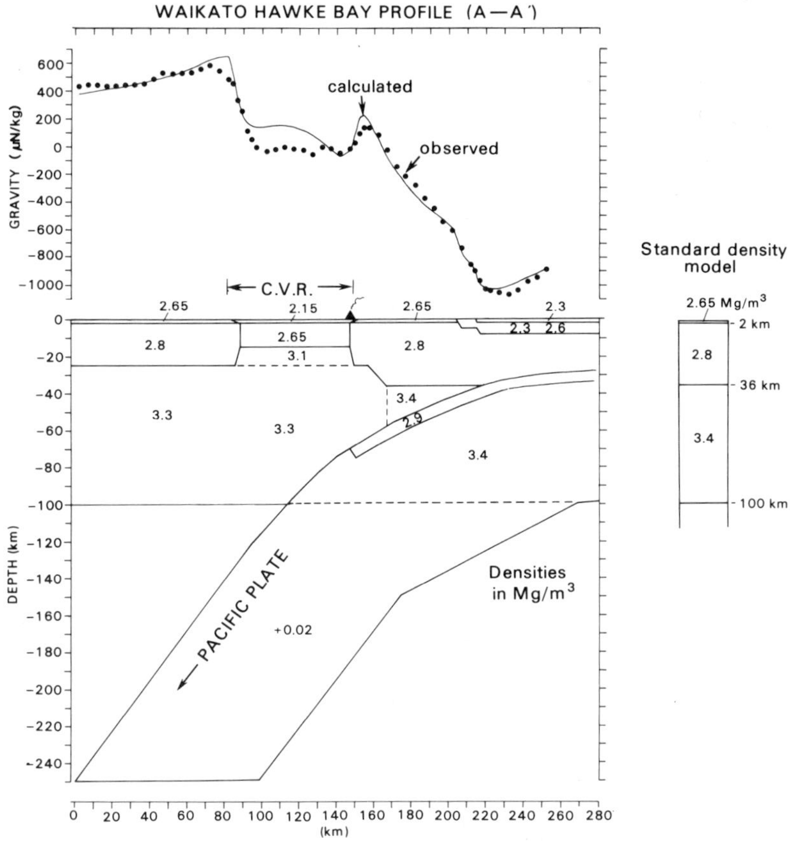

Figure 10
An interpretation of Bouguer gravity anomalies along profile A-A' of Figure 6. Density contrasts are calculated with respect to the standard density model. The portion of the subducted plate in the depth range of 100–250 km is assumed to have a constant density contrast of 0.02 Mg/m^3. 10 $\mu N/kg = 1$ mgal.

crust are modelled as being locally isostatically compensated (compensation depth of 100 km) with respect to the standard density model shown in Figure 10.

Profile B-B' (after STERN *et al.*, 1986) passes through the centre of the South Wanganui Basin where seismic refraction control on crustal structure is less. However, deep multichannel seismic reflection data shows a downwarping of the Moho to about 40 km beneath the basin that appears to roughly parallel the basement-sediment boundary (DAVEY, 1987). Thus a tectonic interpretation of the gravity model in Figure 11 is that there has been a downward flexing of the lithosphere caused by either a mechanical or a hydrodynamical coupling between the Australian and Pacific plates.

The difficulties and ambiguities of using gravity data alone to model subducted plates are well known (WATTS and TALWANI, 1974, 1975; CHAPMAN and TALWANI,

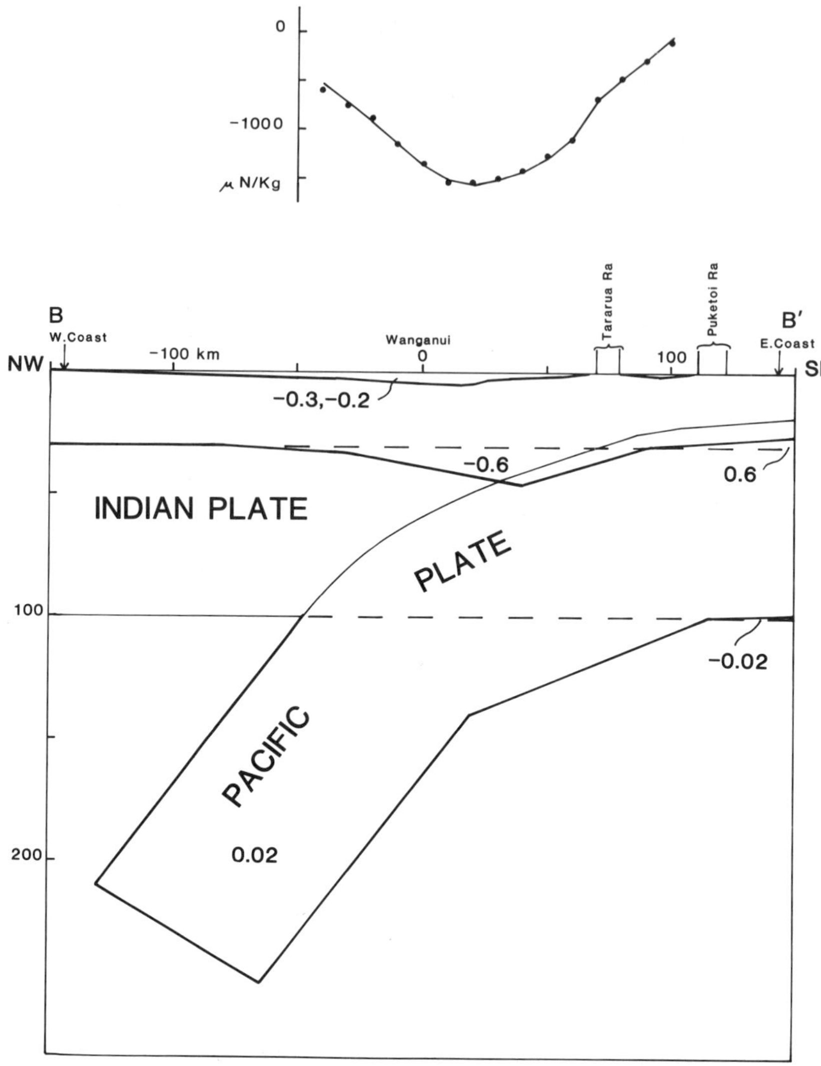

Figure 11
Interpretation of Bouguer anomalies along profile *B-B′* of Figure 6. The same basic model for the subducted Pacific plate is used as in Figure 10. Density contrasts in Mg/m^3 are with respect to the standard density model shown in Figure 10.

1982). In particular, the depth at which oceanic crust melts and changes phase to form denser eclogite is poorly known (RINGWOOD, 1977; ANDERSON *et al.*, 1976; IDA, 1983). Nevertheless, profiles *A-A′* and *B-B′* show that grossly different crust and upper mantle structures occur above adjacent portions of the same subducted plate, and these structures contribute more to the gravity anomaly field than do mass anomalies associated with the subducted plate itself.

Discussion

We shall now attempt to synthesize the results so far described to give a broad description of the Hikurangi Margin and attempt a comparison with other subduction systems after the manner of RUFF and KANAMORI (1980), who neatly characterised subduction zones by their plate convergence rates, subducting plate age, maximum magnitude of shallow earthquakes, and whether or not back-arc basins exist (see Figure 12).

The part of the margin northeast of 39°S has some of the characteristics of a standard subduction zone with active back-arc opening. The wedge-shaped CVR appears to indicate the extent of the opening (Figure 8) in the last 4 Ma. The plate convergence rate is about 55 mm/a which is less than that of a number of subduction zones with back-arc basins (Figure 12) and the largest known earthquake in this region has an M_s of only 7. The age of the subducting lithosphere is not known. Magnetic anomalies at the southern end of the Louisville Ridge indicate

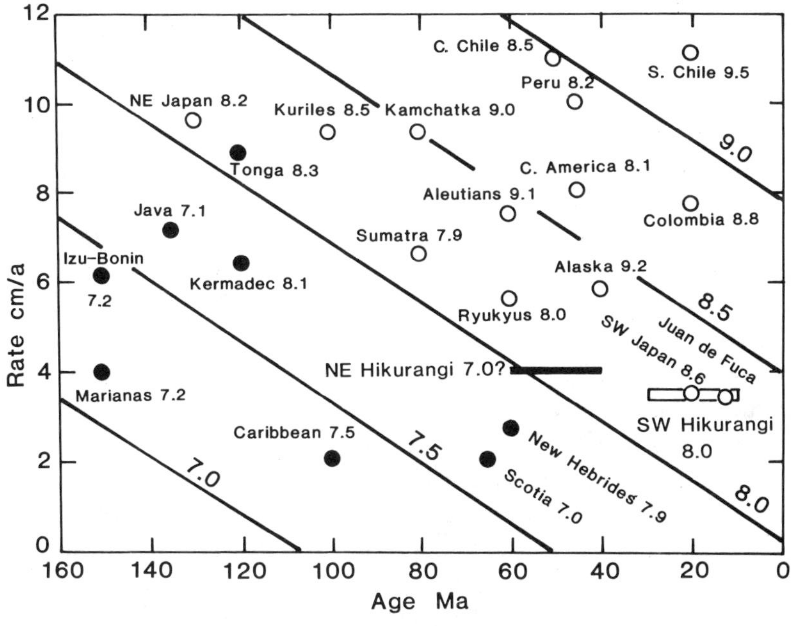

Figure 12

Relationship between age, convergence rate and magnitude at subduction zones, showing possible positions of the NE and SW segments of the Hikurangi Margin. Data other than for Hikurangi Margin, and interpretation reproduced from RUFF and KANAMORI (1980). Solid and open symbols denote subduction zones with and without back-arc basins. Bars indicate range of ages, from SCLATER *et al.* (1980), that would correspond to a water depth east of the Hikurangi Trench corrected for an assumed 3 km of sediment. The rate is taken to be the component of the plate convergence rate perpendicular to the margin.

that the age there is about 72 Ma (CAZENAVE and DOMINH, 1984) and MOLNAR and ATWATER (1978) have estimated the age of Pacific lithosphere adjacent to the Kermadec Trench to be 100 ± 20 Ma. An age in the range of 60–100 Ma would be consistent with the water depth there of 5500 m. However, a water depth of 3500 m, as it is in the northeast part of the Hikurangi Trench, over "normal" oceanic lithosphere would imply an age of only about 10 Ma (SCLATER *et al.*, 1980, or SCHROEDER, 1984, for a treatment of the Pacific Ocean). Since age is used by Ruff and Kanamori as a measurable substitute for negative buoyancy, and negative buoyancy and water depth are related by isostasy, it could be argued that the "effective" age of 10 Ma should be used to plot the northeastern Hikurangi section on Figure 12. This would place the system near SW Japan and Juan de Fuca, away from the group of systems where back-arc opening is or has been occurring.

There are two difficulties with this interpretation. First, the plate convergence is an inappropriate measure of the horizontal component of the subduction velocity when convergence is oblique; the down-dip component of convergence, about 40 mm/a, should be used. Second, it is not known how much of the sediment that contributes to the negative buoyancy of the subducted plate is scraped off to form the accretionary prism or is underplated to the Australian plate crust. To obtain an upper bound on the effective age of the subducting plate in the northeast, assume that all the sediment (3 km) is removed from the crust in Bannister's model (Figure 3d), leaving a residual crustal thickness of 8–9 km, about 2 km more than normal. This crust could have been expected to have lain in about 5000 m of water (500 m shallower than the Kermadec system) and such a water depth is consistent with oceanic crustal ages in the range 40–60 Ma (SCLATER *et al.*, 1980). This would place the northeastern portion of the Hikurangi Margin close to other systems exhibiting back-arc opening (Figure 12).

In the southwest there are none of the symptoms of rifting, although there is a back-arc basin of nonvolcanic, nonextensional character. The southwestern portion of the subducted plate shows evidence of being rather more buoyant: the water depth adjacent to the trench is less, the sediment in the trench is thicker, the coastal uplift rate is greater and the estimate of the subducted plate crustal thickness is greater. At the same time, the rate of convergence and hence component of the rate of convergence down-dip is a little less; about 35 mm/a. This will tend to counterbalance the increased buoyancy. However, it would appear that the southeastern part of the Hikurangi Margin would probably fall outside the boundary of the systems with back-arc opening in Figure 12. Therefore, in view of the likely continuation of very buoyant lithosphere arriving at the trench for future subduction, it must be speculated, as an alternative to the suggestion of STERN *et al.* (1986), that back-arc opening is unlikely to proceed much further south in the Hikurangi system than its current limit—about 39°S.

Concluding Remarks

Although some aspects of the structure and behaviour of the Hikurangi Margin are well understood, a number of problems remain. The single experiment that we feel would most extend our knowledge would be the determination of the crustal structure, age, and possibly anisotropy of seismic velocities of the lithosphere east of the Hikurangi Trench. With this information, a number of uncertainties about the subducted plate should be resolved.

Another offshore region requiring further study occurs between the northeast end of the North Island and the southern termination of the Kermadec Trench. Little is known about this region of transition between the two subduction systems.

The question of why the two back-arc basins are of such different character cannot be regarded as being settled. The change in crustal thickness, uplift and heatflow at about 39°S may result from a combination of a different chemistry of the subducted plate combined with asthenosphere flow induced by the back-arc opening, but detailed modelling is required before this explanation can be accepted.

If the division of the Hikurangi Margin into two zones is accepted, then there are implications for the assessment of seismic potential in the two zones. The position of the NE zone in Figure 12 would suggest a smaller maximum possible earthquake and consequently a smaller potential than for the SW zone. This must be, however, regarded as being highly speculative and does not justify diminishing the current conservative assessment of risk in the NE zone (e.g., SMITH and BERRYMAN, 1986).

Finally, and not least importantly, the very high heatflow observed in the CVR raises the question of whether this is normal for back-arc basins or an abnormality due to the encroachment of rifting into a continental lithosphere. If normal, it could have important implications for heat budget calculations for the earth (e.g., SCLATER *et al.*, 1980).

Acknowledgements

The authors would like to thank their colleagues Helen Anderson, Fred Davey, Russell Robinson and Terry Webb for their comments on the MS.

REFERENCES

ABE, K. (1981), *Magnitudes of large shallow earthquakes from 1904 to 1980*, Phys. Earth Planet. Int. *27*, 72–92.

ABE, K. (1984), *Complements to "Magnitudes of large shallow earthquakes from 1904 to 1980*," Phys. Earth Planet. Int. *34*, 17–23.

ABE, K. and NOGUCHI, S. (1983a), *Determination of magnitude for large shallow earthquakes 1898–1917*, Phys. Earth Planet. Int. *32*, 45–49.

ABE, K. and NOGUCHI, S. (1983b), *Revision of magnitudes of large shallow earthquakes, 1897–1912*, Phys. Earth Planet. Int. *33*, 1–11.

ADAMS, R. D. and FERRIS, B. G. (1976), *A further earthquake at exceptional depth beneath New Zealand (Note)*, N.Z. J. Geol. Geophys. *19*, 269–273.

ADAMS, R. D. and HATHERTON, T. (1973), *Seismological and geothermal research in New Zealand*, Nature *246*, 262–264.

ADAMS, R. D. and WARE, D. E. (1977), *Subcrustal earthquakes beneath New Zealand; locations determined with a laterally inhomogeneous velocity model*, N.Z. J. Geol. Geophys. *20*, 59–83.

ANDERSON, R. N., UYEDA, S., and MIYASHIRO, A. (1976), *Geophysical and geochemical constraints at converging plate boundaries. Pt. 1. Dehydration in the downgoing slab*, Geophys. J. R. Astr. Soc. *44*, 333–357.

ANDERTON, P. W. (1981), *Structure and evolution of the South Wanganui Basin*, New Zealand. N.Z. J. Geol. Geophys. *24*, 39–63.

ANSELL, J. H. and ADAMS, D. A. (1986), *Unfolding the Wadati-Benioff zone in the Kermadec—New Zealand Region*, Phys. Earth Planet. Int. *44*, 274–280.

ANSELL, J. H. and SMITH, E. G. C. (1975), *Detailed structure of a mantle seismic zone using the homogeneous station method*, Nature *253*, 518–520.

ARABASZ, W. J. and LOWRY, M. A. (1980), *Microseismicity in the Tararua-Wairarapa area: depth-varying stresses and shallow seismicity in the southern North Island, New Zealand*, N.Z. J. Geol. Geophys. *23*, 141–154.

ARABASZ, W. J. and ROBINSON, R. (1976), *Microseismicity and geologic structure in the northern South Island, New Zealand*, N.Z. J. Geol. Geophys. *19*, 569–601.

BANNISTER, S. (1986), *Seismicity, stress field and lithospheric structure of the northern Hawkes Bay region, New Zealand*, Ph.D. Thesis, Victoria University of Wellington, New Zealand.

BERRYMAN, K. R. (1979), *Active faulting and derived PHS directions in the South Island, New Zealand*, In *The Origin of the Southern Alps*, Eds. R. I. Walcott and M. M. Cresswell, Roy. Soc. N.Z. Bull. *18*, 29–34.

BIBBY, H. M. (1981), *Geodetically determined strain across the southern end of the Tonga-Kermadec-Hikurangi subduction zone*, Geophys. J. R. Astr. Soc. *66*, 513–533.

BIBBY, H. M., DAWSON, G. B., RAYNER, H. H., STAGPOOLE, V. M., and GRAHAM, D. J. (1984), *The structure of the Mokai geothermal field based on geophysical observations*, J. Volc. Geothm. Res. *20*, 1–20.

BLACK, P. R. and BRAILLE, L. W. (1982), *Pn velocity and cooling of continental lithosphere*, J. Geophys. Res. *87*, 10557–10563.

BROTHERS, R. N. and DELALOYE, M. (1982), *Obducted ophiolites of North Island, New Zealand: origin, age, emplacement and tectonic implications for Tertiary and Quaternary volcanicity*, N.Z. Journ. Geol. and Geophys. *25*, 257–274.

BULLEN, K. E. (1938), *On the epicentre of the 1934 Pahiatua earthquake*, N.Z. J. Sci. Tech. *20*, 61–66.

CALHAEM, I. M. (1973), *Heat flow measurements under some lakes in the North Island, New Zealand*, Ph.D. Thesis, Victoria University of Wellington, New Zealand.

CARTER, L. (1980), *New Zealand Region Bathymetry 1:6,000,000*, (2nd Ed.) New Zealand Oceanographic Inst. Chart. Misc. Series 15.

CAZENAVE, A. and DOMINH, K. (1984), *Geoid heights over the Louisville Ridge (South Pacific)*, J. Geophys. Res. *89*, 11171–11179.

CHAPMAN, M. E. and TALWANI, M. (1982), *Geoid anomalies over deep sea trenches*, Geophys. J. R. Astr. Soc. *68*, 349–369.

CHONG, F. K. L. (1982), *A microearthquake study of the Indian/Pacific plate boundary, central Hawkes Bay, New Zealand*, Ph.D. Thesis, Victoria University of Wellington, New Zealand.

CLOWES, R. M. and AU, D. (1982), *In situ evidence for a low degree of S-wave anisotropy in the oceanic upper mantle*, Geophys. Res. Letters *9*, 13–16.

COLE, J. W. and LEWIS, K. B. (1981), *Evolution of the Taupo-Hikurangi subduction system*, Tectonophysics. *72*, 1–21.

DAVEY, F. J. (1987), *Seismic reflection measurements behind the Hikurangi convergent margin, southern North Island, New Zealand*, Geophys. J. Roy. Astr. Soc. *89*, 443–448.

DAVEY, F. J. and SMITH, E. G. C. (1983a), *The tectonic setting of the Fiordland region, south-west New Zealand*, Geophys. J. R. Astr. Soc. *72*, 23–38.

DAVEY, F. J. and SMITH, E. G. C. (1983b), *A crustal seismic reflection-refraction experiment across the subducted Pacific plate under Wellington, New Zealand*, Phys. Earth Plan. Int. *31*, 327–333.

DAVEY, F. J., HAMPTON, M., CHILDS, J., FISHER, M. A., LEWIS, K., and PETTINGA, J. R. (1986), *Structure of a growing accretionary prism, Hikurangi margin, New Zealand*, Geology *14*, 663–666.

DAWSON, G. B. and DICKINSON, D. J. (1970), *Heat flow studies in thermal areas of the North Island of New Zealand*, Geothermics Spec. Issue *2*, 466–473.

ESHELBY, J. D. (1957), *The determination of the elastic field of an ellipsoidal inclusion and related problems*, Proc. Roy. Soc. London Series A *241*, 376–396.

EVISON, F. F., ROBINSON, R., and ARABASZ, W. J. (1976), *Microearthquakes, geothermal activity and structure, central North Island, New Zealand*, N.Z. J. Geol. Geophys. *19*, 625–637.

FREUND, R. (1971), *The Hope Fault. A strike-slip fault in New Zealand*, N.Z. Geological Surv. Bull. *86*, 49pp and maps. N.Z. Dept. Sci. Ind. Res. Wellington, N.Z.

GHANI, M. A. (1978), *Late Cenozoic vertical crustal movements in the central part of New Zealand*, N.Z. Geol. Geophys. *21*, 117–125.

HAINES, A. J. (1979), *Seismic wave velocities in the uppermost mantle beneath New Zealand*, N.Z. J. Geol. Geophys. *22*, 245–257.

HAINES, A. J. (1981), *A local magnitude scale for New Zealand earthquakes*. Bull. Seis. Soc. Am. *71*, 275–294.

HANKS, T. C. and KANAMORI, H. (1979), *A moment magnitude scale*, J. Geophys. Res. *84*, 2348–2350.

HARRIS, F. (1982a), *Focal mechanisms of subcrustal earthquakes in the North Island, New Zealand*, N.Z. J. Geol. Geophys. *25*, 325–334.

HARRIS, F. (1982b), *Focal mechanism of an intermediate-depth earthquake and its aftershocks (Note)*, N.Z. J. Geol. Geophys. *25*, 109–113.

HATHERTON, T. (1969), *The geophysical significance of calc-alkaline andesites in New Zealand*, N.Z. J. Geol. Geophys. *12*, 436–459.

HATHERTON, T. (1970), *Upper mantle inhomogeneity beneath New Zealand: surface manifestations*, J. Geophys. Res. *75*, 269–284.

HATHERTON, T., MACDONALD, W. J. P., and THOMPSON, G. E. K. (1966), *Geophysical methods in geothermal prospecting in New Zealand*, Bull. Volcanol. *29*, 484–498.

HULL, A. G. and BERRYMAN, K. R. (1986), *Holocene tectonism in the region of the Alpine Fault at Lake McKerrow, Fiordland*, In *Recent Crustal Movements of the Pacific Region*, Eds W. I. Reilly and B. E. Harford. Roy. Soc. N.Z. Bull. *24*, 317–332.

HUNT, T. M. (1980), *Basement structure of the Wanganui Basin, onshore, interpreted from gravity data*, N.Z. J. Geol. Geophys. *23*, 1–16.

IDA, Y. (1983), Convection in the mantle wedge above the slab and tectonic processes in subduction zones. J. Geophys. Res. *88*, 7449–7456.

KANAMORI, H. and ANDERSON, D. L. (1975), *Theoretical basis of some empirical relations in seismology*, Bull. Seism. Soc. Am. *65*, 1073–1095.

KARIG, D. E. (1970), *Kermadec arc—New Zealand tectonic confluence*, N.Z. J. Geol. Geophys. *13*, 21–29.

KATZ, H. R. (1982), *Plate margin transition from oceanic arc-trench to continental system: the Kermadec—New Zealand example*, Tectonophysics *87*, 49–64.

KAYAL, J. R. (1984), *Microseismicity and tectonics at the Indian/Pacific plate boundary: south-east Wellington province, New Zealand*, Geophys. J. R. Astr. Soc. *77*, 567–592.

KAYAL, J. R. and SMITH, E. G. C. (1984), *Upper mantle P-wave velocities in the southeast North Island, New Zealand*, Tectonophysics *104*, 115–125.

KEAR, D. (1959), *Stratigraphy of New Zealand's Cenozoic volcanism northwest of the volcanic belt*, N.Z. J. Geol. Geophys. *2*, 578–589.

KOSMINSKAYA, I. P. and KAPUSTIAN, N. K. (1976), *Generalized seismic model of the typical oceanic crust*, In *Volcanoes and Tectonosphere*. Eds H. Aoki and S. Iizuka: 143–156. Tokai University Press.

LEWIS, K. B. and BENNETT, D. J. (1985), *Structural patterns on the Hikurangi Margin: an interpretation of new seismic data*, In *New Seismic Profiles, Cores, and Dated Rocks from the Hikurangi Margin, New Zealand*, Comp. K. B. Lewis, N.Z. Oceanographic Inst. Field Report *22*, 3–25, Dept. Sci. Ind. Res., Wellington, New Zealand.

MINSTER, J. B. and JORDAN, T. H. (1978), *Present-day plate motions*, J. Geophys. Res. *83*, 5331–5354.

MOLNAR, P. and ATWATER, T. (1978), *Inter-arc spreading and cordilleran tectonics as alternatives related to the age of subducted ocean lithosphere*. Earth Planet. Sci. Lett. *41*, 330–340.

MOONEY, H. M. (1970), *Upper mantle inhomogeneity beneath New Zealand: seismic evidence*, J. Geophys. Res. *75*, 285–309.

OFFICERS OF THE N.Z. GEOLOGICAL SURVEY (1983), Late Quaternary tectonic map of New Zealand. 1:2,000,000. 2nd Ed. N.Z. Geological Survey Series Map 12. Dept. Sci. and Ind. Res., Wellington, New Zealand.

OLIVER, J. and ISACKS, B. (1967), *Deep earthquake zones, anomalous structures in the upper mantle, and the lithosphere*. J. Geophys. Res. *72*, 4259–4275.

PANDEY, O. P. (1981), *Terrestrial heat flow in New Zealand*, Ph.D. Thesis, Victoria University of Wellington, New Zealand.

PILLANS, B. (1983), *Upper Quaternary marine terrace chronology and deformation, South Taranaki, New Zealand*, Geology *11*, 292–297.

PILLANS, B. (1986), *A late Quaternary uplift map for North Island, New Zealand*, In *Recent Crustal Movements of the Pacific Region*, Eds W. I. Reilly and B. E. Harford, Bull. Roy. Soc. N.Z. *24*, 409–418. Wellington, New Zealand.

REILLY, W. I. (1965), *The mean height and hypsographic curve for New Zealand*, N.Z. J. Geol. Geophys. *8*, 128–139.

REYNERS, M. (1980), *A microearthquake study of the plate boundary, North Island, New Zealand*, Geophys. J. R. Astr. Soc. *63*, 1–22.

REYNERS, M. (1983), *Lateral segmentation of the subducted plate at the Hikurangi Margin, New Zealand: seismological evidence*, Tectonophysics *96*, 203–223.

REYNERS, M. (1984), *The Hawke's Bay, New Zealand, earthquake of 1982 September 02 and deformation in the interior of the subducted Pacific plate*, Tectonophysics *106*, 259–273.

RICHTER, C. F. (1958), *Elementary seismology* (W. H. Freeman & Co., San Francisco) 768pp.

RINGWOOD, A. E. (1977), *Petrogenesis in island arc systems*, In *Island Arcs, Deep Sea Trenches and Back-arc Basins* (M. Talwani and W. C. Pitman, eds.) Maurice Ewing series I: 311–324. AGU, Washington.

ROBINSON, R. (1986), *Seismicity, structure, and tectonics of the Wellington region, New Zealand*, Geophys. J. R. Astr. Soc. *87*, 379–409.

ROBINSON, R., SMITH, E. G. C., and LATTER, J. H. (1981), *Seismic studies of the crust under the hydrothermal areas of the Taupo Volcanic Zone, New Zealand*, J. Volc. Geoth. Res. *9*, 253–267.

RUFF, L. and KANAMORI, H. (1980), *Seismicity and the subduction process*, Phys. Earth Planet. Int. *23*, 240–252.

SCHROEDER, W. (1984), *The empirical age-depth relation and depth anomalies in the Pacific Ocean Basin*, J. Geophys. Res. *89*, 9873–9883.

SCLATER, J. G., JAUPART, C., and GALSON, D. (1980), *The heat flow through oceanic and continental crust and heat loss of the earth*, Rev. Geophys. Space Phys. *18*, 269–312.

SMITH, E. G. C. (1977), *The theory of multi-earthquake location by least squares and application to groups of North Island, New Zealand, mantle earthquakes*, Ph.D. Thesis, Victoria University of Wellington, New Zealand.

SMITH, E. G. C. (1981), *Calculation of poles of instantaneous rotation from poles of finite rotation*, Geophys. J. R. Astr. Soc. *65*, 223–227.

SMITH, E. G. C. and DAVEY, F. J. (1984), *Joint Hypocentre Determination of intermediate depth earthquakes in Fiordland, New Zealand*, Tectonophysics *104*, 127–144.

SMITH, E. G. C. and WEBB, T. H. (1986), *The seismicity and related deformation of the Central Volcanic Region, North Island, New Zealand*, In *Late Cenozoic volcanism in New Zealand* (I. E. M. Smith, ed.) Roy. Soc. of N.Z. Bull. *23*, 112–133.

SMITH, W. D. and BERRYMAN, K. R. (1983), *Revised estimates of earthquake hazard in New Zealand*, Bull. N.Z. Nat. Soc. Earthquake Eng. *16*, 259–272.

SMITH, W. D. and BERRYMAN, K. R. (1986), *Earthquake hazard in New Zealand: inferences from seismology and Geology*, In *Recent Crustal Movements of the Pacific Region* (W. I. Reilly and B. E. Harford, Eds.) Bull. Roy. Soc. N.Z. *24*, 223–244, Wellington, New Zealand.

STERN, T. A. (1985), *A back-arc basin formed within continental lithosphere: The Central Volcanic Region of New Zealand*, Tectonophysics *112*, 385–409.

STERN, T. A. (1986), *Geophysical studies of the upper crust within the Central Volcanic Region, New Zealand*, In *Late Cenozoic Volcanism in New Zealand* (I. E. M. Smith, ed.) Roy. Soc. of N.Z. Bull. *23*, 92–111.

STERN, T. A. and DAVEY, F. J. (1987), *A seismic investigation of the crustal and upper mantle structure within the Central Volcanic Region of New Zealand*, N.Z. J. Geol. Geophys. *30*, 217–231.

STERN, T. A., DAVEY, F. J., and SMITH, E. G. C. (1986), *Crustal structure studies in New Zealand*, In *Reflection Seismology: a Global Perspective* (M. Barazangi and L. Brown, eds.) Geodynamic series, Vol. *13*, 121–132. A.G.U., Washington.

STERN, T. A., SMITH, E. G. C., DAVEY, F. J., and MUIRHEAD, K. H. (1987), *Crust and upper mantle structure of the northwestern North Island, New Zealand*, Geophys. J. R. Astr. Soc. *91*, 913–936.

STEVENS, G. R. (1974), *Rugged Landscape. The geology of central New Zealand* (A. H. & A. W. Reed, Wellington, N.Z.) 286pp.

STOCK, J. and MOLNAR, P. (1982), *Uncertainties in the relative positions of the Australia, Antarctica, Lord Howe and Pacific plates since the Late Cretaceous*, J. Geophys. Res. *87*, 4679–4714.

STUDT, F. E. and THOMPSON, G. E. K. (1969), *Geothermal heat flow in the North Island of New Zealand*, N.Z. J. Geol. Geophys. *12*, 673–683.

TAYLOR, B. and KARNER, G. D. (1983), *On the evolution of marginal basins*, Rev. Geophys. and Space Phys. *21*, 1727–1742.

THOMPSON, G. E. K. (1977), *Temperature gradients within and adjacent to the North Island volcanic belt*, N.Z. J. Geol. Geophys. *20*, 85–97.

VAN DER LINGEN, G. J. (1982), *Development of the North Island subduction system, New Zealand*, In *Trench-Forearc Geology: Sedimentation and Tectonics on Modern and Ancient Active Plate Margins: 259–274* (Ed J. K. Leggett) Geol. Soc. London.

WALCOTT, R. I. (1978), *Geodetic strains and large earthquakes in the axial tectonic belt of North Island, New Zealand*, J. Geophys. Res. *83*, 4419–4429.

WALCOTT, R. I. (1984), *The kinematics of the plate boundary zone through New Zealand: a comparison of short and long-term deformation*, Geophys. J. R. Astr. Soc. *79*, 613–633.

WALCOTT, R. I. (1987), *Geodetic strain and the deformation history of the North Island of New Zealand during the late Cenozoic*, Phil. Trans. R. Soc. London Ser. A *321*, 163–181.

WATANABE, T., LANGSETH, M. G., and ANDERSON, R. N. (1977), *Heat flow in back-arc basins of the western Pacific*, In *Island Arcs, Deep Sea Trenches and Back-arc Basins* (M. Talwani and W. Pitman, eds.) A.G.U. Washington, 88–94.

WATTS, A. B. and TALWANI, M. (1974), *Gravity anomalies seaward of deepsea trenches and their tectonic implications*, Geophys. J. R. Astr. Soc. *36*, 57–90.

WATTS, A. B. and TALWANI, M. (1975), *Gravity effects of downgoing lithospheric slabs beneath island arcs*. Geol. Soc. Am. Bull. *86*, 1–4.

WEBB, T. H., WESNOUSKY, S. C. and HELMBERGER, D. V. (1985), *A body-wave analysis of the 1966 Gisborne, New Zealand, earthquake*, Tectonophysics *113*, 271–282.

WEISSEL, J. K. (1981), *Magnetic lineations in marginal basins of the western Pacific*, Phil. Trans. R. Soc. London Ser. A *300*, 223–247.

WELLMAN, H. W. (1972), *Rate of horizontal fault displacement in New Zealand*, Nature *237*, 275–277.

WELLMAN, H. W. (1979), *An uplift map for the South Island of New Zealand and a model for uplift of the Southern Alps*, Bull. R. Soc. N.Z. *18*, 13–20.

WOOD, C. P. (1986), *Stratigraphy and petrology of N.M.4—Ngatamariki geothermal field*. Unpublished report of the N.Z. Geol. Survey, D.S.I.R., Lower Hunt, New Zealand.

ZIAGOS, J. P., BLACKWELL, D. D., and MOOSER, F. (1985), *Heat flow in southern Mexico and the thermal effects of subduction*, J. Geophys. Res. *90*, 5410–5420.

(Received August 5, 1986, revised November 9, 1987, accepted December 29, 1987)

PAGEOPH, Vol. 129, Nos. 1/2 (1989) 0033–4553/89/020233–29\$1.50 + 0.20/0

An Unusual Zone of Seismic Coupling in the Bonin Arc: The 1972 Hachijo-Oki Earthquakes and Related Seismicity

TETSUJI MORIYAMA[1], FUMIKO TAJIMA[2] and TETSUZO SENO[3]

Abstract—The 1972 February and December Hachijo-Oki earthquakes ($M_s = 7.3$ and 7.4), in the northernmost part of the Izu-Bonin subduction zone, are the only major events ($M_s > 7.0$) in the Bonin arc for the past 80 years. Relocation of the hypocenters, using one smaller event having a well-constrained focal depth as a master event, shows that the depth of the February event is 10 km shallower than that of the December event. We have determined the rupture process for both events by minimizing the error in waveform between observed and synthetic seismograms. Although the number of available stations are limited, the depth range of the major energy release for the December event extends deeper than for the February one. The rupture propagated up-dip for both events. It is likely that the rupture zone of the two events overlapped, and that the December event ruptured the deeper part. This suggestion is consistent with the observation that the aftershock zones of both events overlap with that of the December event shifted landward. The waveforms of the December event have a smaller high frequency component than those of the February event, suggesting that the stress at the thrust zone became more uniform or reduced after the February event.

No thrust type smaller event occurred near the rupture zone. Instead, the P-axes of smaller events are parallel to the dip of the slab and their T-axes dip to the southwest. Focal depths of these events estimated by P-wave forward modeling are generally between 40–50 km and located beneath the thrust zone. We thus interpret them as the events within the Pacific slab near the zone ruptured by the two major events. The stress concentration around the rupture zone of the major events is suggested to have triggered these slab events. After the occurrence of the large events, the slab events are concentrated near the deeper portion of the rupture zone. These events may have been caused by the loading of the down-dip compressional stress near the down-dip end of the rupture zone due to the rupture. The occurrence of the doublet of large earthquakes and a number of down-dip compressional events beneath their rupture zones in a shallow portion of the subducting slab indicates an unusual zone of seismic coupling in the Bonin arc, most of which is seismically quiescent.

Key words: Bonin arc, Pacific slab, thrust zone, seismic coupling, Hachijo-Oki.

1. Introduction

The Bonin arc defines the northeastern boundary of the Philippine Sea plate. The seismic activity along this arc is rather low; no great earthquake ($M_s > 7.8$) has

[1] Department of Earth Sciences, Tsukuba University, Sakuramura, Niihari, Ibaraki 305, Japan.

[2] Institute for Geophysics, University of Texas at Austin, Austin, TX 78759–8345, U.S.A.

[3] International Institute of Seismology and Earthquake Engineering, Build. Res. Inst., Tsukuba, Ibaraki 305, Japan.

been recorded for the past 90 years (SENO and EGUCHI, 1983), and the seismic moment release for the past 80 years is small (PETERSON and SENO, 1984). The only significant earthquakes along this arc are the 1972 Hachijo-Oki, off the coast of Hachijo island, earthquakes (doublet; February 29, $M_s = 7.3$ and December 4, $M_s = 7.4$) which occurred in the northernmost part of this subduction zone. Both of these events have thrust mechanisms (YOSHII, 1979b). Their source area south of the triple junction between the Japan and Bonin Trenches and the Sagami Trough thus seems to be a unique place of seismic coupling in the Bonin arc. We studied the source characteristics of these major events and nearby seismicity in the area shown by the box in Figure 1. The two Hachijo-Oki earthquakes occurred within the box and are indicated by the stars in Figure 1.

We have also estimated focal mechanism solutions of smaller earthquakes near the source area and investigated their relationship to the major events and to the stress state of the Pacific slab. We find that the smaller events occur within the slab; no small thrust events are observed. The relationship between the thrust events and the nearby slab seismicity presumably reflects both the stress state of the slab and the effect of the interface. Study of the Hachijo-Oki area, an active seismic region in a generally quiescent subduction zone, provides insight into the interaction between thrust zone coupling and seismicity in the shallow part of the slab.

2. Seismicity

Figure 2 shows the seismicity along the Bonin arc in three different periods; from 1964 to 1970, from 1971 to 1973, and from 1974 to 1981. Only the events listed in the International Seismological Centre (ISC) Bulletins with twenty or more reported P-arrival times are plotted. The star indicates the epicenter of the 1972 earthquakes. A number of aftershocks occurred near the epicenter (Figure 2b). Figure 2 indicates that the activity associated with the 1972 events is the only significant one along this arc.

Figures 3a–e show the details of the seismicity in the study area indicated by the box in Figure 1. Only the events which are reported in the ISC Bulletins with ten or more P-arrival times are plotted. Figure 3a shows the seismicity for the whole period (1964–1983); the others show the seismicity for the periods before the 1972 February event, between the 1972 February and December events, for one year after the December event, and from 1974 through 1983. The upper figure shows the epicentral distribution and the lower one shows a vertical cross-section in the direction perpendicular to the trench axis. Earthquakes with $m_b = 5.5$ and above are marked with numbers which correspond to those studied and listed in Table 1. The aftershock areas of the two large 1972 events overlap each other; however, that of the December event (event 6) is located deeper in the down-dip direction, i.e., more landward of the February event (event 2, Figures 3c and d). In the vertical

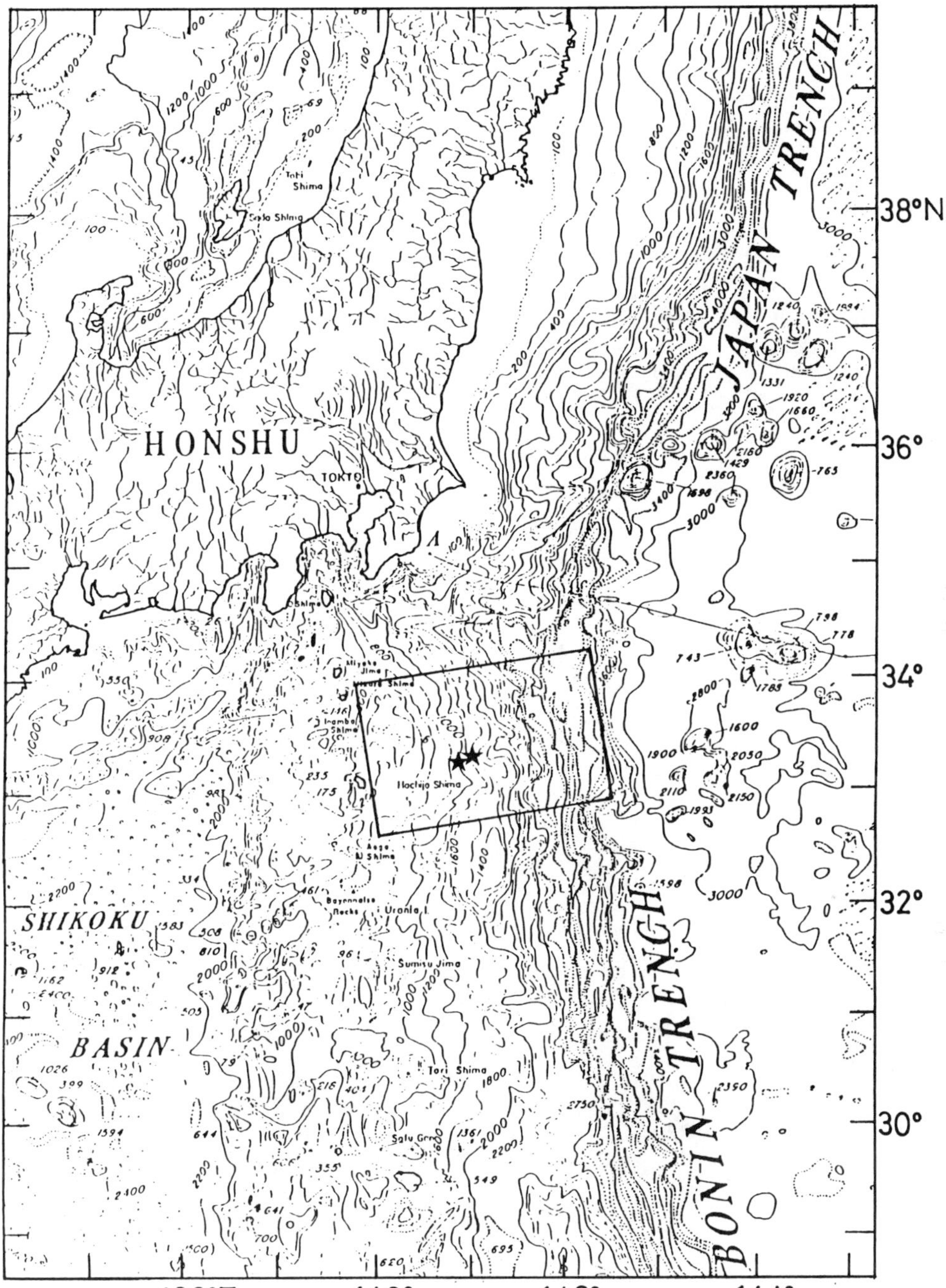

Figure 1
Bathymetric map of the northern Bonin arc (CHASE *et al.*, 1970). The study area indicated by the box is located south of the junction between the Japan and Bonin Trenches. The stars indicate the epicenters of the 1972 Hachijo-Oki earthquakes.

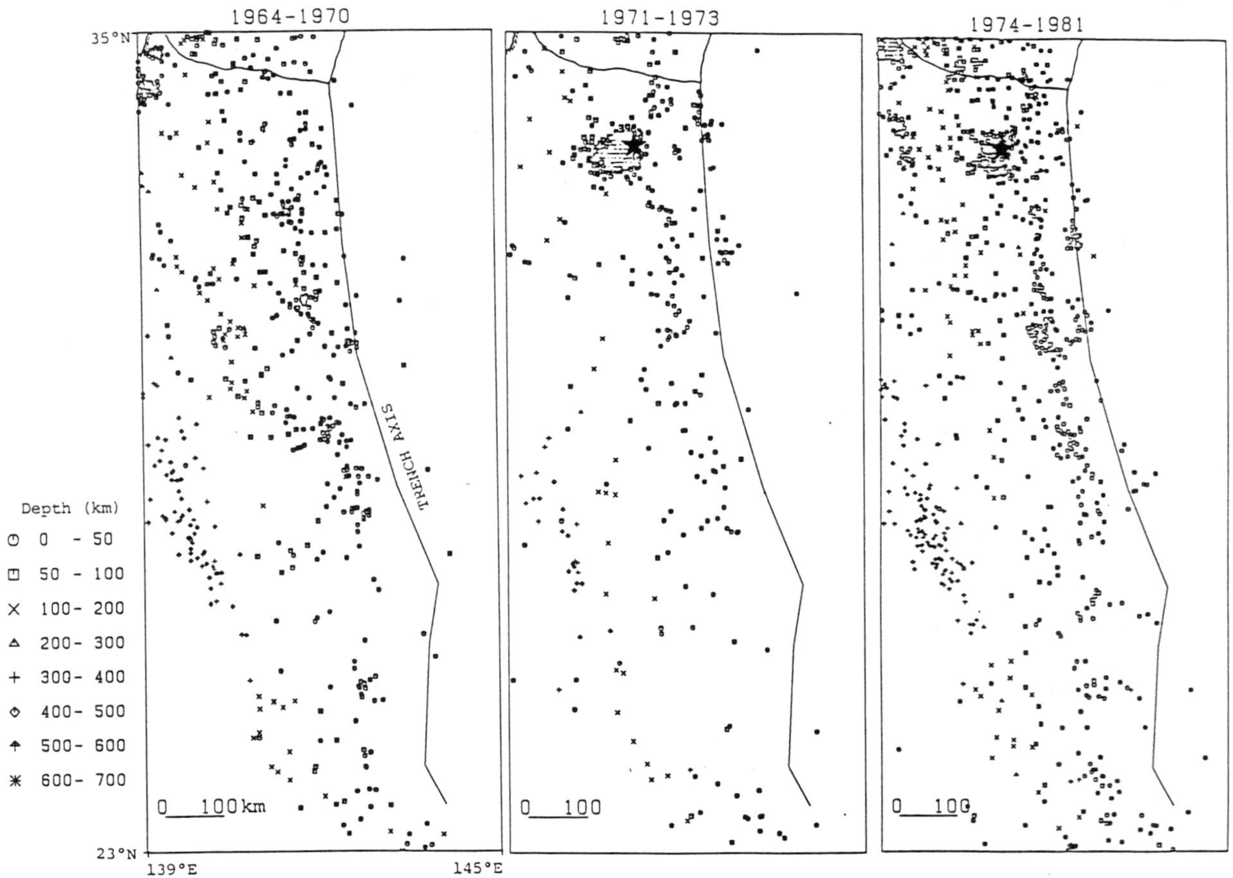

Figure 2
Seismicity along the Bonin arc. Earthquakes with 20 or more *P*-arrival times reported in the ISC Bulletins are plotted. (a): 1964–1970, (b): 1971–1973, (c): 1974–1981. The epicenter of the 1972 large Hachijo-Oki earthquakes is indicated by the star.

Table 1

Earthquakes analyzed in this study.

Event	Date	Origin Time	Lat. (°N)	Lon. (°E)	m_b(ISC)
No. 1	65 Nov. 27	08 42 24.1	33.01	140.94	5.5
No. 2	72 Feb. 29	09 22 59.3	33.38	140.97	6.5
No. 3	72 Mar. 02	20 10 07.8	33.45	141.00	5.8
No. 4	72 Mar. 14	00 47 14.1	33.28	141.02	5.4
No. 5	72 Mar. 18	23 17 39.9	33.60	141.38	5.9
No. 6	72 Dec. 04	10 16 11.5	33.34	140.82	6.7
No. 7	73 Apr. 25	14 21 12.8	33.45	140.88	5.7
No. 8	73 Dec. 26	20 30 05.2	33.46	140.98	5.6
No. 9	74 Sep. 27	03 10 06.3	33.67	141.30	5.7
No. 10	77 Feb. 18	20 51 30.4	33.15	141.06	6.0
No. 11	78 Oct. 11	01 49 00.4	33.47	140.95	5.9
No. 12	80 Oct. 20	03 29 21.6	33.15	140.63	5.5
No. 13	82 Feb. 20	19 18 24.1	33.67	141.17	6.1

The data; origin time, epicentral locations and m_b are obtained from ISC-Bulletins. Events 2 and 6 are the Hachijo-Oki earthquakes.

cross-section, the aftershock zones dip to the west and are centered around 50 km depth for the February event and slightly deeper for the December event (Figures 3c and d). The focal mechanisms of the large 1972 events show thrust faulting with a shallow nodal plane dipping to the west (YOSHII, 1979b). Although the slip vector of the December event is not constrained well, that of the February event, which is fairly well-constrained, is consistent with the predicted Pacific-Philippine Sea relative motion (SENO, 1977). We therefore regard both events as representing underthrusting of the Pacific plate beneath the Philippine Sea plate.

Using an energy contour technique, TAJIMA and KANAMORI (1985) studied the spatial and temporal expansion pattern of the aftershock area of the 1972 Hachijo-Oki earthquakes along with other large subduction zone earthquakes. As shown in Figure 4, the aftershock area expanded significantly for one year in a complex pattern (M_1 indicates the February event and M_2 the December event). The aftershock area expanded slightly into the down-dip direction of the Pacific slab for the first 10 days, and then into the up-dip direction. In the later period the aftershock area also includes expansion to the west, leaving a small quiescent area near the impending December epicenter. For the period of one year after the February event the aftershock area filled the gap area and expanded into the up-dip direction. The ratio of the 100-day aftershock area to that of one day is about 2.5 and grouped into the highest class of expansion ratios among large subduction events. Since the December event and its aftershocks are included in the aftershock sequence of the February event, the one year expansion ratio is as large as about 5. Comparable high expansion ratios are obtained in the Solomon Islands and New Hebrides which reside in a complex tectonics boundary between the Pacific and Indo-Australian plates.

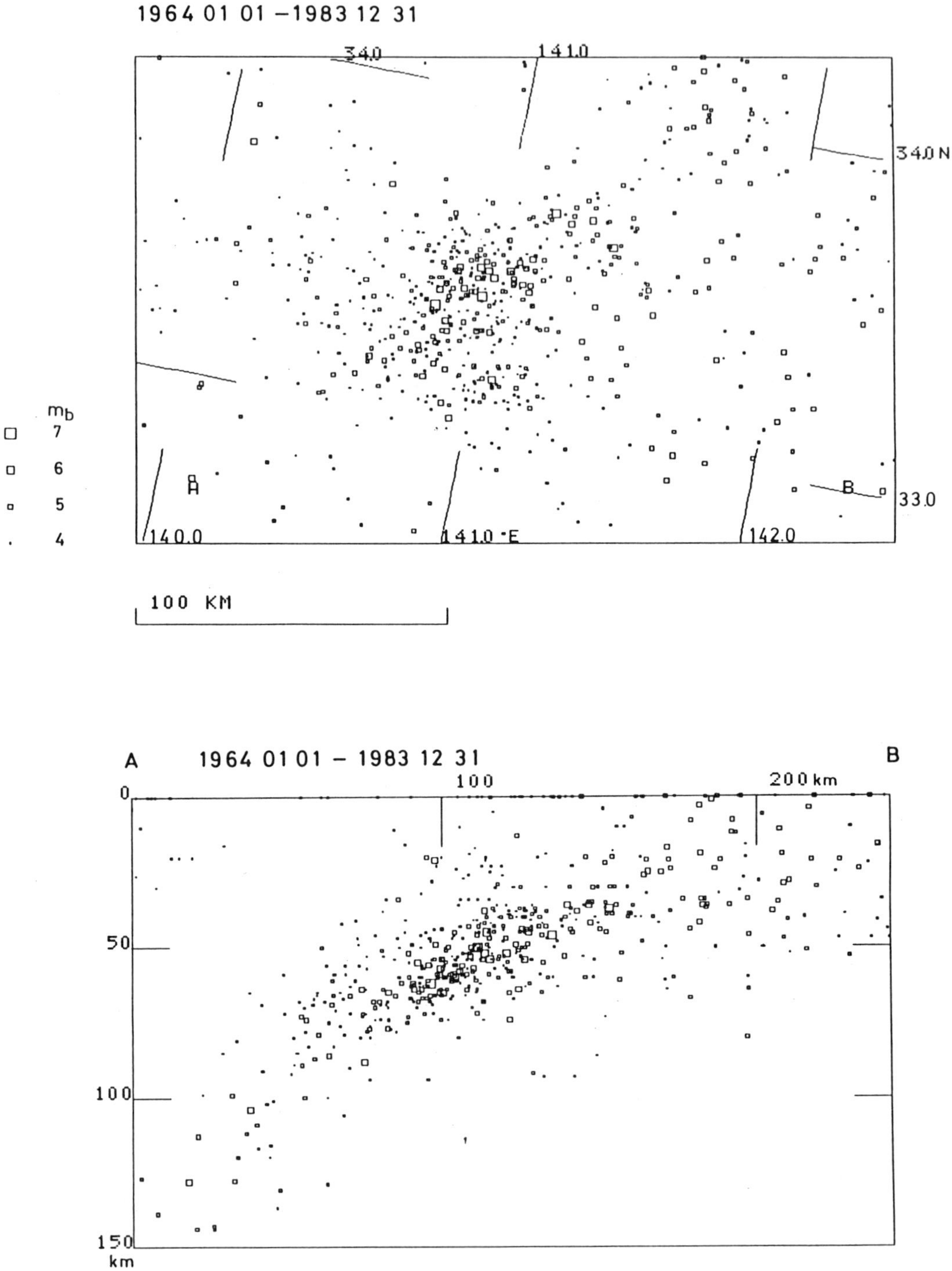
1964 01 01 –1983 12 31
34.0
141.0
34.0 N
mb
7
6
5
4
A
B
33.0
140.0
141.0 E
142.0
100 KM
A
1964 01 01 – 1983 12 31
B
100
200 km
0
50
100
150
km

Fig. 3a.

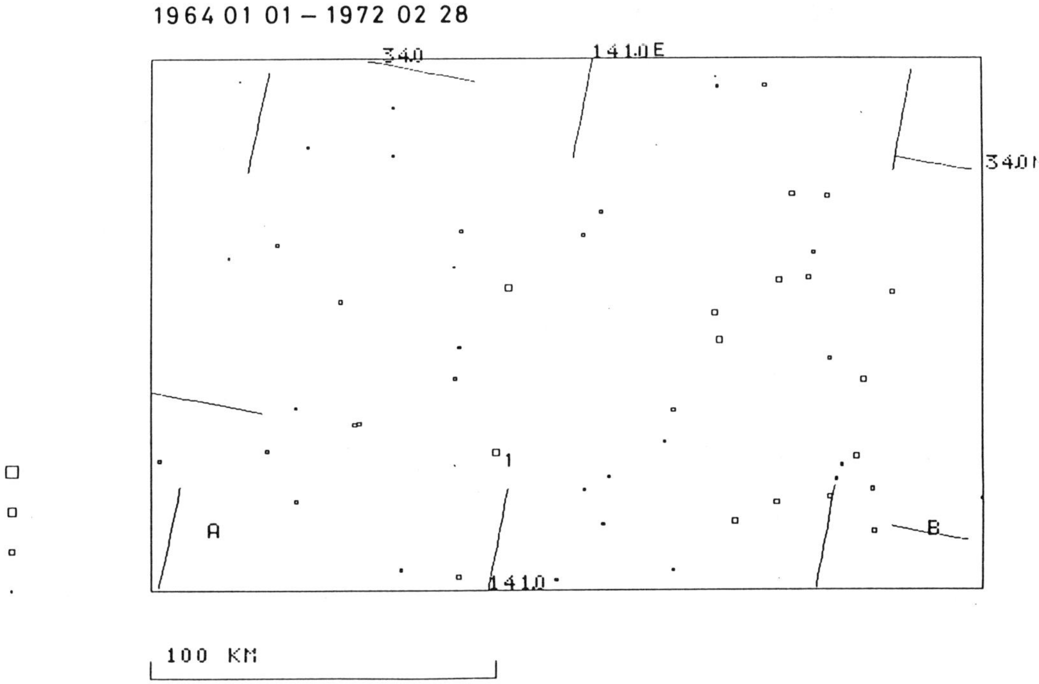
1964 01 01 – 1972 02 28
34.0
141.0E
34.0N
A
B
141.0
100 KM

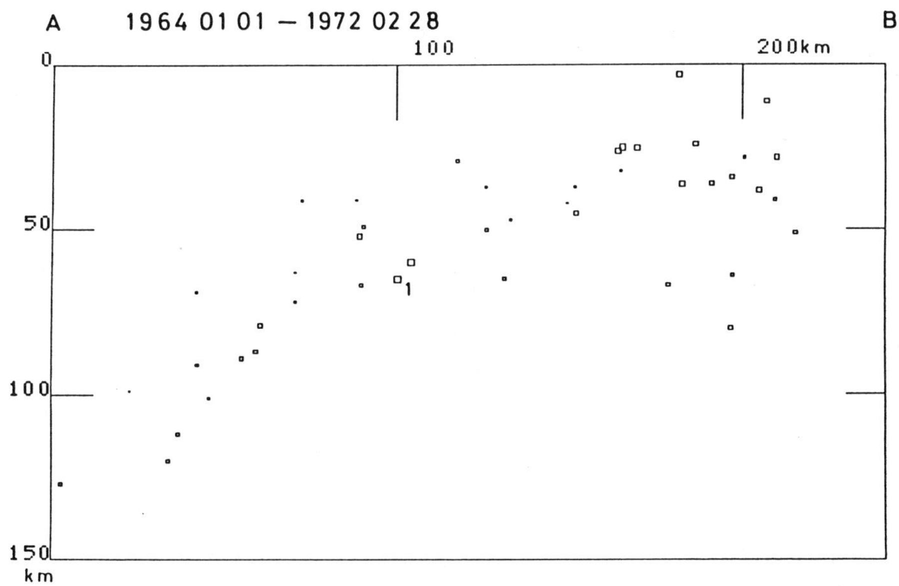
A
1964 01 01 – 1972 02 28
B
0
100
200km
50
100
150
km
1

Fig. 3b.

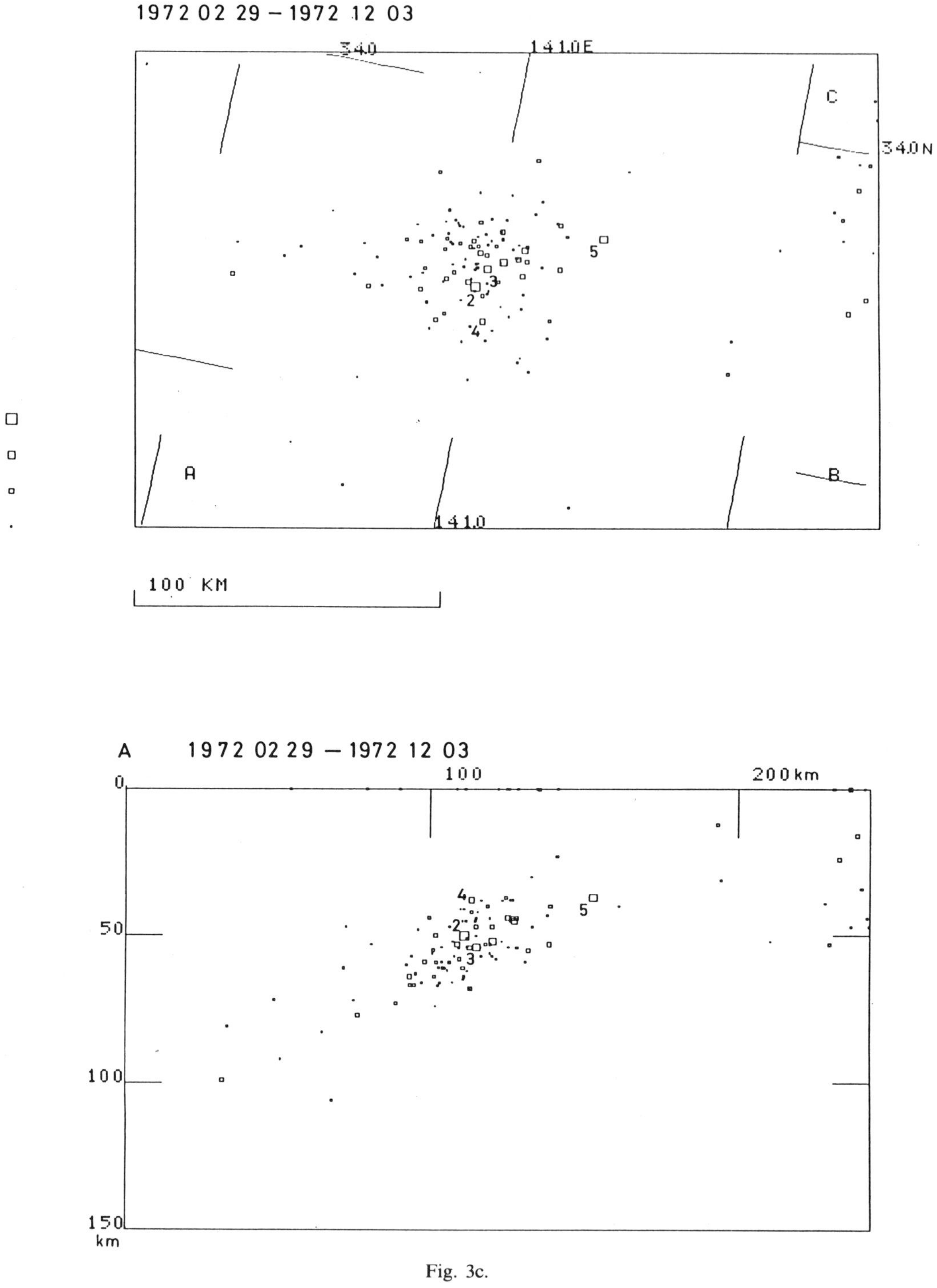
1972 02 29 – 1972 12 03
34.0
141.0E
C
34.0N
5
3
2
4
A
B
141.0
100 KM
A 1972 02 29 – 1972 12 03
0
100
200km
4
5
2
3
50
100
150
km

Fig. 3c.

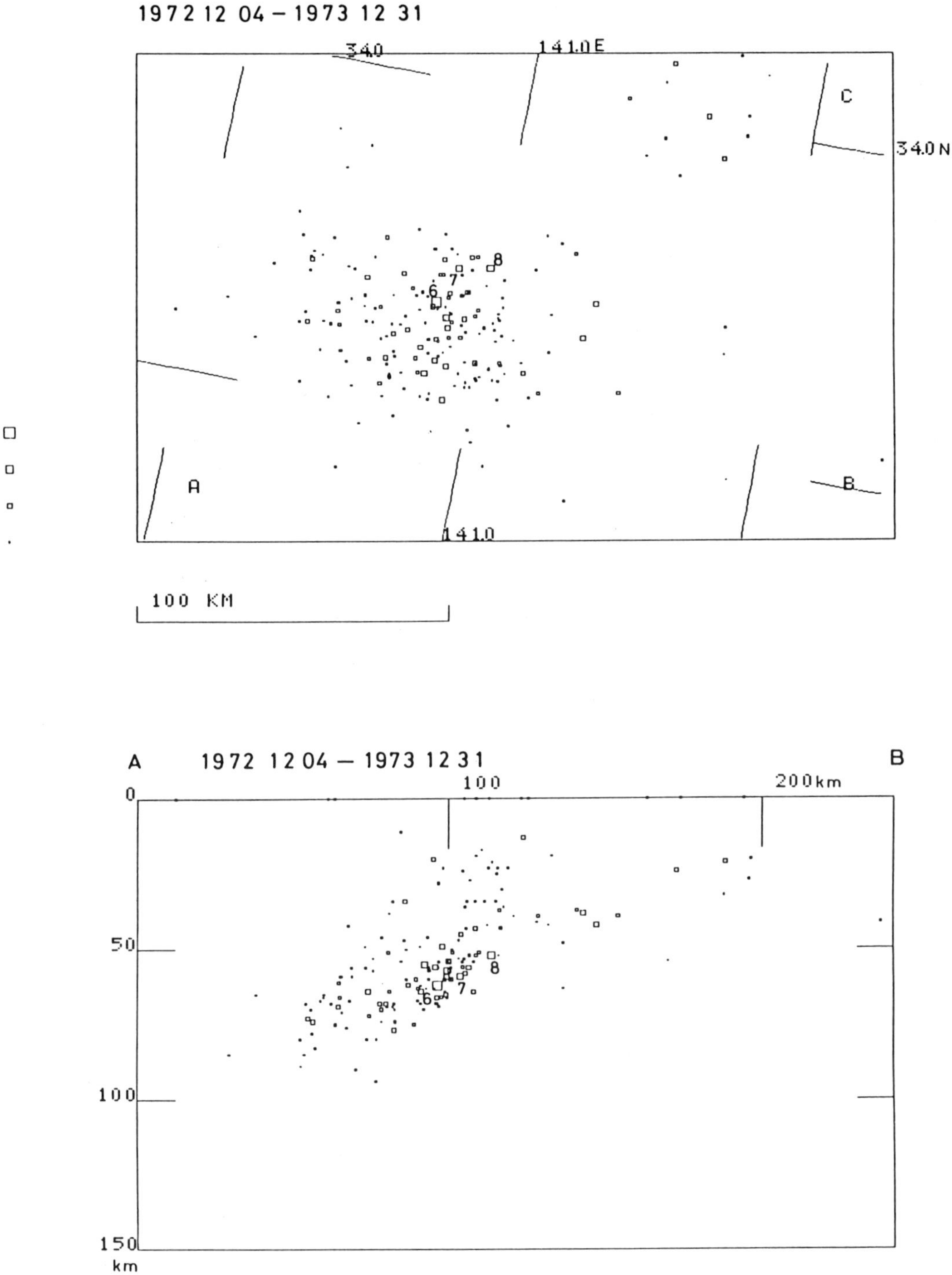
1972 12 04 – 1973 12 31
34.0
141.0E
C
34.0N
8
7
6
A
B
141.0
100 KM
A
1972 12 04 – 1973 12 31
B
100
200km
0
50
100
150
km
8
7
6

Fig. 3d.

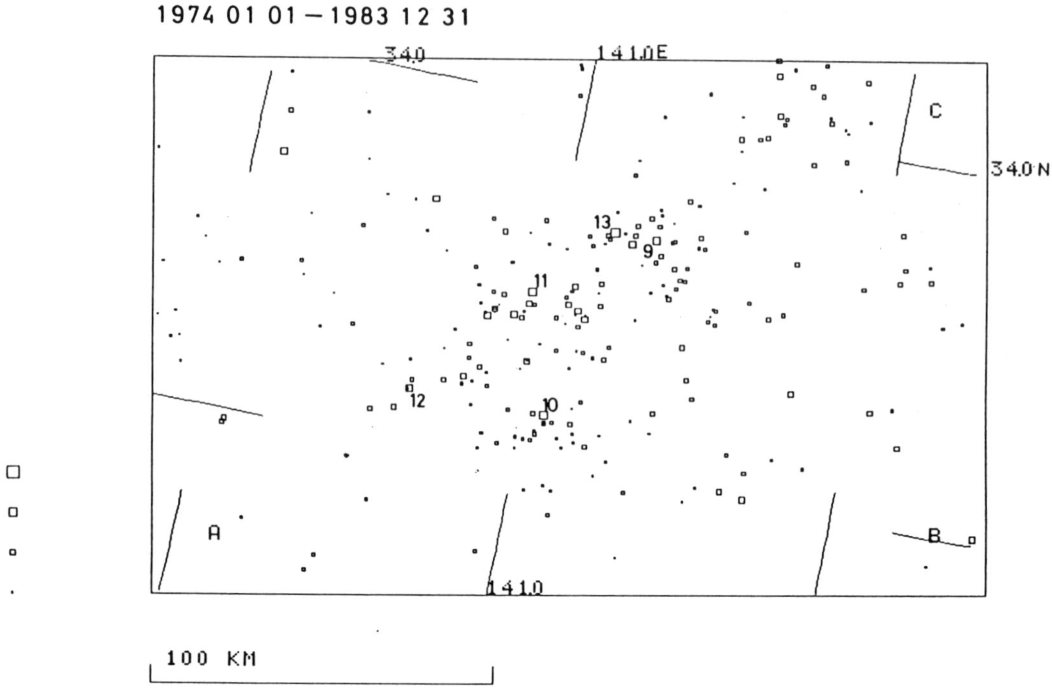
1974 01 01 — 1983 12 31
34.0
141.0E
C
34.0N
13
9
11
12
10
A
B
141.0
100 KM

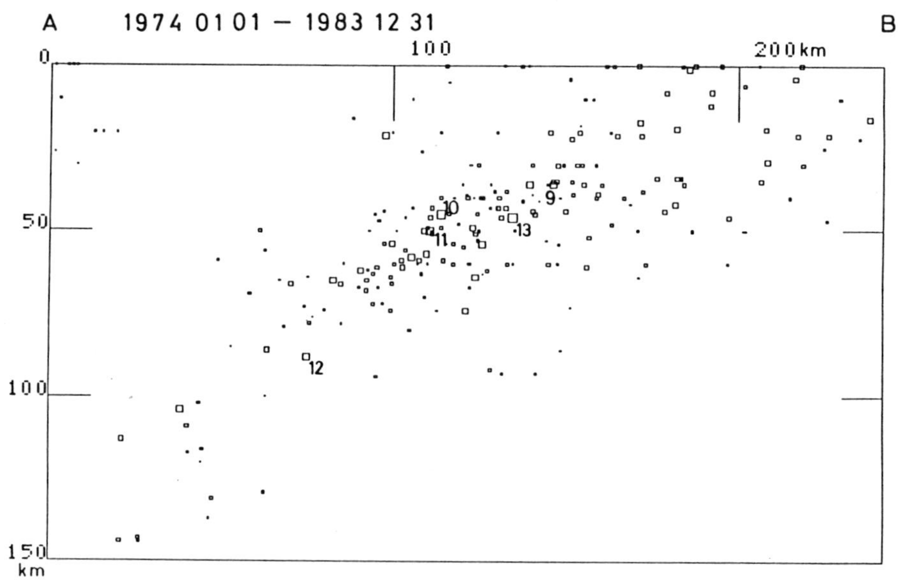
A
1974 01 01 — 1983 12 31
B
0
100
200km
50
100
150
km
10
9
11
13
12

Fig. 3e.

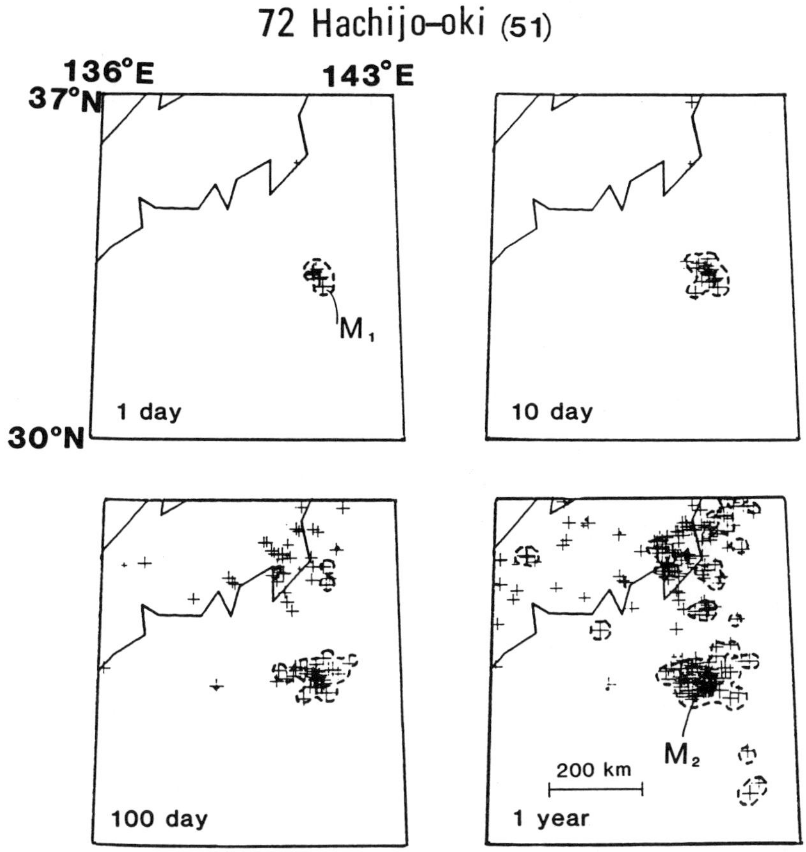

Figure 4
The energy contour map for the aftershock sequence of the 1972 February $M_s = 7.3$ Hachijo-Oki earthquake at four different times: one day, ten days, one hundred days, and one year after the mainshock (TAJIMA and KANAMORI, 1985). M_1 indicates the February event and M_2 indicates the December event. The pluses show smaller events.

Figure 3
Seismicity in the box shown in Fig. 1. Only earthquakes with 10 or more P-arrival times reported in the ISC Bulletins are plotted. The upper shows the epicentral map and the lower shows the cross-section along A-B in the upper figure. The events with $m_b = 5.5$ and above are indicated by the number listed in Table 1. The two large 1972 events are indicated by the solid squares. (a): 1964–1983, (b): 1964–1972 Feb. 28, (c): 1972 Feb. 29–1972 Dec. 3, (d): 1972 Dec. 4–1973, (e): 1974–1983.

3. The 1972 Hachijo-Oki Earthquakes

We relocated the hypocenters of the 1972 Hachijo-Oki events using a master event method. We used the April 25, 1973 $m_b = 5.7$ earthquake (event 7 in Table 1) as a master event because its P-arrival times have a good azimuthal coverage of stations. The epicenter of this event is from the ISC. We determined its depth to be 51 km by forward modeling of P-waveforms, as described in the next section. In the master event relocation, we excluded stations that had P-arrival time residuals of more than 5 sec for the master event and those that had P-arrival time residuals of more than 3 sec for the two main events. The hypocentral parameters obtained from this relocation are listed in Table 2. The depths for both events are several km shallower than the ISC depths. The epicenters of these events moved only a few km from those of ISC. The hypocentral depth of the February event was determined to be 10 km shallower than that of the December event. Although it is not easy to estimate the absolute errors associated with these hypocenters, the deeper depth for the December event is consistent with the difference in aftershock distribution between both events (Figure 3); the aftershock area of the December event is located landward of that of the February event. We will see later that the body-wave analysis of these events also produces the difference in depth range of the rupture zone between these events, consistent with the above results.

Figure 5 shows the long-period P-waveforms of the two events observed at six common stations of World Wide Standardized Seismograph Network (WWSSN). The record of LPS shows diffracted P-waves (epicentral distance $> 90°$); those at KEV and COL are the horizontal components. Except for the LPS records, the waveforms of the February event contain more high-frequency components than those of the December event. The waveforms of the second event look like low-pass filtered ones of the February event. Since core diffraction acts as a natural low-pass filter, neither waveform at LPS contains high-frequency components.

To determine the source processes of these events, we used the deconvolution method of KIKUCHI and FUKAO (1985) which models an earthquake source as a series of double couple point sources, assuming that all the point sources have the same fault mechanism and time history. The inversion algorithm is basically the

Table 2

Relocation of Events 2 and 6 using Event 7 as a master event.

	Date	Origin time	Location		Depth (km)		
No.	Y M D	h m s	(°N)	(°E)	this study	ISC	ISC (pP-P)
2	72 02 29	09 22 59.2	33.35	140.93	45.4	50	43
6	72 12 04	10 16 11.3	33.30	140.80	54.8	62	74

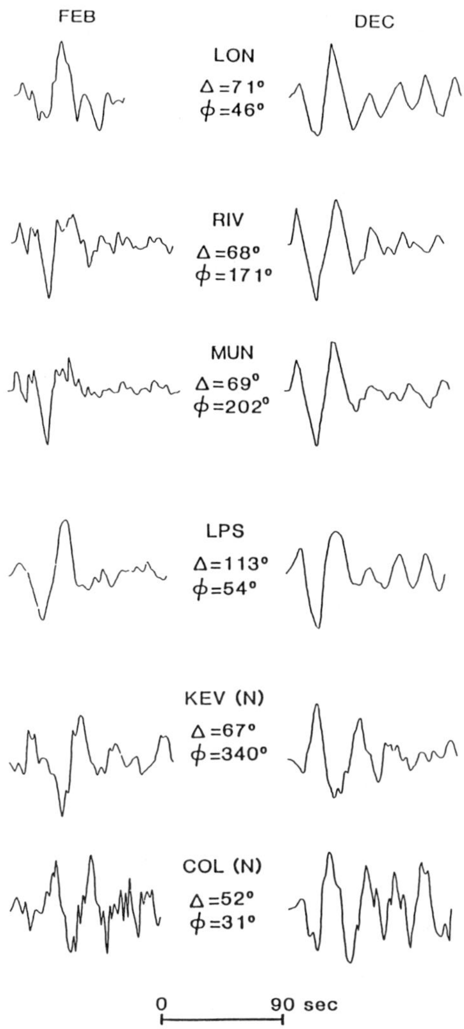

Figure 5

Comparison of the long-period P-wave seismograms of the 1972 February and December events observed at six common stations. The records at LPS are diffracted P-waves and those at KEV and COL are horizontal components.

same as the one developed by KIKUCHI and KANAMORI (1982). This method includes the variation of Green's function over the fault surface for a fixed mechanism. Before the iteration, we set grid points with an interval of 20 km on a rectangular fault surface, which is defined by the strike and dip directions, and generate the synthetic wavelet $w(t, x, y)$ for a point source at each grid point. Surface-reflected phases, pP and sP, are included. After a set of synthetic wavelets are given for all the stations, the point sources are iteratively determined so that the wavelets best fit the residuals of observed records. Eventually, the observed teleseismic P-wave seismogram is approximated by the sum of wavelets:

$$y(t) = \sum_{i=1}^{n} e_i w(t - t_i; x_i, y_i) \tag{1}$$

where e_i is the magnitude of the i-th pulse in units of moment rate. After n iterations, we obtain the far-field source time function as follows:

$$S(t) = \sum_{i=1}^{n} e_i s(t - t_i) \tag{2}$$

where $s(t)$ is the source time function of an individual source with unit seismic moment and defined with two time constants τ_1 and τ_2 as follows:

$$s(t) = \begin{cases} t/\tau_1\tau_2 & (0 \le t \le \tau_1) \\ 1/\tau_2 & (\tau_1 \le t \le \tau_2) \\ (\tau_1 + \tau_2 - t)/\tau_1\tau_2 & (\tau_2 \le t \le \tau_1 + \tau_2) \\ 0 & \text{otherwise} \end{cases} \tag{3}$$

In the analysis, we have adopted the focal mechanism solutions determined by YOSHII (1979b), which are listed in Table 3. We have examined the best-fit hypocentral depths for these events to obtain stable inversion results. Long-period

Table 3

Focal mechanism solution of the earthquakes analyzed in this study.

	Fault Parameters			P-axis		T-axis		Depth (km)			
Event	strike (N°E)	dip (°)	slip (°)	az. (N°E)	pl. (°)	az. (N°E)	pl. (°)	syn.	ISC	ISC (pP-P)	M_s
No. 1	185	80	50	304.9	24.2	57.9	41.0	49	65	—	—
No. 2*	175	34	70	99.4	12.4	142.2	73.3	25	50	43	7.3
No. 3	176	65	135	235.6	8.9	135.3	48.7	45	54	55	—
No. 4	186	80	130	246.9	20.8	135.3	38.7	40	38	38	—
No. 5	118	61	−97	10.7	73.1	213.0	15.7	28	37	41	5.0
No. 6*	163	34	83	78.0	11.2	277.8	78.1	52	62	74	7.4
No. 7	165	72	125	229.5	19.4	114.4	50.3	51	59	61	—
No. 8	192	70	123	258.1	18.3	141.8	53.0	56	52	58	5.1
No. 9	282	90	60	38.6	37.9	165.5	37.7	52	35	43	5.6
No. 10	212	73	135	260.7	15.7	165.3	43.2	47	46	54	5.9
No. 11	210	77	135	268.0	19.2	160.9	40.3	46	50	55	—
No. 12	23	83	−110	271.1	48.2	130.2	35.0	79	89	84	—
No. 13	330	80	40	94.9	19.0	198.7	34.8	47	46	43	—

az: azimuth, pl: plunge.

*Fault parameters of these events are from YOSHII (1979b).

P-waves recorded at only five and three WWSSN stations were available for the February and December events, respectively. The *P*-wave data set for these stations was inverted simultaneously.

Figures 6a, b and c show the source time function and spatial distribution of point source on the fault plane determined for the February event. The optimum depth for this hypocenter is 25 km. This depth is 20 km shallower than that

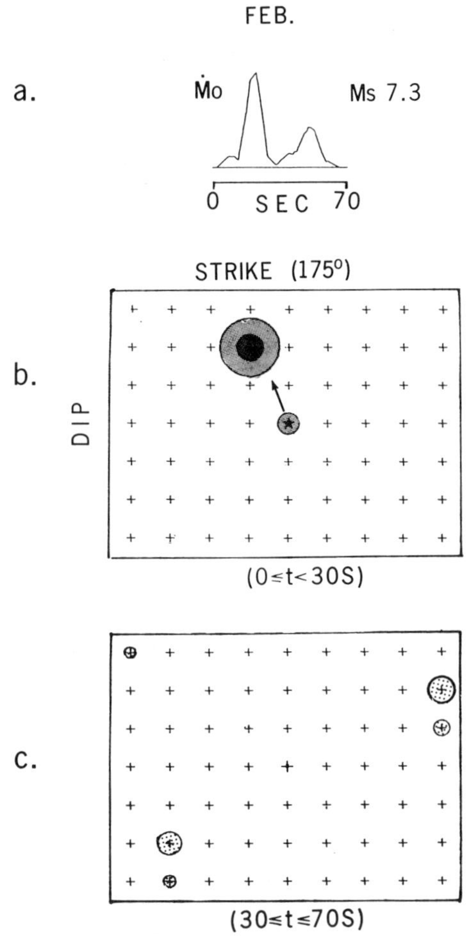

Figure 6

Inversion results for the February event. (a) Source-time function in units of moment-rate. (b) Spatial distribution of the point sources which are located at grids during the 30 sec period after initiation of the rupture. Each grid point represents an area 20 × 20 km on the fixed fault surface which is defined with the strike and dip. The solid star shows the hypocenter at 25 km depth. This pattern indicates the rupture started at the hypocenter with a small amount of energy release and propagated into the up-dip direction with the major energy release. (c) Same as (b) for the period between 30 and 70 sec after the initiation of the rupture. This distribution indicates no coherent rupture pattern. Accordingly the apparent second event in the source-time function seems spurious.

determined by the master event method. The source-time function shows an initial stage of rupture propagation with a small amount of energy release (Figure 6a). This stage is followed by the major energy release. The corresponding point source distribution indicates that the rupture started around the hypocenter and propagated about 50 km into the up-dip direction (Figure 6b). After this major energy release, the located point source distribution is scattered and does not show any coherent pattern of rupture propagation (Figure 6c) even though the source-time function indicates the apparent second event (Figure 6a).

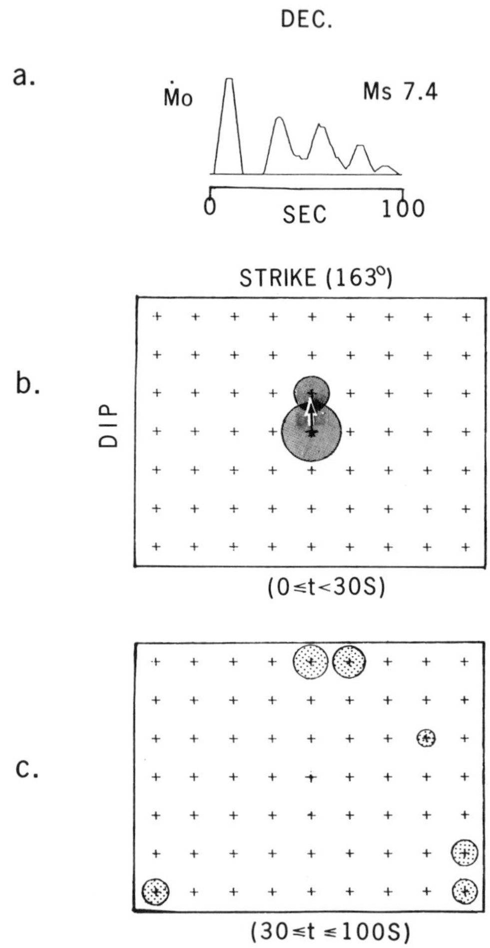

Figure 7
Same as Figure 6 for the December event. (a) Source-time function indicates a sharp rise of the rupture process. (b) Spatial distribution of point sources during the 30 sec period after the initiation of the rupture. The major energy release is located at the hypocenter at 52 km depth and is followed by an up-dip rupture propagation. (c) Same as (b) during the period from 30 and 100 sec. Though two of the point sources are located in the up-dip direction of the hypocenter, the overall distinction does not indicate a coherent pattern.

Figures 7a, b and c show the results for the December event. The optimum hypocentral depth is 52 km. This depth is almost the same as that obtained by the master event method. The source-time function indicates a sharp rise of energy release followed by a few smaller events. The point-source distribution shows that the rupture started with a major energy release around the hypocenter and propagated into the up-dip direction ($0 \leq t \leq 30$ sec, Figure 7b). After this period, the point source distribution is scattered, and does not provide strong evidence for the succeeding rupture pattern (Figure 7c).

The main rupture zone of the February event is located shallower than that of the December event. Although there is some discrepancy in hypocentral depth of the February event by the master event method and by the body-wave analysis, the relationship between the hypocenters of the February and December events, i.e., the deeper hypocentral depth of the December event compared to that of the February one, does not change. Note that this is consistent with the aftershock distribution of the two events, and thus we believe that this is a real feature of the rupture zones of the two events. The above discrepancy between the master event method and the body-wave analysis may be due to the scarceness of the body-wave data and their high noise level and/or inadequacy of the Green's function which assumes a half space to calculate the surface reflections. In addition, the relocation by the master event method may also include some mislocations. Both of the ruptures propagated into the up-dip direction.

If we take into account the depth determination accuracy noted above and the overlapped aftershock distribution of the two events, it is possible that the two events shared a part of the rupture zone. If it is the case, then the rupture of the February event may have smoothed the stress on the fault plane shared by the December event. The seismic moment of the February event was determined to be 1.1×10^{27} dyn-cm and that of the December event to be 1.6×10^{27} dyn-cm. Figures 8a and b show the observed and synthetic waveforms for the February and December events, respectively. The agreement between the two waveforms at each stations is reasonable.

4. *Other Smaller Earthquakes* ($5.5 \leq m_b \leq 6.1$)

We determined focal mechanism solutions for the smaller earthquakes also listed in Table 1 using records of WWSSN. We also redetermined the focal depths of these events by *P*-wave modeling. These earthquakes, which occurred within the box in Figure 1 during the period from 1964 through 1981, are assigned $m_b = 5.5$ or greater by the ISC Bulletins. We added one event, with $m_b = 5.4$, in 1972 and another listed in the Harvard Centroid Moment Tensor Catalogue of 1982 (DZIEWONSKI *et al.*, 1983). Of the eleven smaller earthquakes, one occurred before the February event, three between the February and December events, and seven after the December event. Their hypocentral distribution is shown in Figure 3 with event numbers.

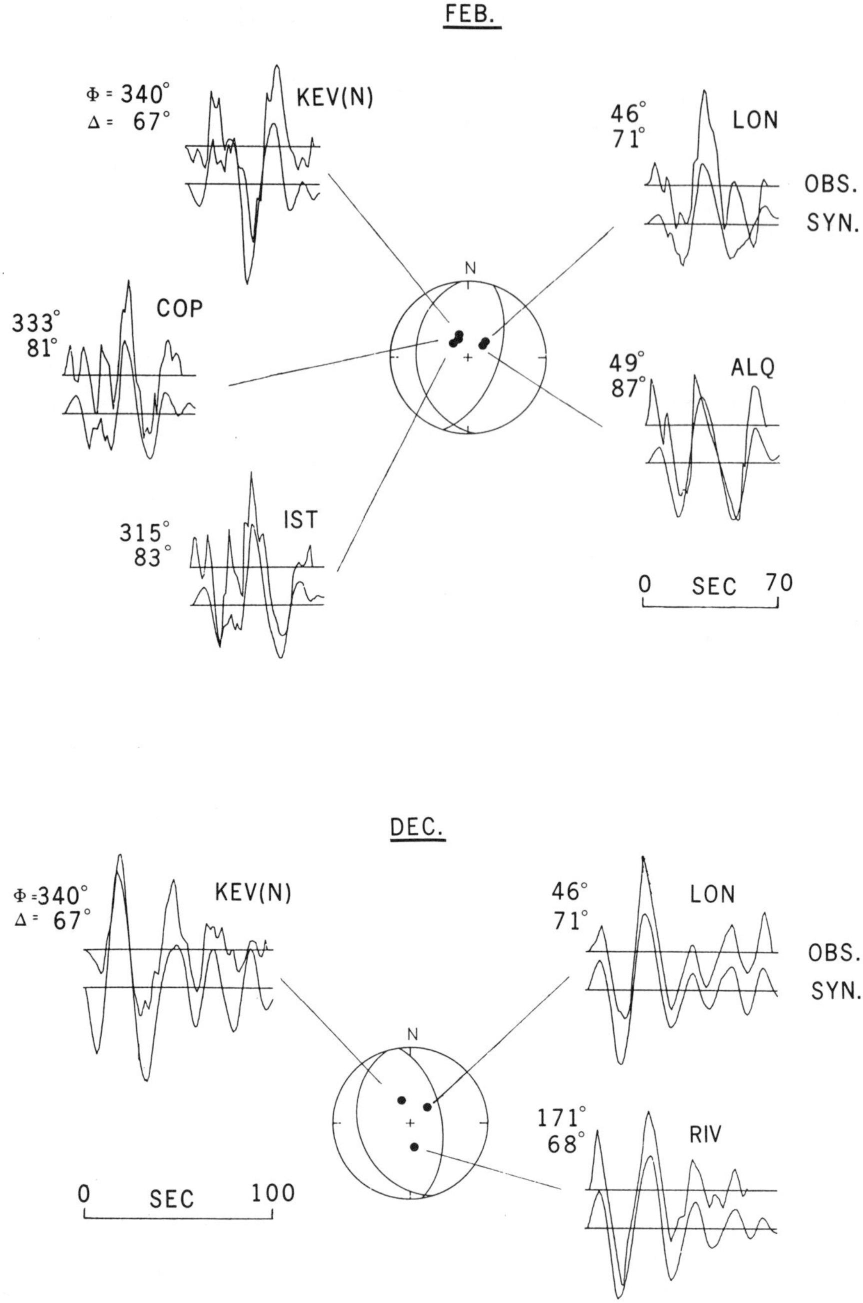

Figure 8
Comparison between the observed and synthetic waveforms. (a) February event. (b) December event.

Figure 9 shows the focal mechanisms of these events. The fault parameters are listed in Table 3. One of the nodal planes is generally not well constrained. We estimated this plane using S-wave polarization angles, although there are some apparently erroneous data. These focal mechanisms and those of the two large 1972 events on the bathymetric map of the area (HYDROGRAPHIC DEPARTMENT, 1982) are plotted in Figure 10. Only the two large 1972 events show a thrust type

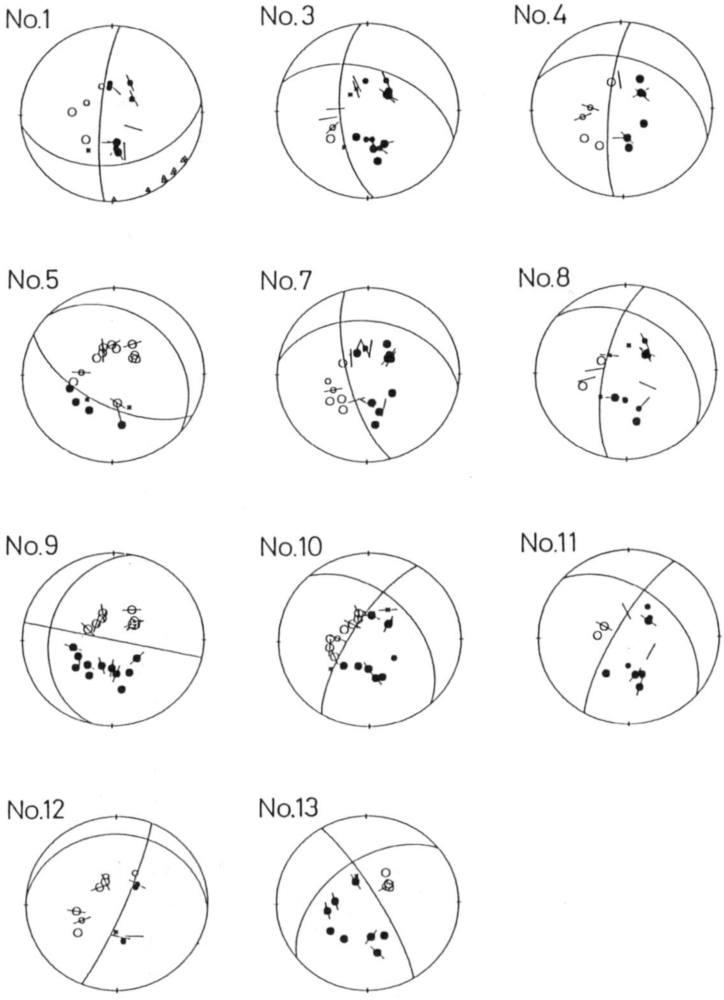

Figure 9

Focal mechanism solutions plotted in the equal area net (lower hemisphere). The solid and open circles indicate compression and dilatation first motions of P-waves, respectively. The larger symbols are from long-period records. The stars indicate nodes of the P-wave first motion. The open triangles indicate dilatation first motions of P-wave from the local stations in Japan reported in the ISC Bulletins. The arrows indicate the S-wave polarization angles.

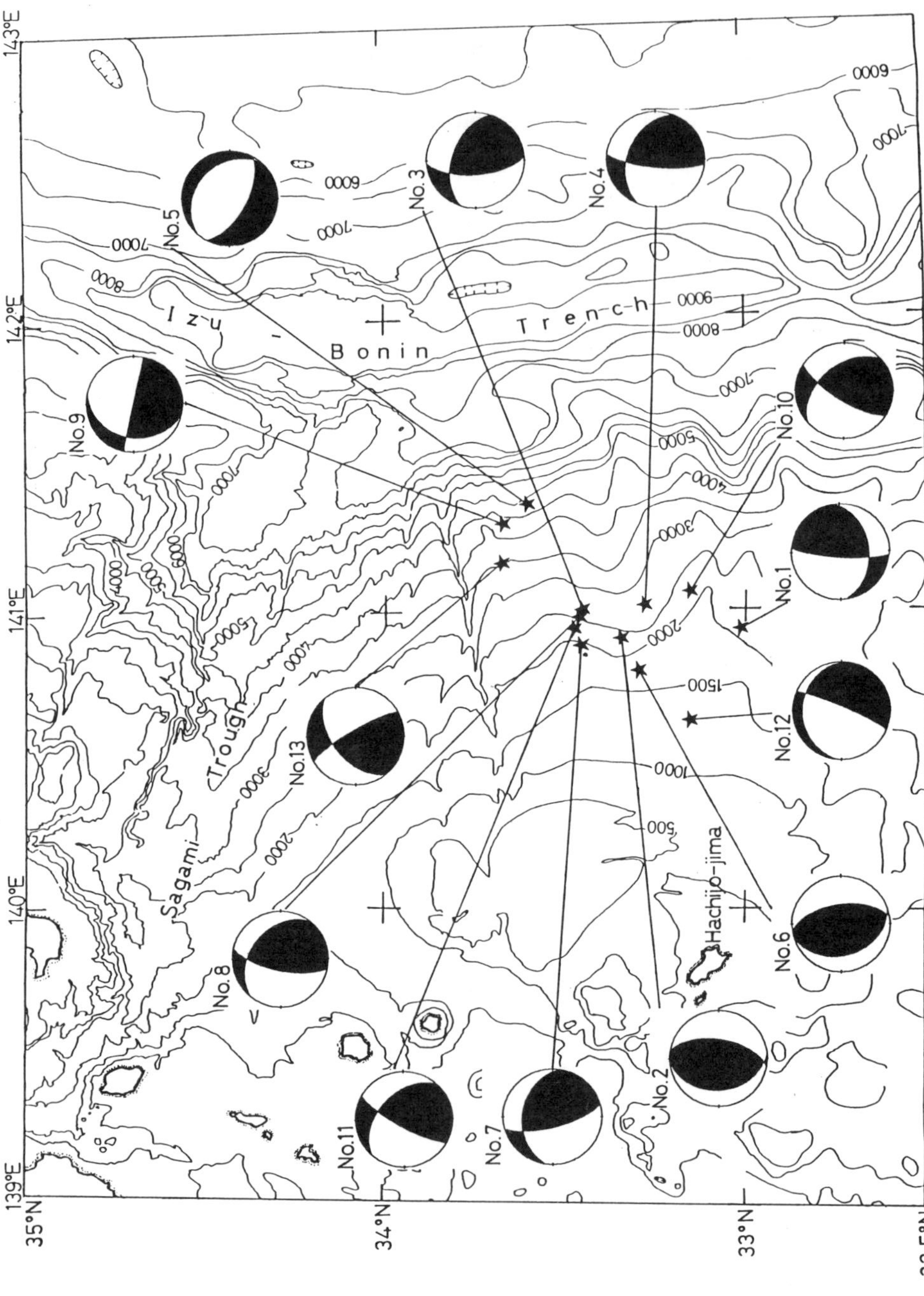

Figure 10
Focal mechanisms plotted in the bathymetric map (HYDROGRAPHIC DEPARTMENT, 1982). The solutions for the two large 1972 events are from YOSHII (1979b).

mechanism. The mechanisms of the smaller events form two groups. One composed of events in the northernmost part of the study area has a nearly vertical fault plane striking northwest-west (events 5, 9 and 13). The other composed of the events near the rupture zones of the two large 1972 events has a P-axis dipping to the west.

We estimated the focal depths of these events by comparing P-waveforms to synthetic waveforms for various focal depths generated, using the algorithm of KROEGER and GELLER (1986). We assumed a point source, and used the crustal structure across the Bonin arc, obtained by seismic refraction (HONZA and TAMAKI, 1985) to calculate reflections and conversions near the source. The thickness of the water layer was estimated at each epicenter from the bathymetric map (HYDROGRAPHIC DEPARTMENT, 1982). The pulses are convolved with the Q operator (CARPENTER, 1966), instrument response and far-field trapezoidal source time function. The attenuation factor t_p^* is assumed to be 1.0 sec. Because the earthquakes treated in this section have short durations ($M_s < 6.0$), there is no serious trade-off between the source time function and the depth. The source time function can be estimated from the first half-cycle of the P-wave and the depth from the overall waveform. Figure 11 shows examples of a comparison between the synthetic and observed seismograms at COL for event 8 and at ADE for event 9. The best fit depth for these stations is indicated by the arrow at the right. We obtained the depth for each event by averaging the best fit depth at several stations; the resulting depth, number of stations used and standard deviations are listed in Table 4. The maximum of the standard deviation is 6.1 km. The comparison with the ISC depths is given in Table 3. For some events, the difference between the depth in this study and the ISC depth amounts to more than 20 km.

We plotted the foci thus obtained and the zones of the major energy release of the two main events in a cross-section perpendicular to the trench in Figure 12a; the map view is shown in Figure 12b. The dotted line approximates the fault plane of

Table 4

The depths of the events determined in this study.

Event	Number of Stations	Average (km)	S.D.
No. 1	4 (1)	48.9	4.1
No. 3	5 (1)	45.0	0
No. 4	6 (3)	40.0	4.1
No. 5	8	27.5	5.0
No. 7	8	51.3	4.1
No. 8	4 (1)	56.3	4.1
No. 9	15	52.3	5.1
No. 10	11	46.7	6.3
No. 11	4	45.8	5.6
No. 12	5 (2)	79.0	3.7
No. 13	8	46.9	6.1

Number in parentheses in this column of "Number of stations" is the number of short period P-waves used in the comparison in waveforms. S.D. denotes standard deviation of depth determination.

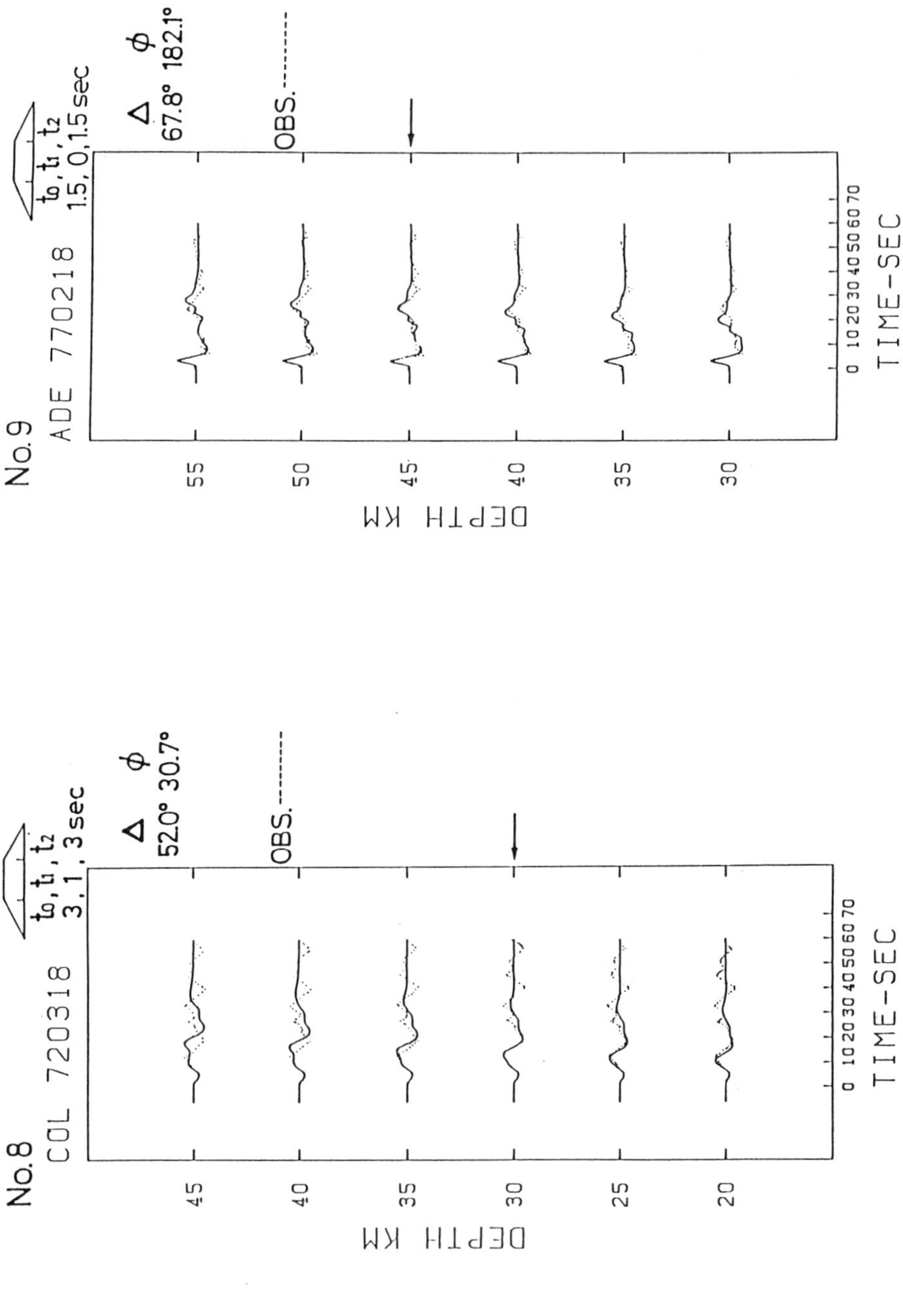

Figure 11

Examples of the comparison between synthetic seismograms for various focal depths and the observed seismogram at COL for event 8 and ADE for event 9. The assumed source time function is indicated at the upper right.

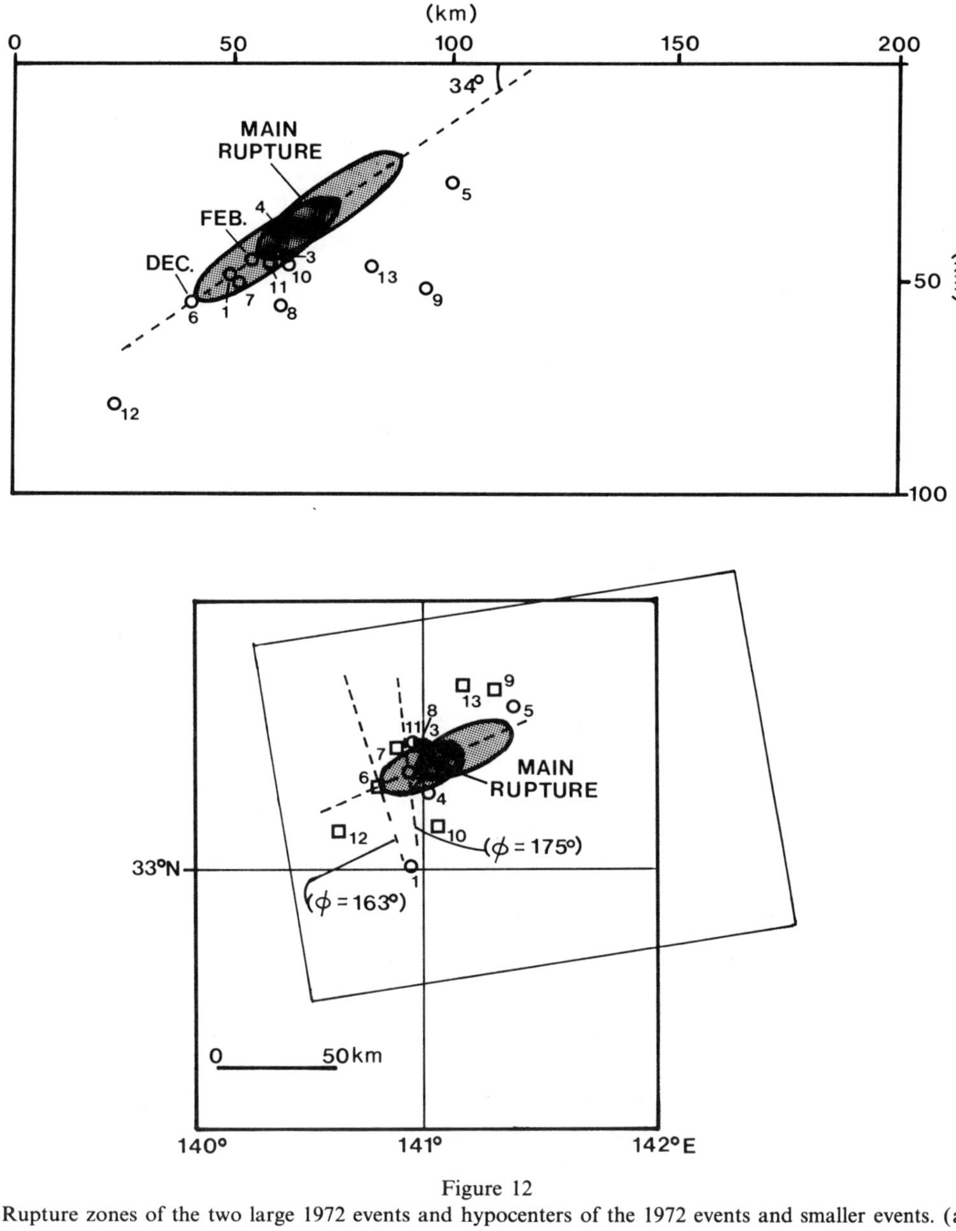

Figure 12

Rupture zones of the two large 1972 events and hypocenters of the 1972 events and smaller events. (a) Cross-section along the dip direction of the fault planes of the two large events. Foci of the smaller events and the two large events are determined by the P-wave modeling and the master event method, respectively. (b) The map view. The epicenters of the smaller events are from ISC and those of the two large events are from the relocation using a master event method.

the two large 1972 events; it goes through their hypocenters with the dip of the mechanism solutions. We used the hypocenters of the two main events determined by the master event method. This is because we regarded that the hypocentral depths by the body-wave analysis may include larger errors due to the scarceness of the data and also because the hypocentral depths by the master event method are more consistent with the aftershock distribution. The dimensions of the rupture zones, 55 km for the February event and 40 km for the December event, were estimated from the body-wave analysis. The epicenters of the smaller events are from the ISC. Assuming that the two large events occurred in the thrust zone, we regard all the other smaller events occurring within the Pacific slab based on the fact that they are generally located beneath the thrust zone and do not have a thrust type mechanism. It is rather surprising that the mechanisms of the smaller events in the vicinity of the rupture zone differ from those of the two large events. In the next section, we will discuss the meaning of these smaller events in terms of the interaction with the large thrust events.

1. Discussion

5.1. Stress State of the Slab

In order to examine the stress generating the earthquakes, we plotted the P and T-axes in the lower hemisphere using an equal area net (Figures 13a and b). In Figure 13a, we also plotted the upper surface of the slab defined by the seismicity in this area (YOSHII, 1979b). We can see that except for the events in the north (events 5, 9 and 13) and the large 1972 events (events 2 and 6), all the events have P-axes parallel to the dip of the Pacific slab. In contrast, events 5, 9 and 13 have horizontal T-axes directing southward.

The solutions of events, 5, 9 and 13 have a sense of motion with the southwest block downthrown with respect to the northeast block, if we assume the more vertical nodal plane is the fault plane. To the north near the triple junction (34°N, 142°E), two large earthquakes are known to have the similar type of mechanism solution as that of these events; they are the 1953 $M_s = 7.9$ and 1984 $M_s = 6.7$ earthquakes (ANDO, 1971; SENO and TAKANO, 1988). Two smaller similar type of earthquakes also occurred 80 and 120 km northwest of the triple junction (SENO and TAKANO, 1988). Further to the west beneath central Honshu, a few similar type of events have also occurred (ISACKS and MOLNAR, 1971; ITO and ANNAKA, 1977). Thus this mode of deformation, illustrated schematically in Figure 14, is common in the slab near the junction between the northeast Japan and Bonin arcs. The sense of motion of this mechanism solution indicates that the slab near the junction is pulled to the south. Presumably, the greater dip of the slab beneath the Bonin-Mariana arc, compared to that beneath northern Honshu (e.g., YOSHII, 1979b), causes this deformation of the slab near the junction (SENO and TAKANO, 1988).

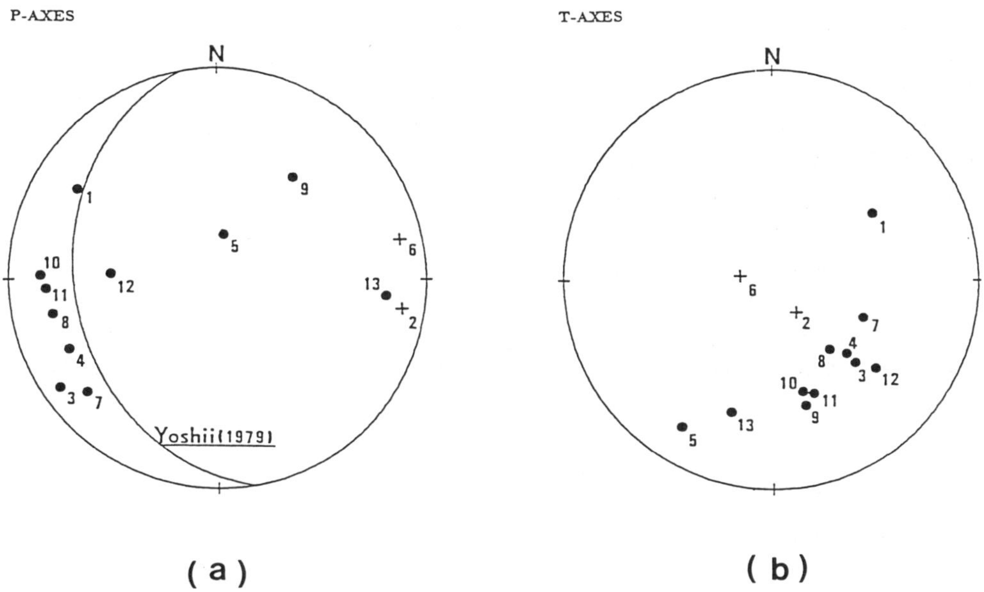

Figure 13
(a) Equal area protection of the P-axes of the events listed in Table 1. The upper surface of the Pacific slab estimated from the regional seismicity (YOSHII, 1979b) is shown. (b) Equal area projection of the T-axes of the events listed in Table 1.

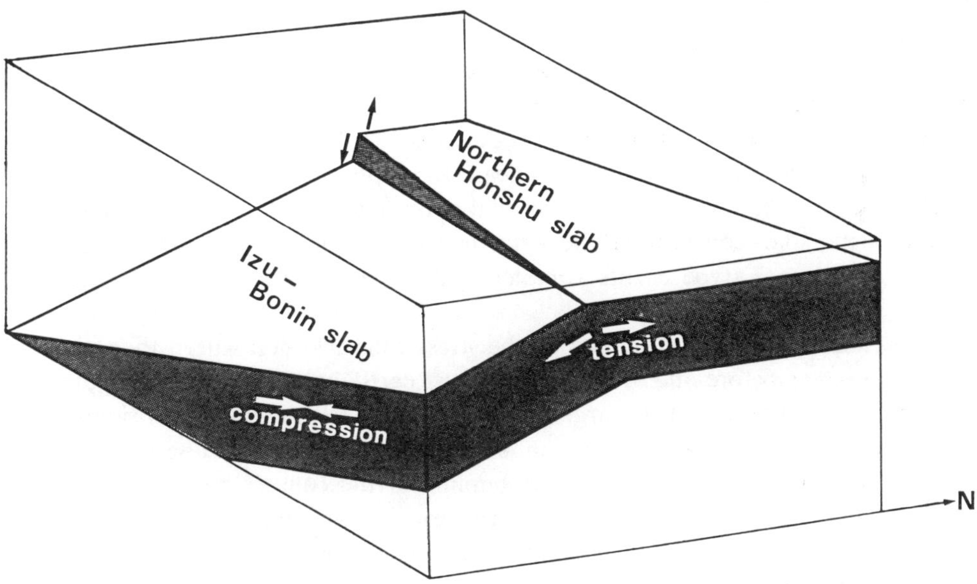

Figure 14
A schematic illustration showing the stress state and the mode of deformation of the slab near the northern Honshu-Bonin arc junction.

The others events except 5, 9 and 13 have down-dip P-axes. Previous studies of mechanism solutions of deep and intermediate-depth earthquakes beneath the Bonin arc show that the slab is in down-dip compression at the depth larger than 300 km (ANNAKA, 1977; KATSUMATA and SYKES, 1969) and at intermediate depth between 70–250 km (FUJITA and KANAMORI, 1981). The down-dip compression of the earthquakes in this study is thus concordant with the compressional stresses in the deeper part. The stress state of the slab near the study area is also illustrated in Figure 14. However, the depth of the earthquakes in the present study is mostly around 50 km, shallower than those of the down-dip compressional events usually seen in the down-going slab (YOSHII, 1979a; KAWAKATSU and SENO, 1983). They may then reflect an unusual mode of deformation caused by the interaction with the large 1972 thrust events.

5.2. Interaction Between the Thrust Zone Events and the Slab Events

There are only few studies of the interaction between a large thrust earthquake in a subduction zone and smaller slab earthquakes near its rupture zone. MALGRANGE and MADARIAGA (1983) studied Chilean subduction zone earthquakes and found that moderate size slab earthquakes are distributed around the rupture zone of large thrust earthquakes before and after their occurrence. There was a characteristic feature in the location and timing of the occurrence of down-dip compressional type (C-type) and down-dip tensional type (T-type) earthquakes. Before the occurrence of a large thrust earthquake, T-type end C-type earthquakes tend to occur landward of and beneath the rupture zone, respectively. In contrast, after the occurrence of the thrust earthquake, C-type and T-type events tend to occur landward and seaward of the rupture zone, respectively. KAWAKATSU and SENO (1983) studied the northern Honshu subduction zone and found that the double seismic zone extended to just beneath the rupture zone of a large thrust type earthquake, the 1978 Miyagi-Oki earthquake, prior to its occurrence. In that case, C-type earthquakes concentrated beneath the rupture zone, and to the landward of the rupture zone, C-type events completely disappeared prior to the Miyagi-Oki event.

These features can be explained by the stress concentration within the slab near the rupture zone before and after a large thrust earthquake as shown in Figure 15. Before the occurrence of a large thrust earthquake, down-dip tensional and compressional stresses within the slab will appear landward and seaward of the rupture zone, respectively, due to the coupling at the rupture zone (Figure 15a). After the occurrence of the large thrust earthquake, the stress pattern in the slab will be reversed as shown in Figure 15b. Recently a similar stress change within the slab at intermediate depth and at the outer rise before and after large thrust type earthquakes is proposed, based on a large amount of focal mechanism data by LAY *et al.* (1988) and ASTIZ and KANAMORI (1986).

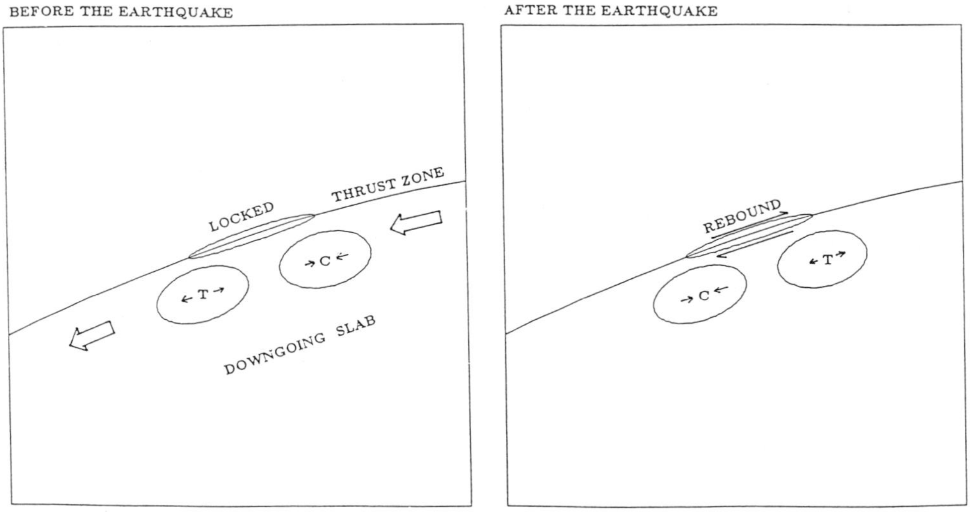

Figure 15
A schematic illustration of the pertubation to the stress of the slab (a) due to seismic coupling at the thrust zone and (b) due to faulting at the time of thrust earthquake.

In the case of the Hachijo-Oki rupture zone, the situation seems more complex, possibly because of the twin events with a nine month interval; nevertheless, there is a systematic feature in the location and timing of the occurrence of smaller events beneath this rupture zone. During the period between the two large thrust events, C-type events (events 3 and 4) occurred beneath the deeper portion of the rupture zone of the February event (Figure 12). Within about one year after the occurrence of the December event, some C-type events (events 7 and 8) occurred beneath the deeper portion of the rupture zone of the December event. Other events (events 1, 5, 9, 10 and 13) are not located close to the rupture zone and are not easily related to the occurrence of the large events. The occurrence of the February event, which ruptured the shallower portion of the thrust zone, would have loaded the slab near the end of its rupture zone in a down-dip direction and might have caused events 3 and 4. Similarly, the occurrence of the December event would have further loaded the deeper portion of the slab and might have caused events 7 and 8. These stresses are perturbations to the average tectonic stress in the slab. The average tectonic stress in the slab beneath the Bonin arc seems to be down-dip compressional as discussed in the former subsection. The perturbing stresses may thus trigger only the down-dip compressional events and not down-dip tensional events.

It is interesting whether seismicity and, by implication, stresses are concentrated near the thrust interface only before and after large events. KAWAKATSU and SENO (1983) suggested that for the large thrust Miyagi-Oki earthquake the stress concentration near the rupture zone will be released sometime after the event. A recent study (SENO and KAWAKATSU, 1985), including the five years after the event, shows that activity within the slab near the rupture zone is still continuing. In the case of the Hachijo-Oki events, a few slab events occurred even 5 to 8 years after the rupture (events 10–12). It is thus possible that stress concentration near the thrust interface in subduction zones with seismic coupling may be a permanent feature. Alternatively, the Hachijo-Oki case may be special in that the stress concentration within the slab almost always exists and causes smaller slab events. Resolution of this question requires data on seismicity for the longer period after the Miyagi-Oki and Hachijo-Oki events and other large thrust earthquakes.

6. Conclusion

The source area of the 1972 Hachijo-Oki earthquakes in the northern Bonin arc is the only site of significant seismic energy release along the arc for the past 90 years. The rupture zones of the February and December Hachijo-Oki events probably overlapped partially, with that of the December event deeper than that of the February event. The smaller events near the area do not show thrusting along the plate interface but rather occurred within the descending Pacific slab. They have down-dip P-axes and, along with the similar mechanism solutions of deeper events, indicate that the Pacific slab beneath the forearc of the northern Bonin arc is in down-dip compression. The unusually shallow activity within the slab, however, may have been caused by the stress disturbance due to the seismic coupling and succeeding seismic faulting of the zone.

Further to the north near the triple junction, normal fault events within the Pacific slab show steeply dipping nodal planes perpendicular to the trench axis with the southwest block downthrown with respect to the northeast block; possibly these events occurred due to the change in slab dip between the northern Honshu and Bonin arcs.

Acknowledgements

We thank Seth Stein and Paul Silver for critical review of the manuscript. We also thank Yoji Kobayashi, Tadashi Sato and the members of his laboratory for encouragement and discussion throughout this work. This research was supported by the National Science Foundation under Grant EAR–8518649.

REFERENCES

ANDO, M. (1971), *A fault model of the 1953 Boso-Oki earthquake*, Abstr. Seismol. Soc. Jpn *2*, 49 (in Japanese).

ANNAKA, T. (1977), *Spatial distribution and focal mechanisms of deep earthquakes in the Izu-Bonin and northeastern Japan arcs*, Zisin *30*, 213–225 (in Japanese).

ASTIZ, L. and H. KANAMORI (1986), *Interplate coupling and temporal variation of mechanisms of intermediate-depth earthquakes in Chile*, Bull. Seismol. Soc. Am. *76*, 1614–1622.

CHASE, T. H., W. MENARD, and MAMMERICKS, J., *Bathymetry of the North Pacific* (Scripps Inst. of Oceanography and Inst. of Marine Resources, 1970).

CARPENTER, E. W., *Absorption of elastic waves—an operator for constant Q mechanism*, Report 0–43166 (Atomic Weapons Research Establishment, London, 1966).

DZIEWONSKI, A. M., A. FRIEDMAN, D. GIARDINI, and J. H. WOODHOUSE (1983), *Global seismicity of 1982: Centroid moment tensor solutions for 308 earthquakes*, Phys. Earth Planet. Inter. *33*, 76–90.

FUJITA, K. and H. KANAMORI (1981), *Double seismic zones and stresses of intermediate depth earthquakes*, Geophys. J. R. Astron. Soc. *66*, 131–156.

HONZA, E. and K. TAMAKI, *The Bonin Arc*, in *The Ocean Basins and Margins*, 7A (eds. A. E. M. Narin, F. G. Stehli and S. Uyeda) pp. 459–502, 1985.

HYDROGRAPHIC DEPARTMENT, Maritime Safety Agency, Bathymetric Chart, Central Japan, no. 6313, 1982.

ISACKS, B. and P. MOLNAR (1971), *Distribution of stresses in the descending lithosphere from a global survey of focal-mechanism solutions of mantle earthquakes*, Rev. Geophys. Space Phys. *9*, 103–174.

ITO, K. and M. ANNAKA (1977), *Spatial clustering and focal mechanisms of the deep earthquakes in central Japan*, Zisin *30*, 201–212 (in Japanese).

KATSUMATA, M. and L. R. SYKES (1969), *Seismicity and tectonics of the western Pacific: Izu-Mariana-Caroline and Ryukyu-Taiwan regions*, J. Geophys. Res. *74*, 5923–5948.

KAWAKATSU, H. and T. SENO (1983), *Triple seismic zone and the regional variation of seismicity along the northern Honshu arc*, J. Geophys. Res. *88*, 4215–4230.

KIKUCHI, M. and Y. FUKAO (1985), *Iterative deconvolution of complex body waves from great earthquakes—the Tokachi-Oki earthquake of 1968*, Phys. Earth Planet. Inter. *37*, 235–248.

KIKUCHI, M. and H. KANAMORI (1982), *Inversion of complex body waves*, J. Bull. Seismol. Soc. Am. *72*, 491–506.

KROEGER, G. C. and R. J. GELLER, *An efficient method for synthesizing teleseismic body waves for shallow sources in a vertically stratified medium*, J. Geophys. Res., submitted, 1986.

LAY, T., L. ASTIZ, H. KANAMORI, and D. H. CHRISTENSEN, *Temporal variation of large intraplate earthquakes in coupled subduction zones*, Phys. Earth Planet. Inter., in press, 1988.

MALGRANGE, M. and R. MADARIAGA (1983), *Complex distribution of large thrust and normal fault earthquakes in the Chilean subduction zone*, Geophys. J. Roy. Astron. Soc. *83*, 489–505.

PETERSON, E. T. and T. SENO (1984), *Factors affecting seismic moment release rates in subduction zones*, J. Geophys. Res. *89*, 10233–10248.

SENO, T. (1977), *The instantaneous rotation vector of the Philippine Sea plate relative to the Eurasian plate*, Tectonophysics *42*, 209–226.

SENO, T. and T. EGUCHI, *Seismotectonics of the western Pacific region*, *Geodynamics of the Western Pacific—Indonesian Region*, Geodynamics Series, 11 (AGU, Washington, D.C., 1983).

SENO, T. and H. KAWAKATSU, *Monitoring seismicity within the slab to predict large earthquakes at the deep thrust zone, northern Honshu*, *Abstr.*, 23rd IASPEI meeting, Tokyo, 31, 1985.

SENO, T. and T. TAKANO (1988), *Seismotectonics at the trench-trench-trench triple junction off central Honshu*, Pure and Appl. Geophys. *129*, *112*, 27–41.

STEIN, S. and D. WIENS (1986), *Depth determination for shallow teleseismic earthquakes: Methods and results*, Rev. Geophys. *24*, 806–832.

TAJIMA, F. and H. KANAMORI (1985), *Global survey of aftershock area expansion patterns*, Phys. Earth Planet. Interiors *40*, 77–134.

YOSHII, T. (1979a), *A detailed cross-section of the deep seismic zone beneath northern Honshu, Japan*, Tectonophysics *55*, 349–360.

YOSHII, T. (1979b), *Compilation of geophysical data around the Japanese islands*, Bull. Earthq. Res. Inst. *54*, 75–117 (in Japanese).

(Received September 26, 1986, revised April 2, 1988, accepted April 15, 1988)

PAGEOPH, Vol. 129, Nos. 1/2 (1989) 0033-4553/89/020263-20$1.50 + 0.20/0

Do Trench Sediments Affect Great Earthquake Occurrence in Subduction Zones?

LARRY J. RUFF[1]

Abstract—Seismic energy release is dominated by the underthrusting earthquakes in subduction zones, and this energy release is further concentrated in a few subduction zones. While some subduction zones are characterized by the occurrence of great earthquakes, others are relatively aseismic. This variation in maximum earthquake size between subduction zones is one of the most important features of global seismicity. Previous work has shown that the variation in maximum earthquake size is correlated with the variation in two other subduction zone properties: age of the subducting lithosphere and convergence rate. These two properties do not explain all the variance in maximum earthquake size. I propose that a third subduction zone property, "trench sediments", explains part of the remaining variance in maximum earthquake size. Subduction zones are divided into two groups: (1) those with excess trench sediments, and (2) those with horst and graben structure at the trench. Thirteen of the 19 largest subduction zone events, including the three largest, occur in zones with excess trench sediments. About half the zones with excess trench sediments are characterized by great earthquake occurrence. Most of the other zones with excess trench sediments but without great earthquakes are predicted to have small earthquakes by the age-rate correlation. Two notable exceptions are the Oregon-Washington and Middle America zones. Overall, the presence of excess trench sediments appears to enhance great earthquake occurrence. One speculative physical mechanism that connects trench sediments and earthquake size is that excess trench sediments are associated with the subduction of a coherent sedimentary layer, which at elevated temperature and pressure, forms a homogeneous and strong contact zone between the plates.

Key words: Subduction zones, great earthquakes, trench sediments.

1. Introduction

Most of the largest earthquakes occur in subduction zones and represent the underthrusting of the oceanic lithosphere. Beyond this basic observation, subduction zone seismicity presents many puzzling features. One rather remarkable feature is the variation in maximum earthquake size between subduction zones: this variation is more than three orders of magnitude as measured by the seismic moment. To emphasize this range in characteristic earthquake size, the Chilean

[1] Department of Geological Sciences, University of Michigan, Ann Arbor, MI 48109, U.S.A.

subduction zone frequently generates giant earthquakes with fault area dimensions reaching hundreds of kilometers, widespread strong shaking, and destructive tsunamis. On the other hand, underthrusting in the Marianas arc will occasionally generate a magnitude ≈ 7 earthquake. As another illustration, since the seismic energy release is dominated by the largest events (see, for example, GUTENBERG and RICHTER, 1954), a map of the total seismic energy release in the twentieth century would show a high concentration in some subduction zones while others would be virtually "aseismic". Understanding this variation in large earthquake occurrence is a fundamental pursuit in seismology.

Two lines of inquiry into this subject will be followed: (1) the relationship between maximum earthquake size and other subduction zone "parameters", (2) observations and models of the earthquake rupture process. The above two approaches are not the only lines of study into the subject, but several investigators have made progress using these approaches. The first approach can explain much of the variation in maximum earthquake size, though significant deviations remain. On the other hand, the second approach explains the occurrence of great earthquakes in terms of the strength distribution of the plate interface. In this paper, I propose that the presence or absence of trench sediments forms a link between the two approaches. In particular, trench sediments seem to enhance seismic coupling. I will first briefly review some of the relevant results on seismic coupling and the rupture process, then proceed with a short discussion of trench sediments, and finish with a global survey on the association of excess trench sediments with great earthquakes.

2. Global Variation in Seismic Coupling

KANAMORI (1971a) noted that interplate seismic coupling varies systematically in the subduction zones of the northwestern Pacific. He equated the maximum earthquake size in a region with seismic coupling. At that time, KANAMORI explained the variation in seismic coupling by an evolutionary process, whereby the upper plate contact surface degrades. UYEDA and KANAMORI (1979) presented a global perspective and discussed the possible interconnection between several subduction zone parameters. Their paper promoted the approach of "comparative subductology": using global comparisons of the essential characteristics of subduction zones to uncover important aspects of the subduction process. UYEDA and KANAMORI included seismic coupling as an essential characteristic, and suggested that the kinematic parameter of upper-plate absolute velocity exerts a strong influence on coupling. RUFF and KANAMORI (1980) then focussed on seismic coupling, and established that maximum earthquake size is significantly correlated with two parameters: convergence rate and age of the subducting oceanic lithosphere. In particular, great earthquakes tend to occur in subduction zones with faster convergence and younger lithosphere, while the relatively aseismic subduction zones are characterized by slower convergence and older lithosphere (see Figure 1). JARRARD (1986) then updated and extended the data set, both in terms of the number of subduction zones and number of parameters, and further confirmed that maximum earthquake size is correlated with oceanic lithosphere age and rate. Although much of the variance in maximum earthquake size is explained by age and rate, there are a few subduction zones that deviate from the overall trend. Thus, there are other factors that influence earthquake size.

Statistical correlations do not necessarily indicate a direct cause and effect relationship. Nevertheless, RUFF and KANAMORI (1980) introduced preferred trajectory as a simple physical mechanism that might explain the above correlation. Plate age and convergence rate directly translate into the preferred vertical and horizontal velocities of the down-going slab. Note that a long-term effect of a steep preferred trajectory, i.e., old lithosphere and slow rate, could be back-arc spreading. Indeed, the subduction zones with recent back-arc spreading activity are characterized by small earthquake size (Figure 1, see RUFF and KANAMORI, 1980). Shallow preferred trajectory might cause large earthquakes by either increasing the tectonic compressive stress or by increasing the fault width as a consequence of a shallow dip angle. Are these mechanisms consistent with our knowledge of the source process of great earthquakes? In the next section, we will see that the current physical model used to explain the earthquake rupture process is not easily related to the above mechanisms.

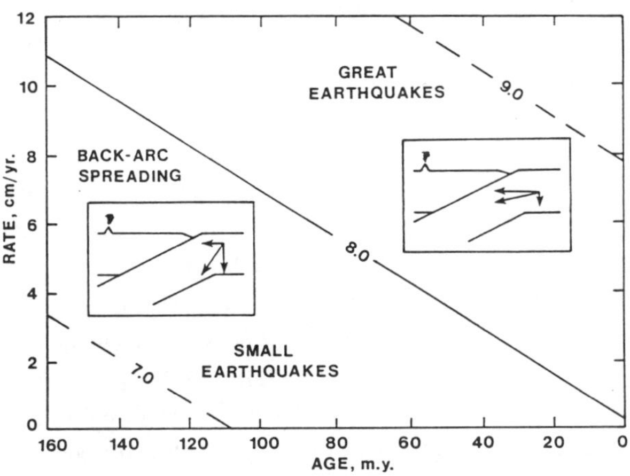

Figure 1

Seismic coupling as a function of lithosphere age and convergence rate (after RUFF and KANAMORI, 1980). Linear regression analysis shows that maximum earthquake size depends on age and rate; the best-fit plane is indicated by the contours of predicted M_W. Subduction zones that are subducting young lithosphere at a high rate would plot in the upper-right corner, and these zones are characterized by great earthquake occurrence (i.e., $M_W > 8$). The lower-left corner contains subduction zones with smaller earthquakes and back-arc spreading. The insets depict the mechanism of preferred trajectory, whereby age and rate determine the preferred vertical and horizontal velocities of the slab.

3. *Large Subduction Zone Earthquakes*

3.1. Seismic Gaps

Subduction at shallow depths is accommodated by slip across a narrow zone, the plate boundary interface. As subduction proceeds, all segments along the plate boundary must eventually slip, either seismically or aseismically. Plate boundary segments that have previously slipped as large earthquakes are likely to slip again as large earthquakes. Observations indicate that the recurrence intervals between large earthquakes in the same segment vary from less than one to a few hundred years. Hence, future earthquakes are more likely to occur sooner in plate boundary segments with a history of large earthquakes that have not experienced a large earthquake in many decades. This simple yet powerful concept of seismic gaps has been developed and applied to the major plate boundaries, and has successfully forecast the locations of a few large earthquakes (FEDOTOV, 1965; MOGI, 1969; SYKES, 1971; KELLEHER, 1972; KELLEHER *et al.*, 1973; MCCANN *et al.*, 1979; NISHENKO and MCCANN, 1981; NISHENKO, 1985). Plate boundary segmentation is an essential component of seismic gaps which in part determines the size of the largest earthquakes.

3.2. Plate Boundary Segmentation

Underthrusting can occur as seismic slip in the shallow part of the subduction zone, while slip apparently occurs only aseismically below a depth of about 40 km (see RUFF and KANAMORI, 1983a). For the typical shallow dip angle of 10° to 20°, the potential fault width is one to two hundred kilometers. Major subduction zones extend to thousands of kilometers along the arc system, thus the seismically active part of a plate boundary can be viewed as a long narrow strip that is composed of seismic segments. One of the primary tasks of seismic gap research has been to determine the rupture length of the large plate boundary earthquakes. It is usually assumed that the one-day aftershock area delimits the mainshock rupture area. The one-day aftershock area frequently coincides with the region of uniform and intense shaking. Hence, observations of aftershocks or shaking can identify the seismic segments of a subduction zone. In several cases, the seismic segmentation corresponds to the physiographic segmentation, e.g., major fracture zones or ridges in the oceanic plate form the boundary between two segments.

The overall size of an earthquake is best measured by the seismic moment, defined as $M_0 = \mu AD$, where μ is the source region shear modulus, A is the fault area, and D is the average seismic displacement. The moment magnitude scale is directly dependent on the seismic moment ($M_W = (\log M_0)/1.5 - 10.7$, with M_0 in dyn cm, KANAMORI, 1978). As discussed in KANAMORI and ANDERSON (1975), the average displacement tends to increase as the fault area increases. Thus, in general, a larger fault area is equivalent to a larger seismic moment and M_W. From this perspective, maximum earthquake size depends on those factors that control fault area.

Fault area obviously depends on fault width and length. KELLEHER *et al.* (1974) claimed that variations in fault width play a dominant role in seismic coupling. They noted that some of the largest earthquakes have occurred in subduction zones with a wide contact zone. They offered support for this claim by measuring the width between the trench and the 70 km depth contour of the Benioff zone. While their definition of "width" is attractive in that it can be globally measured, unfortunately it does not directly yield the width of the seismogenic zone. To further explore this idea, we can look at the seismogenic width as displayed in the aftershock zones of large earthquakes. The rupture width and length of selected large earthquakes are shown in Figure 2. Aftershock areas are usually determined subjectively by drawing a rectangle or oval around the scattered aftershocks. To obtain a more objective determination, TAJIMA and KANAMORI (1985) devised an energy contour method. I have simply measured the maximum widths and lengths, perpendicular and parallel to the trench axes, of the one-day aftershock regions in TAJIMA and KANAMORI. Two events not in their paper are included in Figure 2, the dimensions for the 1960 Chile event are from PLAFKER and SAVAGE (1970), and the 1957 Aleutian values are taken from HOUSE *et al.* (1981). For the examples presented in Figure 2, it would seem that the variation in fault length is more

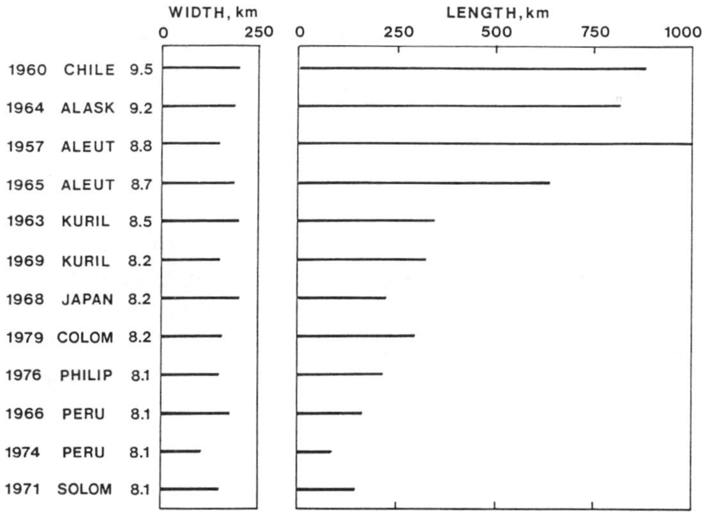

Figure 2
Width and length of the one-day aftershock areas of recent great earthquakes, identified by year of occurrence, region, and M_W. Except for two exceptions noted in the text, the measurements are based on the aftershock areas of TAJIMA and KANAMORI (1985). Rupture length varies more than rupture width.

important than the variation in fault width as an explanation for variation in maximum earthquake size.

3.3. Rupture Length and the Asperity/Barrier Model

Fault length is a consequence of where the earthquake rupture front starts and stops. The development of the asperity/barrier model gives us a simple physical characterization of earthquake rupture (for example, AKI, 1979; DAS and AKI, 1977; KANAMORI and STEWART, 1978; NUR and ISRAEL, 1980; KANAMORI, 1981; LAY *et al.*, 1982; AKI, 1983; RUFF, 1983; BECK and RUFF, 1984, 1987; SCHWARTZ and RUFF, 1987). The essence of this model is that a fault zone can be characterized by strong and weak regions. The strong regions are called asperities and slip seismically. Slip in the weak regions concentrates the stress at the asperities. The largest earthquake in a seismic segment results from the failure of the largest asperity, or the sequential failure of the largest asperities. The rupture front of the maximum size earthquake can start in either a weak or strong region, but must break the dominant asperity. The rupture front stops either at a major geometric barrier or in a weak region that has previously slipped and therefore is not highly stressed. Major geometric barriers include drastic kinks in the plate boundary strike or changes in plate boundary type.

The asperity/barrier model provides a qualitative physical connection between seismic segmentation and the earthquake rupture process. This connection leads to some interesting consequences. Figure 3 shows two hypothetical strength distributions along a subduction zone. The distribution with small isolated strong points would cause small localized rupture areas, while the distribution with a relatively constant strength would result in a large rupture length. Thus, strong seismic coupling depends more on the "smoothness" of the strength distribution than on the peak value of the strength. This view is supported by the apparent smoothness of the rupture process of the 1964 Great Alaskan earthquake (see RUFF and KANAMORI, 1983b).

The above discussion is not a comprehensive review of all the research on fault heterogeneity, nor does it present all viewpoints on the rupture process. Given this qualifier, I state the following conclusion: stronger seismic coupling is caused by a smoother strength distribution along the seismogenic zone. How does this conclusion relate to our earlier considerations on the global variation in seismic coupling? Are there any mechanisms that relate lithosphere age and convergence rate to fault strength distribution? RUFF and KANAMORI (1983a) noted two facts relevant to the above questions: (1) the old lithosphere subducting in the Marianas is characterized by rough ocean-floor topography, (2) the two largest earthquakes (1960 Chilean and 1964 Alaskan events) occurred in segments that seem to have "excess trench sediments". If a rough ocean-floor topography translates into highly heterogeneous strength distribution, then the first fact would explain the relatively aseismic nature

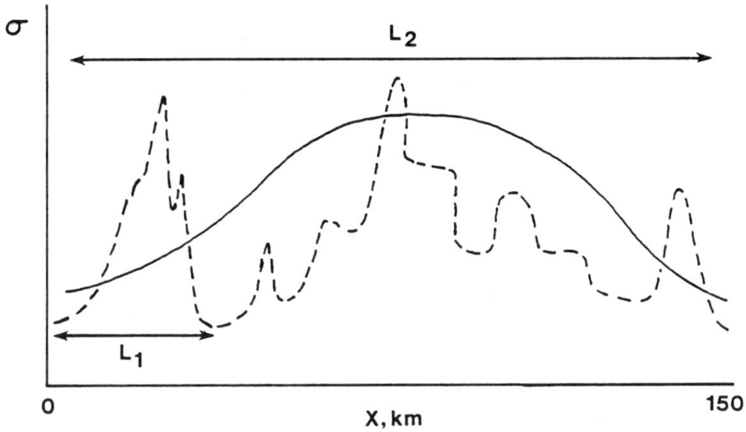

Figure 3
Two hypothetical strength distributions along a fault: smooth (solid line) and irregular (dashed). Each distribution is plotted as failure stress (σ) *versus* distance along the fault zone. Slip occurs when the tectonic stress level reaches the failure stress. Slip in the weaker regions transfers stress to the strong regions. Since the strongest regions in the irregular distribution are isolated, the earthquakes that eventually break these regions have a rather small rupture dimension (L_1). For the relatively smooth distribution, the dimension of the strong region is larger, hence the earthquake will have a greater rupture length (L_2).

of the Marianas subduction zone. If we allow the interpretative leap that the presence of excess trench sediments is related to a smoother strength distribution, then the second fact would explain the two giant earthquakes. The main thrust of this paper is to pursue the possible contribution of excess trench sediments. Rather than analyze just one or two trench segments, I will adopt the *comparative subductology* attitude and present a global survey. The term "excess trench sediments" will first be defined and developed.

4. Excess Trench Sediments

Even in their pioneering plate tectonics paper, ISACKS *et al.* (1968) made some provision for sediment subduction. At first glance, the tectonic role of sediments appears to be a minor one. Sediments do however significantly impact geology, geochemistry, and geophysics: many geological terranes are thought to be exumed trench sediments; recycling of crustal material via sediment subduction affects the geochemical evolution of the mantle; and as promoted in this paper, sediments may influence subduction zone seismicity.

Sediments are transported into the trench region from both the oceanic plate and island arc sides. There are two apparently contradictory views on the fate of these sediments: (1) the sediments are subducted along with the down-going plate,

(2) sediments riding on the oceanic plate are scraped off and deposited; the accretionary prism contains the accumulated oceanic sediments along with continent-derived sediments. We now know that both views are correct. That is, some subduction zones are presently subducting all available sediments while others seem to be actively constructing accretionary prisms (see e.g., KLUM and FOWLER, 1974; KARIG and SHARMAN, 1975; SCHOLL *et al.*, 1977; SCHOLL *et al.*, 1980; VON HUENE *et al.*, 1982; HILDE, 1983). In some subduction zones, not only are the available sediments being subducted, but the sediment subduction process is apparently eroding the upper-plate in the trench region. In many of these subduction zones, the down-going oceanic plate develops a horst and graben structure between the outer-rise and the trench. This structure frequently strikes parallel to the trench axis, and is thought to result from tensional stresses associated with plate bending (see CHAPPLE and FORSYTH, 1979). One proposed mechanism for sediment subduction is that the horst and graben structure provides "buckets" that fill up with sediments, which are then carried down the subduction zone (Figure 4, for a discussion of this mechanism, see HILDE, 1983). If the available sediments do not fill the graben "buckets", then the leading edge of the upper plate erodes to provide

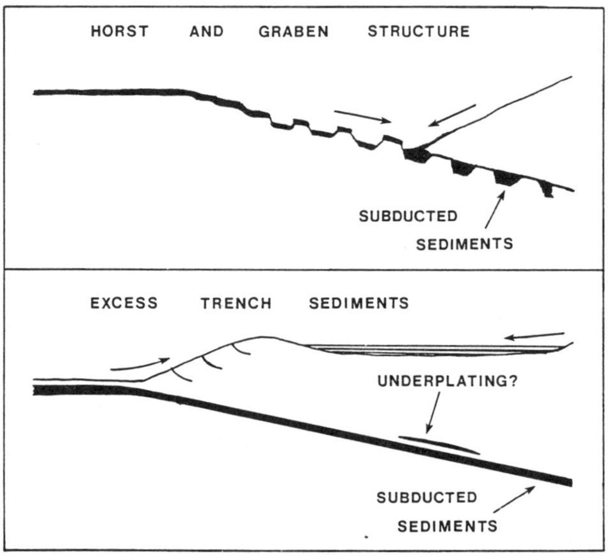

Figure 4
Two trench morphologies and possible mechanisms of sediment subduction. These graphs are schematic and not to scale, but the depth of the grabens is typically < 1 km, and the horizontal dimension is approximately 10 km, hence the seismogenic zone is off to the right. A well-developed horst and graben structure provides "buckets" which can erode the upper-plate and carry down sediments. In contrast, one of the characteristics of excess trench sediments is that sediments are added to the upper-plate and form an accretionary prism. The role of underplating is not known. At this time, it is only speculation that zones with excess trench sediments subduct a coherent sedimentary layer that serves as a smooth contact zone between the plates.

the requisite material. At the other end of the spectrum are those subduction zones that accumulate sediments. This can happen in three ways: (1) the oceanic sediments are scraped off and are added to the front of the rapidly deforming accretionary prism, (2) continent-derived sediments are deposited on top of the accretionary prism, or they spill into the trench and are added to the highly deformed front, (3) sediments that are initially subducted then underplate the bottom of the accretionary prism. Note that sediment subduction may occur even if the accretionary prism is growing. Accretionary prisms form where the sediment supply rate is greater than the sediment subduction rate. There is no complete global assessment of the relative rates of sediment supply and subduction. Although it is thought that most oceanic sediments are subducted, there are dissenting opinions (for discussion, see ARMSTRONG, 1981; KARIG and KAY, 1981).

The horst and graben structure is not found in subduction zones with extensive accretionary prisms (HILDE, 1983). Perhaps the horsts and grabens are buried by the excess sediments, or maybe the horst and graben structure does not exist at all. Studies of the southern Caribbean arc indicate that the incoming oceanic sediments are split, the upper portion is added to the accretionary prism while the lower portion is apparently subducted undeformed (e.g., BIJU-DUVAL *et al.*, 1982). If we assume that the horst and graben formation is suppressed, then the lower sedimentary portion could remain undeformed. The contact between the plates would then be composed of coherent layered sediments, as opposed to the "bucket" mechanism where the contact material alternates between basaltic crust and chaotic sediments (Figure 4). If coherent layered sediments enhance seismic coupling at deeper depths, then the metamorphosed oceanic sediments must be capable of seismic failure at elevated temperatures and pressures. There are some indications that this may be possible (WANG, 1980). Thus, the mechanism of sediment subduction might significantly affect the seismic strength distribution along the contact zone between the plates.

My use of the term "excess trench sediments" refers to the presence of an accretionary prism and, generally, the absence of the horst and graben structure. Excess trench sediments then closely coincides with HILDE'S (1983) classification of thick trench deposits. The main distinction between these two terms is the speculative implication that I attach to excess trench sediments. In future discussion, excess trench sediments will be equated with a smooth strength distribution. I emphasize that this identification should be treated with some skepticism, but I assert that it is at least plausible.

5. Global Comparison of Trench Sediments and Large Earthquakes

The division of subduction zones into the binary classification of "excess trench sediments" and "horst-graben structure" certainly has a subjective element. The global survey by HILDE (1983) at least provides an independent classification. As previously noted, I equate HILDE'S thick trench sediments with excess trench sediments (ETS). The major subduction zones with ETS are shown in Figure 5. More than half of the surveyed subduction zones are classified as zones with ETS. We shall test the hypothesis that excess trench sediments promote the occurrence of large and great earthquakes. If trench sediments were quantified by a continuous variable, then we could simply test the correlation between trench sediments and maximum earthquake size for all subduction zones. Given our binary classification scheme, we seek significant differences between the two populations. There are two types of tests: (1) do great earthquakes occur in subduction zones with or without excess trench sediments; (2) are subduction zones with excess trench sediments characterized by great or small earthquakes.

The data for the first test are listed in Table 1. The primary source for the earthquake data is the list in KANAMORI (1977), with two modifications from the updated list in KANAMORI (1983) (specifically, the 1923 Kamchatka and 1979 Colombia events). There is one other change, the M_W value for the 1957 Great

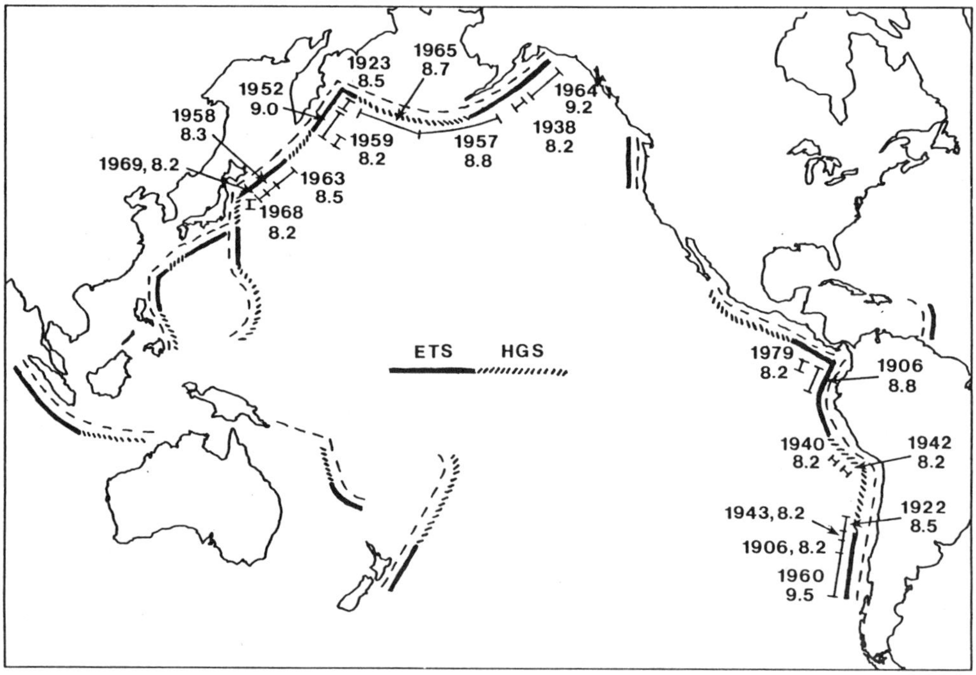

Figure 5
World map showing the trench sediment classification of the major subduction zones. Also shown are the year, M_W, and approximate rupture length of the largest underthrusting earthquakes. Zones are classified either as ETS (excess trench sediments) or HGS (horst and graben structure). The trench sediment classification is largely based on the map in HILDE (1983).

Table 1

Great subduction zone underthrusting earthquakes with $M_W \geq 8.2$. A cross in the ETS column indicates excess trench sediments.

Region	Date	M_W	ETS
Southern Chile	May 22, 1960	9.5	X
Alaska	March 28, 1964	9.2	X
Kamchatka	Nov. 4, 1952	9.0	X
Aleutians	March 9, 1957	8.8	
Colombia-Ecuador	Jan. 31, 1906	8.8	X
Aleutians	Feb. 4, 1965	8.7	
Kurile Islands	Oct 13, 1963	8.5	X
Central Chile	Nov. 11, 1922	8.5	
Kamchatka	Feb. 3, 1923	8.5	X
Kurile Islands	Nov. 6, 1958	8.3	X
Chile	Aug. 17, 1906	8.2	X
Alaska	Nov. 10, 1938	8.2	X
Kamchatka	May 4, 1959	8.2	X
Colombia	Dec. 12, 1979	8.2	X
Northern Honshu	May 16, 1968	8.2	
Peru	May 24, 1940	8.2	
Peru	Aug. 24, 1942	8.2	
Central Chile	April 6, 1943	8.2	X
Kurile Islands	Aug. 11, 1969	8.2	X

Aleutian event has been reduced from 9.1 to 8 ¾ (RUFF *et al.*, 1985). There are several great earthquakes listed in KANAMORI (1983) that do not appear in Table 1, three earthquakes are intraplate events in central Asia, two are thought to be strike-slip events (1958 Alaska, and 1938 Banda Sea, see BEN-MENAHEM, 1977), and two subduction zone events represent intraplate normal faulting in the oceanic lithosphere (1933 Sanriku, and 1977 Java, see KANAMORI, 1971; FITCH *et al.*, 1981; SPENCE, 1986). The nineteen earthquakes in Table 1 are thought to be all of the underthrusting subduction events in the twentieth century with a documented M_W of 8.2 or larger. While the sediment classification for most of the earthquakes is straightforward, a few events need some discussion. A close examination of Figure 5 reveals that the rupture zones of the two great Aleutian events include segments with excess sediments. However, most of the moment release for both events occurs outside these regions, thus these events do not receive the ETS classification. Sediments in southern Chile are apparently transported northward along the trench axis (KULM *et al.*, 1977), and consequently thin to the north. Hence there is some ambiguity where to end the ETS classification in central Chile. Using additional information given in KULM *et al.* (1977), I have placed the dividing line between the 1943 and 1922 rupture zones, broadly consistent with HILDE's map. Figure 6 compares the distribution of earthquake size for zones with and without excess sediments. More than half of the earthquakes, 13 out of 19, are associated with

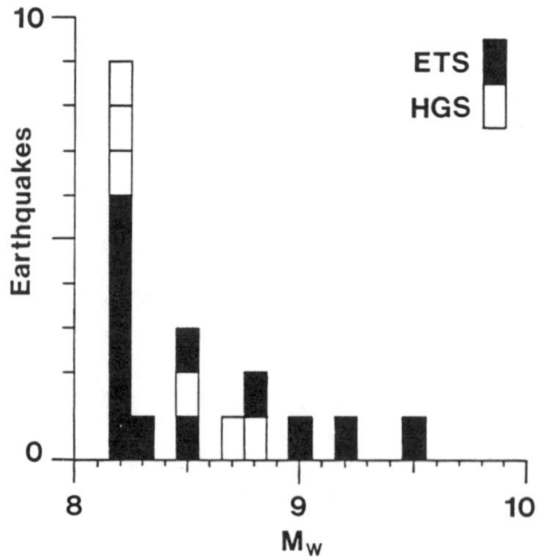

Figure 6
Histogram of the 19 largest underthrusting subduction zone earthquakes, plotted as a function of M_W. There are two groups of earthquakes: those that occur in zones with excess trench sediments (ETS), and those that occur in zones with horst and graben structure (HGS). Most of the largest earthquakes occur in ETS zones.

excess trench sediments. It seems that most of the largest earthquakes occur in regions of excess sediments with the strong exception of the two great Aleutian earthquakes. Due in part to the large number of $M_W = 8.2$ events, the mean M_W values for the two groups are not significantly different. A preliminary conclusion is that great earthquakes tend to occur in zones with excess sediments, but not exclusively.

The other test is to examine the maximum earthquake size of all zones with excess trench sediments. There are subduction zones in Figure 5 with ETS that are not represented by any earthquakes in Table 1. These zones are Southern Honshu, Sumatra, Caribbean, New Zealand-Kermadec, Izu-Bonin, Taiwan-Philippines, New Hebrides, Middle America (south of Mexico), and Oregon-Washington (USA). If these zones have had no great earthquakes in the twentieth century, can we immediately claim that these zones are characterized by the lack of great earthquakes? This question opens the difficult issue of whether the twentieth century catalog of great earthquakes can be regarded as characteristic. If the recurrence interval between the largest earthquakes in some subduction zone is greater than 100 years, then the maximum size earthquake may be missing from the twentieth century catalog. While this problem may also apply to zones that have registered a great earthquake, we will only examine the historical record for the zones with ETS that lack a great earthquake.

It appears that at least two of the above zones, the Southern Honshu and Sumatra zones, are characterized by great earthquakes. Two earthquakes with $M_W = 8.1$ have occurred in the Southern Honshu zone, in 1944 and 1946. Hence, the exclusion of Southern Honshu from Table 1 results from the arbitrary magnitude threshold. There is an excellent historical record in this region; it is likely that one or both of the great earthquakes in 1854 would be above the 8.2 magnitude threshold, and presumably the great earthquake in 1707 would also be 8.2 or larger (ANDO, 1975). Turning to the Sumatra zone, NEWCOMB and MCCANN (1987) document the occurrence of two great earthquakes in the nineteenth century, the 1833 and 1861 events with probable rupture lengths of 500 and 300 km, respectively. On the basis of rupture length, they assign a M_W on the order of 8 ¾ to the 1833 event. This M_W value is consistent with the rupture length versus M_W plot in Figure 2. The lack of Sumatran great earthquakes in the twentieth century can be explained by a recurrence interval greater than 150 years.

The historical record for the Caribbean indicates that a large event occurred in the northern part of the Lesser Antilles arc in 1843 (see KELLEHER *et al.*, 1973). With a rupture length of 200–250 km, this event should be close to the 8.2 magnitude threshold of Table 1. Thus, the Caribbean arc seems capable of generating at least a magnitude 8 event. Although the accretionary prism is less developed at the location of the 1843 event, the ETS classification is given to the entire zone. NISHENKO and MCCANN (1981) suggest that the New Zealand earthquake that occurred in 1833 has a rupture length of approximately 200 km. Thus, New Zealand can also generate an earthquake close to the 8.2 magnitude threshold.

There is no evidence of a great subduction event in the Izu-Bonin arc. There are no earthquakes along the Taiwan-Eastern Philippines trench with a documented M_W greater than 8. Likewise, no great earthquakes in the New Hebrides or Middle America regions are known.

The seismic hazard of the Oregon-Washington region is currently an active issue (e.g., HEATON and KANAMORI, 1984; and see HEATON and HARTZELL, this volume). There are no great, or even large, earthquakes in this region in the twentieth century. Assessment of the previous activity is hampered by a limited historical record and the long recurrence interval for a great earthquake. Although this zone may be capable of generating a great earthquake, for the present concerns of a global survey, Oregon-Washington zone is listed as aseismic.

In summary, of the 14 distinct regions with excess sediments, 7 zones appear to generate great earthquakes, 2 appear to be borderline cases (Caribbean and New Zealand-Kermadec), and 5 do not generate great earthquakes. Thus, the subduction zones with great earthquakes do not form an overwhelming majority of the ETS subduction zones.

5.1. Effect of Plate Age and Convergence Rate

To review our findings: most great earthquakes occur in zones with excess sediments, and about half of the subduction zones with excess sediments have great earthquakes. Clearly there is not a one-to-one correspondence between excess trench sediments and great earthquakes. The lack of a one-to-one correspondence implies that other factors influence earthquake size. At this point, it is instructive to place trench sediments within the framework of the established global correlation for maximum earthquake size.

Those zones with ETS but no great earthquakes have another feature in common: they plot in the lower-left corner of Figure 1 (except for Oregon-Washington and Middle America). This association implies that age and rate determine the *potential* for a great earthquake, hence a subduction zone that falls in the lower-left corner of Figure 1 may not generate a great earthquake regardless of excess trench sediments. The best-fit plane from the linear regression in RUFF and KANAMORI (1980) is used to calculate a predicted M_W from the age and rate values for each subduction zone (Table 2). The observed maximum M_W is plotted versus the

Table 2

Maximum earthquake size in subduction zones: Observed (Obs), Predicted (Pre), Residual (Obs − Pred), Zones with Excess Trench Sediments (X).

Zone	M_W (Obs)	M_W (Pre)	Residual	ETS
Southern Chile	9.5	9.3	0.2	X
Central Chile	8.5	9.0	−0.5	
Peru	8.2	8.9	−0.7	
Ecuador-Colombia	8.8	8.8	0.0	X
Central Am. (Mexico)	8.1	8.6	−0.5	
Kamchatka	9.0	8.5	0.5	X
Aleutians	8.8	8.4	0.4	
Alaska	9.2	8.4	0.8	
Kurile Islands	8.5	8.3	0.2	X
Taiwan-Philippines	8.0	8.2	−0.2	X
Sumatra	8.7	8.2	0.5	X
Northern Honshu	8.2	8.1	0.1	
Tonga	8.3	8.1	0.2	
Java	7.1	7.8	−0.7	
Kermadec	8.1	7.8	0.3	X
New Zealand	8.0	7.6	0.4	X
New Hebrides	7.9	7.8	0.1	X
Scotia	7.1	7.7	−0.6	
Izu-Bonin Islands	7.2	7.4	−0.2	X
Caribbean	8.0	7.3	0.7	X
Marianas	7.2	7.2	0.0	
Southern Honshu	8.3	8.3	0.0	X
Middle America	7.5	8.4	−0.9	X
Oregon-Washington	<7.0	8.3	−1.3	X

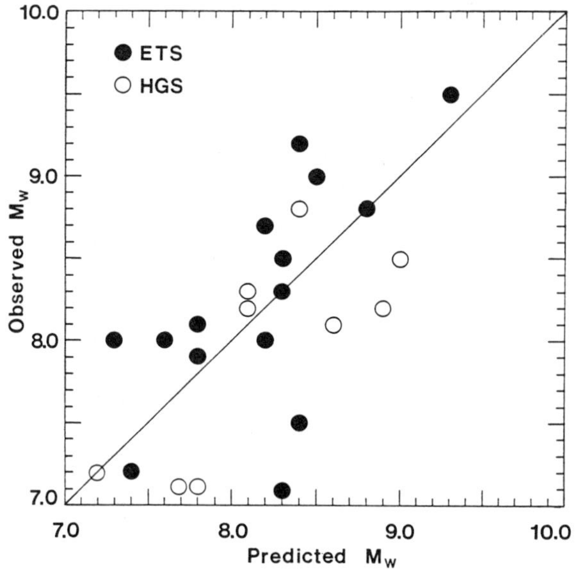

Figure 7
Observed maximum earthquake size versus the size predicted by the linear combination of convergence rate and lithosphere age (see Figure 1, and RUFF and KANAMORI, 1980). If age and rate exactly predicted the observed M_W, then all subduction zones would plot on the diagonal line. The two groups of subduction zones are: excess trench sediments (ETS) and horst and graben structure (HGS).

predicted value in Figure 7. KANAMORI (1983) shows a similar plot for the time/space normalized M_W. The predicted values for the maximum and normalized M_W are nearly identical (compare Figures 2 and A1 in RUFF and KANAMORI, 1980). The observed values in Figure 7 are the same as in RUFF and KANAMORI (1980) except for the revised values for the Aleutians, Sumatra, Caribbean, and New Zealand, and the addition of Southern Honshu, Middle America (south of Mexico), and Oregon-Washington. Figure 8 plots the observed M_W values from Figure 7 and shows the large range in maximum earthquake size for zones with sediments, as previously concluded. We now "correct" for the effect of age and rate by plotting the variation from the predicted values (Figure 9). A positive value corresponds to a maximum earthquake size larger than the predicted number. The two populations in Figure 9 appear to be distinct, though the standard errors about the means overlap.

The Oregon-Washington and Middle America zones violate the trend that ETS zones with no great earthquakes plot in the lower-left corner of Figure 1. Are they true exceptions, or have great earthquakes occurred before the historical records began? This important question cannot be answered here, except that to say that this study supports the concerns of HEATON and KANAMORI (1984).

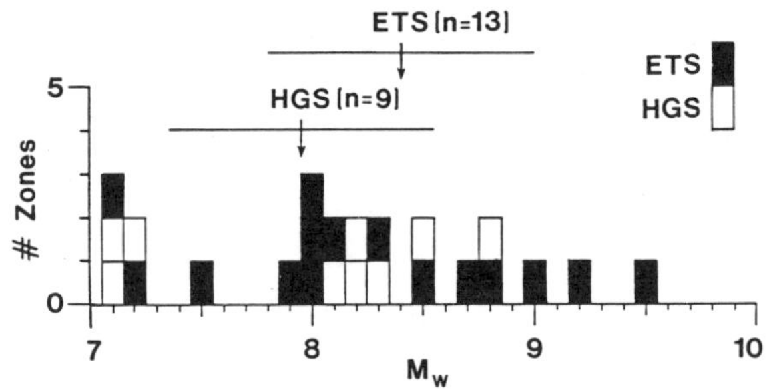

Figure 8
Histogram of the maximum earthquake size of subduction zones. Also plotted are the means and standard errors of the ETS and HGS groups. Of the 24 subduction zones, 15 are classified as ETS zones. The mean and standard error for the ETS zones are based on 13 zones, Oregon-Washington and Middle America are excluded (see text).

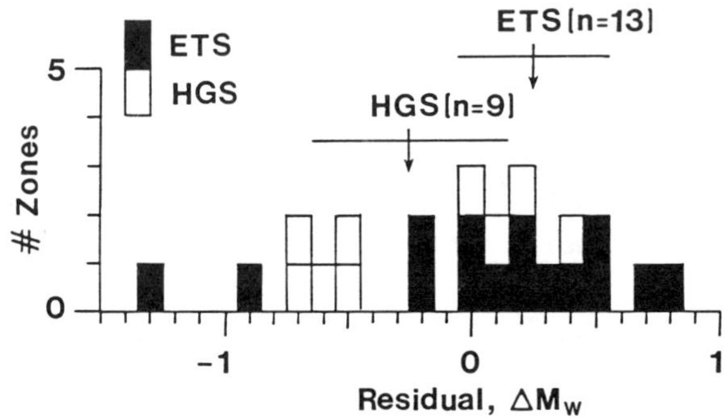

Figure 9
Histogram of the residual M_W of subduction zones. The residuals are the variation from the predicted trend in Figure 7. The means and standard errors of the residuals are plotted for the ETS and HGS zones. As in Figure 8, the Oregon-Washington and Middle America zones are excluded from the mean. The ETS and HGS means are better separated than in Figure 8; this suggests that trench sediments may explain part of the variation in earthquake size about the age-rate prediction.

6. Discussion and Conclusions

The global survey of great earthquakes and subduction zones shows that most great earthquakes occur in subduction zones with excess trench sediments (ETS), and about half of the zones with ETS are characterized by great earthquake occurrence. Also, the mean M_W value is higher for the population of great earthquakes in zones with ETS. If we were only concerned with the global distribution of the 19 largest earthquakes, we might conclude that excess trench sediments explain great earthquake occurrence, with just a few exceptions. This conclusion is weakened by the observation that many zones with ETS have no great earthquakes. Hence, excess trench sediments do not always cause great earthquakes, neither are they the sole cause of great earthquakes. There must be other factors. The established correlation of maximum earthquake size with age and rate provides the complimentary factor: the zones with ETS but without great earthquakes have low M_W values predicted from age and rate (except for two zones). The quantitative significance of age and rate can be assessed by comparing Figures 8 and 9. Application of the age-rate "correlation" better separates the two populations. Therefore, the overall conclusion is that the variation in maximum earthquake size depends on at least three variables: plate age, convergence rate, and excess trench sediments. The association between the largest earthquakes and excess trench sediments can then be attributed to the enhancement of M_W in those ETS zones with high seismic potential.

Recall the proposed chain of cause-and-effect mechanisms that connects trench sediments to earthquake size: excess trench sediments → subduction of a coherent sedimentary layer → smooth seismic strength distribution → large earthquake rupture area. Although this chain of mechanisms is presently untested, future studies can critically examine these speculations.

RUFF and KANAMORI (1980) suggested that age and rate affect seismic coupling through preferred trajectory. Now that sediments and smooth strength distributions are invoked to cause great earthquakes, a new possibility is that age and rate influence the sediment subduction process. For example, suppression of the horst and graben structure may be an important component in achieving a smooth strength distribution; perhaps the horst and graben structure is more easily suppressed in rapidly subducting young lithosphere.

Plate tectonics explains why there are underthrusting earthquakes in subduction zones, but it cannot explain the global variation in the size of these earthquakes. This fundamental problem is still unsolved. Indeed, given the sketchy nature of the physical mechanisms, I will *not* claim that the three factors of age, rate, and sediments *control* earthquake size. I will also admit that the statistical correlation between excess sediments and great earthquake occurrence is less than compelling. However, at the very least, the association between excess trench sediments and great earthquakes should dispel the commonly expressed notion that sediments cause aseismic subduction. It is my hope that the assumptions and speculations in this paper will be probed and tested.

Acknowledgments

The program in earthquake studies at the University of Michigan is supported by the National Science Foundation (Grants EAR–8407786 and P.Y.I./EAR–8351515 to LJR). My thanks to the Maurice Ewing Symposium program for the opportunity to present these ideas to rock mechanics.

References

AKI, K. (1979), *Characterization of barriers on an earthquake fault*, J. Geophys. Res. *84*, 6140–6148.

AKI, K., *Strong-motion seismology*, in *Earthquakes: Observation, Theory, and Interpretation* (edited by H. Kanamori and E. Boschi) (North Holland, New York 1983) pp. 223–250.

ANDO, M. (1975), *Source mechanisms and tectonic significance of historical earthquakes along the Nankai trough, Japan*, Tectonophysics *27*, 119–140.

ARMSTRONG, R. C. (1981), *Radiogenic isotopes: the case for crustal recycling on a near-steady-state no-continental-growth earth*, Phil. Trans. Roy. Soc. London *A301*, 443–472.

BECK, S. L. and L. J. RUFF (1984), *The rupture process of the great 1979 Colombia earthquake: evidence for the asperity model*, J. Geophys. Res. *89*, 9281–9291.

BECK, S. L. and L. J. RUFF (1987), *Rupture process of the great 1963 Kurile Islands earthquake sequence: asperity interaction and multiple event rupture*, J. Geophys. Res. *92*, 14,123–14,138.

BEN-MENAHEM, A. (1977), *Renormalization of the magnitude scale*, Phys. Earth Planet. Int. *15*, 315–340.

BIJU-DUVAL, B., P. LE QUELLEC, A. MASCLE, V. RENARD, and P. VALERY (1982), *Multibeam bathymetric survey and high resolution seismic investigations on the Barbados Ridge complex (eastern Caribbean): A key to the knowledge and interpretation of an accretionary wedge*, Tectonophysics *86*, 275–304.

CHAPPLE, W. M. and D. W. FORSYTH (1979), *Earthquakes and bending of plates at trenches*, J. Geophys. Res. *84*, 6729–6749.

DAS, S. and K. AKI (1977), *Fault planes with barriers: A versatile earthquake model*, J. Geophys. Res. *82*, 5658–5670.

FEDOTOV, S. A. (1965), *Regularities of the distribution of strong earthquakes of Kamchatka, the Kurile Islands, and northeastern Japan*, Trudy Inst. Fiz. Zemli Akad. Nauk. SSSR *36*, 66–93.

FITCH, T. J., R. G. NORTH, and M. W. SHIELDS (1981), *Focal depths and moment tensor representations associated with the great Sumba earthquake*, J. Geophys. Res. *86*, 9357–9374.

GUTENGERG, B. and C. F. RICHTER, *Seismicity of the Earth* (Princeton University Press, Princeton, N.J., 2nd ed., 1954) 310 pp.

HEATON, T. H. and S. H. HARTZELL (1988), *Estimation of strong ground motions from hypothetical earthquakes on the Cascadia subduction zone, Pacific Northwest*, Pure and Appl. Geophys. *129*, *1/2*, 133–203.

HEATON, T. H. and H. KANAMORI (1984), *Seismic potential associated with subduction in the northwestern United States*, Bull. Seism. Soc. Am. *74*, 933–941.

HILDE, T. W. C. (1983), *Sediment subduction versus accretion around the Pacific*, Tectonophysics *99*, 381–397.

HOUSE, L. S., L. R. SYKES, J. N. DAVIES, and K. H. JACOB, *Identification of a possible seismic gap near Unalaska Island, eastern Aleutians, Alaska*, in *Earthquake Prediction, an International Review* (edited by D. W. Simpson and P. G. Richards) (American Geophysical Union, Washington, D.C. 1981) pp. 81–92.

ISACKS, B., J. OLIVER, and L. SYKES (1968), *Seismology and the new global tectonics*, J. Geophys. Res. *73*, 5855–5899.

JARRARD, R. D. (1986), *Relations among subduction parameters*, Rev. Geophys. *24*, 217–284.

KANAMORI, H. (1971a), *Great earthquakes at island arcs and the lithosphere*, Tectonophysics *12*, 187–198.

KANAMORI, H. (1971b), *Seismological evidence for a lithospheric normal faulting—the Sanriku earthquake of 1933*, Phys. Earth Planet. Int. *4*, 289–300.

KANAMORI, H. (1977), *The energy release in great earthquakes*, J. Geophys. Res. *82*, 2981–2987.

KANAMORI, H. (1978), *Quantification of earthquakes*, Nature *271*, 411–414.

KANAMORI, H., *The nature of seismicity patterns before large earthquakes*, in *Earthquake Prediction, an International Review* (edited by D. W. Simpson and P. G. Richards) (American Geophysical Union, Washington D.C., 1981) pp. 1–19.

KANAMORI, H., *Global seismicity*, in *Earthquakes: Observation, Theory, and Interpretation* (edited by H. Kanamori and E. Boschi) (North Holland, New York, 1983) pp. 596–608.

KANAMORI, H. and D. L. ANDERSON (1975), *Theoretical basis of some empirical relations in seismology*, Bull. Seism. Soc. Am. *65*, 1073–1095.

KANAMORI, H. and G. S. STEWART (1978), *Seismological aspects of the Guatemala earthquake of February 4, 1976*, J. Geophys. Res. *83*, 3427–3434.

KARIG, D. E. and R. W. KAY (1981), *Fate of sediments on the descending plate at convergent margins*, Phil. Trans. Roy. Soc. London *A301*, 443–472.

KARIG, D. E. and G. F. SHARMAN (1975), *Subduction and accretion in trenches*, Geol. Soc. Am. Bull. *86*, 377–389.

KELLEHER, J. (1972), *Rupture zones of large South American earthquakes and some predictions*, J. Geophys. Res. *77*, 2087–2103.

KELLEHER, J., J. SAVINO, H. ROWLETT, and W. MCCANN (1974), *Why and where great thrust earthquakes occur along island arcs*, J. Geophys. Res. *79*, 4889–4899.

KELLEHER, J., L. SYKES, and J. OLIVER (1973), *Possible criteria for predicting earthquake locations and their application to major plate boundaries of the Pacific and the Caribbean*, J. Geophys. Res. *78*, 2547–2585.

KULM, L. D. and G. A. FOWLER, *Oregon continental margin structure and stratigraphy—a test of the imbricate thrust model*, in *The Geology of Continental Margins* (edited by C. A. Burk and C. L. Drake) (Springer, New York, 1974) pp. 261–283.

KULM, L. D., W. J. SCHWELLER, and A. MASIAS, *A preliminary analysis of the subduction processes along the Andean continental margin, 6° to 45°S*, in *Island Arcs, Deep Sea Trenches, and Back-Arc Basins* (edited by M. Talwani and W. C. Pitman III) (American Geophysical Union, Washington D.C., 1977) pp. 285–302.

LAY, T., H. KANAMORI, and L. RUFF (1982), *The asperity model and the nature of large subduction zone earthquakes*, Earthquake Pred. Res. *1*, 3–71.

MCCANN, W. R., S. P. NISHENKO, L. R. SYKES, and J. KRAUS (1979), *Seismic gaps and plate tectonics: Seismic potential for major plate boundaries*, PAGEOPH *117*, 1087–1147.

MOGI, K. (1969), *Relationship between the occurrence of great earthquakes and tectonic structures*, Bull. Earthquake Res. Inst., Tokyo Univ. *47*, 429–451.

NEWCOMB, K. R. and W. R. MCCANN (1987), *Seismic history and seismotectonics of the Sunda arc*, J. Geophys. Res. *92*, 421–439.

NISHENKO, S. P. (1985), *Seismic potential for large and great interplate earthquakes along the Chilean and southern Peruvian margins of South America: A quantitative reappraisal*, J. Geophys. Res. *90*, 3589–3616.

NISHENKO, S. P. and W. R. MCCANN, *Seismic potential for the world's major plate boundaries: 1981*, in *Earthquake Prediction, an International Review* (edited by D. W. Simpson and P. G. Richards) (American Geophysical Union, Washington D.C., 1981) pp. 20–28.

NUR, A. and M. ISRAEL (1980), *The role of heterogeneities in faulting*, Phys. Earth Planet. Int. *21*, 225–236.

PLAFKER, G. and J. C. SAVAGE (1970), *Mechanism of the Chilean earthquakes of May 21 and 22, 1960*, Geol. Soc. Am. Bull. *81*, 1001–1030.

RUFF, L. J., *Fault asperities inferred from seismic body waves*, in *Earthquakes: Observation, Theory, and Observation* (edited by H. Kanamori and E. Boschi) (North Holland, New York, 1983) pp. 251–276.

RUFF, L. and H. KANAMORI (1980), *Seismicity and the subduction process*, Phys. Earth Planet. Int. *23*, 240–252.

RUFF, L. and H. KANAMORI (1983a), *Seismic coupling and uncoupling at subduction zones*, Tectonophysics *99*, 99–117.

RUFF, L. and H. KANAMORI (1983b), *The rupture process and asperity distribution of three great earthquakes from long-period diffracted P-waves*, Phys. Earth Planet. Int. *31*, 202–230.

RUFF, L., H. KANAMORI, and L. SYKES (1985), *The 1957 great Aleutian earthquake*, EOS *66*, 298 (abstract).

SCHOLL, D. W., M. S. MARLOW, and A. K. COOPER, *Sediment subduction and offscraping at Pacific margins*, in *Island Arcs, Deep Sea Trenches, and Back-arc, Basins* (edited by M. Talwani and W. C. Pitman III) (American Geophysical Union, Washington D.C., 1977) pp. 199–210.

SCHOLL, D. W., R. VON HUENE, T. L. VALLIER, and D. G. HOWELL (1980), *Sedimentary masses and concepts about tectonic processes at underthrust margins*, Geology *8*, 564–568.

SCHWARTZ, S. Y. and L. J. RUFF (1987), *Asperity distribution and earthquake occurrence in the southern Kurile Islands arc*, Phys. Earth Planet. Int. *49*, 54–77.

SYKES, L. R. (1971), *Aftershock zones of great earthquakes, seismicity gaps, and earthquake prediction for Alaska and the Aleutians*, J. Geophys. Res. *76*, 8021–8041.

SPENCE, W. (1986), *The 1977 Sumba earthquake series: evidence for slab pull force acting at a subduction zone*, J. Geophys. Res. *91*, 7225–7239.

TAJIMA, F. and H. KANAMORI (1985), *Global survey of aftershock area expansion patterns*, Phys. Earth Planet. Int. *40*, 77–134.

UYEDA, S. and H. KANAMORI (1979), *Back-arc opening and the mode of subduction*, J. Geophys. Res. *84*, 1049–1061.

VON HUENE, R., M. LANGSETH, N. NASU, and H. OKADA (1982), *A summary of the Cenozoic tectonic history along the IPOD Japan Trench transect*, Geol. Soc. Am. Bull. *93*, 829–846.

WANG, C. (1980), *Sediment subduction and frictional sliding in a subduction zone*, Geology *8*, 530–533.

(Received August 13, 1986, revised February 29, 1988, accepted April 11, 1988)